Rethinking the Scientific Revolution

This book challenges the traditional historiography of the Scientific Revolution, probably the single most important unifying concept in the history of science. Usually referring to the period from Copernicus to Newton (roughly 1500 to 1700), the Scientific Revolution is considered to be the central episode in the history of science, the historical moment at which that unique way of looking at the world that we call "modern science" and its attendant institutions emerged.

Reexamination of the preoccupations of early modern natural philosophers undermines many of the assumptions underlying standard accounts of the Scientific Revolution. Starting with a dialogue between Betty Jo Teeter Dobbs and Richard S. Westfall, whose understanding of the Scientific Revolution differed in important ways, the chapters in this volume reconsider canonical figures, their areas of study, and the formation of disciplinary boundaries during this seminal period of European intellectual history.

Margaret J. Osler is Professor of History and Adjunct Professor of Philosophy at the University of Calgary. She has also taught at Wake Forest University, Harvey Mudd College, and Oregon State University. Professor Osler is the author of *Divine Will and the Mechanical Philosophy: Gassendi and Descartes on Contingency and Necessity in the Created World*. Her articles have appeared in *Journal of the History of Ideas*, *Isis*, *Studies in History and Philosophy of Science*, and many other publications.

Rethinking the
Scientific Revolution

Edited by
MARGARET J. OSLER
University of Calgary

CAMBRIDGE UNIVERSITY PRESS
Cambridge, New York, Melbourne, Madrid, Cape Town, Singapore, São Paulo

Cambridge University Press
The Edinburgh Building, Cambridge CB2 8RU, UK

Published in the United States of America by Cambridge University Press, New York

www.cambridge.org
Information on this title: www.cambridge.org/9780521661010

© Margaret J. Osler 2000

This publication is in copyright. Subject to statutory exception
and to the provisions of relevant collective licensing agreements,
no reproduction of any part may take place without the written
permission of Cambridge University Press.

First published 2000

A catalogue record for this publication is available from the British Library

Library of Congress Cataloguing in Publication data

Rethinking the scientific revolution / Margaret J. Osler.
p.   cm.
Includes index.
ISBN 0-521-66101-3 (hc.). – ISBN 0-521-66790-9 (pbk.)
1. Science, Renaissance. 2. Science – France – History –
17th century. 3. Science – England – History – 17th century.
I. Osler, Margaret J., 1942– .
Q125.2.R48   1999
509'.03 – dc21   99-21576
                                                           CIP

ISBN 978-0-521-66101-0 hardback
ISBN 978-0-521-66790-6 paperback

Transferred to digital printing 2008

In memory of
Betty Jo Teeter Dobbs and Richard S. Westfall,
mentors, colleagues, and friends

CONTENTS

List of Contributors   *page* ix

Preface   xi

Introduction

1   The Canonical Imperative: Rethinking the Scientific Revolution   *Margaret J. Osler*   3

Part I   The Canon in Question

2   Newton as Final Cause and First Mover   *B. J. T. Dobbs*   25

3   The Scientific Revolution Reasserted *Richard S. Westfall*   41

Part II   Canonical Disciplines Re-Formed

4   The Role of Religion in the Lutheran Response to Copernicus   *Peter Barker*   59

5   Catholic Natural Philosophy: Alchemy and the Revivification of Sir Kenelm Digby   *Bruce Janacek*   89

6   Vital Spirits: Redemption, Artisanship, and the New Philosophy in Early Modern Europe   *Pamela H. Smith*   119

7   "The Terriblest Eclipse That Hath Been Seen in Our Days": Black Monday and the Debate on Astrology during the Interregnum   *William E. Burns*   137

8   Arguing about Nothing: Henry More and Robert Boyle on the Theological Implications of the Void *Jane E. Jenkins*   153

Part III Canonical Figures Reconsidered

9 Pursuing Knowledge: Robert Boyle and Isaac Newton
   Jan W. Wojcik — 183

10 The Alchemies of Robert Boyle and Isaac Newton: Alternate Approaches and Divergent Deployments
   Lawrence M. Principe — 201

11 The Janus Faces of Science in the Seventeenth Century: Athanasius Kircher and Isaac Newton   *Paula Findlen* — 221

12 The Nature of Newton's "Holy Alliance" between Science and Religion: From the Scientific Revolution to Newton (and Back Again)   *James E. Force* — 247

13 The Fate of the Date: The Theology of Newton's *Principia* Revisited   *J. E. McGuire* — 271

14 Newton and Spinoza and the Bible Scholarship of the Day   *Richard H. Popkin* — 297

Part IV The Canon Constructed

15 The Truth of Newton's Science and the Truth of Science's History: Heroic Science at Its Eighteenth-Century Formulation   *Margaret C. Jacob* — 315

Index — 333

CONTRIBUTORS

PETER BARKER is Professor of the History of Science at the University of Oklahoma.

WILLIAM E. BURNS received a doctorate in history in 1994. He is currently working on a study of the astrologer John Gadbury (1627–1704).

B. J. T. DOBBS was Professor of History at the University of California at Davis.

PAULA FINDLEN teaches history of science at Stanford University, where she directs the Science, Technology and Society Program.

JAMES E. FORCE is Professor of Philosophy at the University of Kentucky.

MARGARET C. JACOB is Professor of History at the University of California at Los Angeles.

BRUCE JANACEK is Assistant Professor of History at North Central College, Naperville, Illinois.

JANE E. JENKINS teaches the history of science at the University of New Brunswick.

J. E. McGUIRE is Professor in the Department of History and Philosophy of Science at the University of Pittsburgh.

MARGARET J. OSLER is Professor of History and Adjunct Professor of Philosophy at the University of Calgary.

RICHARD H. POPKIN is Professor Emeritus at Washington University, St. Louis. He is also Adjunct Professor of History and Philosophy at the University of California at Los Angeles.

LAWRENCE M. PRINCIPE is Associate Professor in the Department of History of Science, Medicine and Technology and the Department of Chemistry at The Johns Hopkins University.

PAMELA H. SMITH is Associate Professor of History at Pomona College.

RICHARD S. WESTFALL was Distinguished Professor in the Department of the History and Philosophy of Science at Indiana University.

JAN W. WOJCIK is Associate Professor of Philosophy at Auburn University.

PREFACE

The most satisfying tribute a scholar can receive is serious consideration of his or her work by other scholars. When Betty Jo Teeter Dobbs died suddenly and prematurely on March 29, 1994, I decided to invite a number of her colleagues, students, and friends to contribute papers to a volume in memory of her scholarly interests and the impact of her work. Dobbs's groundbreaking studies of the meaning of Newton's alchemy irrevocably altered our understanding of the Scientific Revolution and early modern natural philosophy. The scholarship of her associates reflects this impact – in spirit as well as detail. This volume stands as a tribute to her contributions.

Dobbs herself articulated some of the far-reaching ramifications of her work on Newton in her History of Science Society Distinguished Lecture, "Newton as Final Cause and First Mover," in which she challenged the received understanding of the Scientific Revolution. This essay, which was originally published in *Isis*, opens the volume and sets the themes for the chapters that follow. Richard S. Westfall contributed an essay that went head-to-head with Dobbs's and which provides an eloquent defense of the utility – indeed necessity – of thinking in traditional terms about the Scientific Revolution. The debate between these two giants about the central concept in our field provides the broader context for the chapters in the volume. Subsequent events altered the direction of the volume after it was well underway. Westfall's sudden death on August 21, 1996, reinforced my decision to construct the volume in terms of their debate and, at the same time, to honor Westfall's memory along with Dobbs's.

Readers will observe that most of the chapters in the volume lean towards Dobbs's revisionism rather than Westfall's reassertion of the received view of the Scientific Revolution. Westfall was aware that his essay was going to serve as something of a foil for the volume. We had a long discussion about this fact some months before his death, and he understood what the tilt of the book would be. Despite the fact that his views receive serious criticism in many of these chapters, the outlook of the volume itself was something he understood and accepted.

I am indebted to many people who contributed to the conception, content, and final production of this volume. Early conversations with Peter Barker, Paula Findlen, and Deborah Harkness helped formulate the shape of the volume and the list of contributors. Margaret C. Jacob read drafts of all the chapters and rewrote her own essay to make it serve as an epilogue to the volume. Margaret G. Cook, Andrew Cunningham, and Lawrence M. Principe read and reread various incarnations of the Introduction, making numerous suggestions for improving it. I am especially grateful to Alex Holzman, History of Science editor at Cambridge University Press, who supported this project from the beginning and provided wise counsel during difficult times. Brian MacDonald provided efficient and sensitive editing and advice as he guided the book through the production process. Jeff Wigelsworth assisted me with proofreading and other chores. My friend Betty Flagler was a constant source of good advice and emotional support.

Fellowships from the National Endowment for the Humanities and the Calgary Institute for the Humanities provided time to work on this project. Research grants from the Social Sciences and Humanities Research Council of Canada and the University of Calgary Research Grants Committee provided material support.

INTRODUCTION

1

The Canonical Imperative: Rethinking the Scientific Revolution

MARGARET J. OSLER

The Scientific Revolution is probably the single most important unifying concept in the history of science. Usually referring to the period from Copernicus to Newton (roughly 1500 to 1700), it is considered to be the central episode in the history of science, the historical moment at which that unique way of looking at the world that we call "modern science" and its attendant institutions emerged. It has been taken as the terminus ad quem of classical and medieval science and the terminus a quo of all that followed. Not itself an explanatory concept, the Scientific Revolution has become the reference point for questions that guide historians of science, questions about what it was, what exactly happened, why it happened, and why it happened when and where it did.[1]

Traditional histories of the Scientific Revolution have customarily focused on a list of canonical individuals who explored a canonical set of subjects. The individuals usually include Copernicus, Tycho, Kepler, Galileo, Vesalius, Harvey, Descartes, Boyle, and Newton. The subjects are astronomy, physics, mathematics, anatomy, physiology, and chemistry.[2] This book reflects the problematization of the canon in recent scholarship. The traditional canonical figures often devoted

---

I am grateful to Andrew Cunningham, Margaret G. Cook, Margaret C. Jacob, Pamela McCallum, Lawrence M. Principe, J'nan Morse Sellery, Jan W. Wojcik, and two anonymous readers for Cambridge University Press for suggestions that greatly improved this chapter. The title was Betty Flagler's idea.

1 Floris Cohen organizes his book around this set of questions. See H. Floris Cohen, *The Scientific Revolution: A Historiographical Inquiry* (Chicago: University of Chicago Press, 1994). See also Toby E. Huff, *The Rise of Early Modern Science: Islam, China, and the West* (Cambridge: Cambridge University Press, 1993).

2 Consider, for example, the following old standards: Herbert Butterfield, *The Origins of Modern Science, 1300–1800*, rev. ed. 1949: reprint, (New York: Free Press, 1957); A. Rupert Hall, *The Revolution in Science, 1500–1750* (New York: Longman, 1983; first published 1954); and Richard S. Westfall, *The Construction of Modern Science: Mechanisms and Mechanics* (New York: John Wiley, 1971).

themselves to noncanonical subjects and frequently resembled many of their contemporaries who have not found a place in the Scientific Revolution's hall of fame. Moreover, the subjects that engaged their attention do not readily map onto the canonical list of modern sciences. Questioning the canon leads us to inquire why and how it was formed. And this inquiry, in turn, causes us to interrogate our own presuppositions as historians and how those presuppositions affect what we see in the past.

The chapters in this volume engage in a critical dialogue with the traditional understanding of the Scientific Revolution. The late Betty Jo Teeter Dobbs opens the discussion by stating her intention "to undermine one of our most hallowed explanatory frameworks, that of the Scientific Revolution." Starting from I. B. Cohen's definition of "revolution" as "a change that is sudden, radical, and complete,"[3] she argues that the Scientific Revolution had none of these characteristics. She observes that the reception of Copernicanism was slow, that traditional histories extend the process to one that took anywhere from 150 to 500 years, that the break with Aristotelianism was not complete, and that even the heroes of the traditional accounts – most notably Newton – did not think in the same way as modern scientists. Indeed, Newton's intellectual preoccupations had more to do with theology and alchemy than with the physics and mathematics on which his modern reputation rests.

In a resounding defense of the historiography for which he was one of the most distinguished spokesmen, the late Richard S. Westfall responds with an eloquent reassertion of "our central organizing idea ... [because] without it our discipline will lose its coherence, and what is more, the cause of historical understanding take a significant step backward." Arguing for the undeniable impact of science on the modern world and for the central role of the Scientific Revolution as the one idea that has brought coherence to the entire history of science, Westfall reaffirms the importance of the traditional historiography that he helped to create. He argues that the break with Aristotelianism was a major discontinuity in the history of ideas and that the new science that developed was qualitatively different from traditional natural philosophy. As for the canon, Westfall states that Newton is remembered today for his contributions to physics, optics, and mathematics – not for his studies of alchemy and theology.

This debate between Dobbs and Westfall signals a deep rupture in

---

3 I. Bernard Cohen, *The Revolution in Science* (Cambridge, Mass.: Harvard University Press, 1985), pp. 51–90.

the present understanding of the historical development of science. Westfall's analysis is fundamentally forward-looking, based on the assumption that what is interesting in the past are those developments that led to our present understanding of the world. In contrast to Westfall, Dobbs seeks to understand the presuppositions and assumptions of her historical actors rather than searching for anticipations of modern ideas in their thought.

In considering intellectual developments in the period between 1500 and 1700 and in considering the figure of Isaac Newton, Dobbs and Westfall start from different assumptions, which inevitably lead them to different conclusions. Dobbs challenges a traditional assumption about the heroes of the Scientific Revolution, namely, "that their thought patterns were fundamentally just like ours." It is only because they make this assumption that historians have found it difficult to explain Kepler's Pythagoreanism or Newton's devotion to alchemy. By making a different assumption, namely that people have not always viewed the world in the same way that we do, Dobbs is able to argue that we can make sense of their diverse interests and preoccupations. This is the crux of Westfall's disagreement with Dobbs. He assumes that thinkers in the past are similar to us and that what is important for the historian is that aspect of a thinker's work that has survived until the present or that has led to our present way of looking at things.

Despite the polar differences between Westfall and Dobbs about the existence of the Scientific Revolution and its utility as a historiographical concept, they are talking about the same revolution, the one that either did or did not occur during the period bounded by 1500 and 1700. As Margaret Jacob argues in the concluding chapter, however, their disagreement is badly posed: it rests on the heroic assumption of who and what made the revolution. The reconsideration of the canonical heroes of the revolution in this volume leaves the question – Was there or was there not a scientific revolution? – dangling. One possible answer, suggested in Jacob's chapter, is that there was a scientific revolution, but not the one that both Dobbs and Westfall tacitly assume to be at stake. Rather, the revolution was constructed in the eighteenth century when natural philosophers selectively took up Newton's physics and mathematics while ignoring his alchemical and theological views. In addition to understanding the intellectual developments of this period, then, future research must address the interests and concerns of subsequent generations, which created the perception that a scientific revolution occurred in the sixteenth and seventeenth centuries and then bequeathed it to us. If Jacob's interpretation is valid, its power derives from the kinds of questions, reassessments, and challenges posed by the scholarship represented in this volume.

These differences bring the question of the nature of intellectual change into bold relief. Because contemporary science makes claims about the world that scientists and historians take to be true, historians of science have sometimes succumbed to the Whiggish tendency to understand the history of science as the unfolding of ideas by the force of their own, internal logic. This tendency explains why they have left certain critical developments incompletely explained. Examples include the acceptance of Copernican astronomy, the rise of the mechanical philosophy, the decline of astrology and transmutational alchemy, and the acceptance of Newtonian physics.[4] Such developments seem not to require explanation, since they are presumed to be *right*. Less anachronistic historians understand that all historical developments demand explanation, since there is no preordained or *right* way for ideas to develop. Interrogating the developments that seem to require no explanation yields insight into the assumptions guiding our historical actors. Peter Barker asks how we can account for the fact that some people were more receptive to heliocentric astronomy than others. He argues that Lutheran theology with its emphasis on order and design predisposed certain astronomers to be especially receptive to Copernicanism.

Ideas do not influence subsequent ideas; nor do they develop by their own intrinsic power. Rather they are deployed or developed by particular individuals in real historical contexts to solve problems of their own. Taking questions of agency seriously means using actors' categories to account for the development of ideas.[5] What, then, do actors do in the context of intellectual history? Borrowing a metaphor that Peter Barker appropriated from A. I. Sabra, I maintain that thinkers appropriate ideas from the traditions within which they work and use them in their own contexts to solve the particular problems that concern them.[6] The metaphor of appropriation gives agency to the historical actors who work within their own particular social, ideolog-

---

4 Westfall, *Construction of Modern Science*, pp. 30–1, 159. On the decline of transmutational alchemy, see William R. Newman and Lawrence M. Principe, "Alchemy vs. Chemistry: The Etymological Origins of a Historiographical Mistake," *Early Science and Medicine* 3 (1998): 32–65.
5 My analysis here is indebted to Quentin Skinner, "Meaning and Understanding in the History of Ideas," *History and Theory* 8 (1969): 3–53.
6 Peter Barker develops this point in detail in "Understanding Change and Continuity: Transmission and Appropriation in Sixteenth Century Natural Philosophy," in *Tradition, Transmission, Transformation*, ed. F. J. Ragep and S. Ragep, with Steven Livesey (Leiden: Brill, 1996), pp. 527–50.

ical, and intellectual contexts. "By speaking of 'appropriation,' we acknowledge the change in a previously established idea, theory, technique, or practice as it enters a new historical (and perhaps geographical) location. If the practice, idea, or whatever takes root, it is because it serves the continuing needs of the appropriators."[7] The act of borrowing may be more or less deliberate; nonetheless, natural philosophers draw on earlier thought to address problems in their own context.

Thinking in terms of appropriation leads us to consider different questions from those asked by traditional historiography. The contexts in which ideas are used become of paramount importance, as do detailed histories of particular concepts within those contexts. Why were particular figures attracted to one tradition or another? How does the way they ask their questions affect the use they make of the ideas they borrow? Why do they ask questions in the particular way that they do? The new historiography is characterized by an increasing awareness of the importance of the intellectual and social context within which ideas have developed, along with a renewed respect for the presuppositions and concepts of the historical actors rather than those of historians.[8] It takes the history of science to places where it has not usually been seen before: into the courts, into the streets, into the countryside, and into local societies.[9] Accordingly, Pamela Smith

---

7 Ibid., p. 21.
8 Westman has shown how this problem affects Thomas S. Kuhn's analysis in *The Copernican Revolution: Planetary Astronomy in the Development of Western Thought* (Cambridge, Mass.: Harvard University Press, 1957). He comments that "the narrative is historical, but not historicist." See Robert S. Westman, "Two Cultures or One? A Second Look at Kuhn's *The Copernican Revolution*," *Isis* 85 (1994): 88.
9 For science in court culture, see Robert S. Westman, "Proof, Poetics, and Patronage: Copernicus' Preface to *De revolutionibus*," in *Reappraisals of the Scientific Revolution*, ed. David C. Lindberg and Robert S. Westman (Cambridge: Cambridge University Press, 1990), pp. 167–206; Mario Biagioli, *Galileo Courtier: The Practice of Science in the Culture of Absolutism* (Chicago: University of Chicago Press, 1993); Bruce T. Moran, *The Alchemical World of the German Court: Occult Philosophy and Chemical Medicine in the Circle of Moritz of Hessen (1572–1632)* (Stuttgart: Franz Steiner Verlag, 1991); David S. Lux, *Patronage and Science in Seventeenth-Century France* (Ithaca, N.Y.: Cornell University Press, 1989); and Pamela H. Smith, *The Business of Alchemy: Science and Culture in the Holy Roman Empire* (Princeton: Princeton University Press, 1994). For science in the streets, see William Eamon, *Science and the Secrets of Nature: Books of Secrets in Medieval and Early Modern Culture* (Princeton: Princeton University Press, 1993); and Patrick Curry, *Prophecy and Power: Astrology in Early Modern England* (Ithaca, N.Y.: Cornell University Press, 1989). For science in the countryside, see Keith Thomas, *Man and the Natural World: Changing Attitudes in England, 1500–1800* (London: Allen Lane, 1983); Andrew Cunningham, "The Culture of Gardens" and Paula Findlen, "Courting Nature," in *Cultures of Natural History*, ed. N. Jardine, J. A. Secord, and E. C. Spary (Cambridge: Cambridge University Press, 1996), pp. 38–56, 57–74. For science in mu-

takes us to the artisan's workshop and to various German courts; William Burns, to the streets of London during the Interregnum; and Margaret Jacob, to the schools and local philosophical societies of the eighteenth century in which Newtonianism was taught to diverse and often nonprofessional audiences.

Once we think in terms of appropriation, the traditional canons of historical figures and disciplines become problematized because we must ask why and how the canon was constructed. By elucidating the similarities between Athanasius Kircher and Isaac Newton, Paula Findlen raises the question of why Newton was incorporated into the canon and Kircher was not. It is only the judgment of later generations that forged our distinction between genius and crackpot. In the new historiography, questions of when and why certain figures become canonical replace the practice of reading history backward from the present. Thus, it becomes evident that the concerns of the appropriators determine which past figures are useful, seminal, or important rather than the intrinsic or timeless merit of their ideas. Returning to the example of Newton, it was the interests of eighteenth-century natural philosophers that put his mathematical physics in the foreground, as Margaret Jacob argues, and it was the concerns of twentieth-century historians of science who valued mathematical sciences above all others that gave him his ruling position in the canon. Moreover, different commentators may find different aspects of a single thinker's ideas significant. Twentieth-century scientists and historians may value Newton's contributions to mathematics and physics, but, as Richard Popkin demonstrates, religious fundamentalists are more impressed by his approach to biblical scholarship.

Nevertheless, to contextualize the canon is not to deny the reality of historical change. Despite historical continuities and the appropriation of ideas from ancient, medieval, and Renaissance sources, the period from 1500 through 1700 witnessed major changes in natural philosophy. These changes become evident if we examine ideas separated by a sufficiently long temporal gap. As Westfall wisely notes, "Scientists of today can read and recognize works done after 1687. It takes a historian to comprehend those written before 1543." Unlike Steven Shapin who opens his book by proclaiming, "There is no such thing as the Scientific Revolution," I think it is crucial to acknowledge that there were major changes in thinking about the natural world during the early modern period, changes in cosmology, metaphysics, episte-

seums, see Paula Findlen, *Possessing Nature: Museums, Collecting, and Scientific Culture in Early Modern Italy* (Berkeley: University of California Press, 1994). For an attempt to incorporate some of these approaches into a new synthesis, see Steven Shapin, *The Scientific Revolution* (Chicago: University of Chicago Press, 1996).

mology, matter theory, physics, and optics.[10] Likewise, the practices and social locations in which these subjects were supported and pursued underwent considerable change. However all these changes cannot be embraced in a single formulation as implied by a phrase like "*The* Scientific Revolution."[11] The trick is to recognize that intellectual change occurred while at the same time recognizing that change is not necessarily linear or self-evident progress toward our modern way of thinking.

Avoiding presentism or Whiggish historiography, uncontroversial though such a strategy may appear, raises a further conundrum. If we assert that historiographical sophistication is increasing as we learn to take actors' categories into account, are we unwittingly giving a Whiggish history of our own historiographical practice?[12] Such an infinite regress of Whiggism can be avoided as long as we do not claim progress for historical method itself. Historians need to recognize the role that their own assumptions play in their constructions of the past. There is no escaping them, but consciously acknowledging them staves off the temptation of claiming objectivity and progress.

10 On cosmology, see Alexandre Koyré, *From the Closed World to the Infinite Universe* (Baltimore: Johns Hopkins Press, 1957). On metaphysics and epistemology, see E. A. Burtt, *The Metaphysical Foundations of Modern Science* (reprint, Garden City, N.Y.: Doubleday Anchor, 1954; first published 1924); Richard H. Popkin, *The History of Scepticism from Erasmus to Spinoza*, rev. and exp. ed. (Berkeley: University of California Press, 1979; first published 1960); Margaret J. Osler, *Divine Will and the Mechanical Philosophy: Gassendi and Descartes on Contingency and Necessity in the Created World* (Cambridge: Cambridge University Press, 1994); Barbara Shapiro, *Probability and Certainty in Seventeenth-Century England: A Study of the Relationships between Natural Science, Religion, History, Law, and Literature* (Princeton: Princeton University Press, 1983); and Jan W. Wojcik, *Robert Boyle and the Limits of Reason* (Cambridge: Cambridge University Press, 1997). On matter theory, see Dennis Des Chene, *Physiologia: Natural Philosophy in Late Aristotelian and Cartesian Thought* (Ithaca, N.Y.. Cornell University Press, 1996); *The Fate of Hylomorphism: "Matter" and "Form" in Early Modern Science*, ed. C. H. Lüthy and William R. Newman, *Early Science and Medicine* 2 (1997): 216–352; and Norma E. Emerton, *The Scientific Reinterpretation of Form* (Ithaca, N.Y.: Cornell University Press, 1984). On physics, see Daniel Garber, *Descartes' Metaphysical Physics* (Chicago: University of Chicago Press, 1992), and Richard S. Westfall, *Force in Newton's Physics: The Science of Dynamics in the Seventeenth Century* (London: MacDonald and American Elsevier, 1971). On optics, see A. I. Sabra, *Theories of Light from Descartes to Newton* (London: Oldbourne, 1967), and Alan E. Shapiro, *Fits, Passions, and Paroxysms: Physics, Method, and Chemistry and Newton's Theories of Colored Bodies and Fits of Easy Reflection* (Cambridge: Cambridge University Press, 1993).
11 This kind of essentialism marks Cohen's attempt to formulate a general account of the Scientific Revolution. See *The Scientific Revolution*, part 3.
12 See Thomas Nickles, "Philosophy of Science and the History of Science," in *Constructing Knowledge in the History of Science*, ed. Arnold Thackray, *Osiris* 2 ser., 10 (1995): 139–63.

This historical approach is at odds with traditional accounts of the Scientific Revolution. In the scholarly tradition stemming from nineteenth-century positivist Ernst Mach, historians have told a story that stresses the radical discontinuity of the Scientific Revolution from what came before and locates the point of rupture in the mind of Galileo.[13] Subsequent historians of science – whatever their historiographical predilection[14] – all tended to accept Mach's identification of the revolutionary moment as Galileo's transition from the theory of impetus in the early *De motu* to his mature science of inertial motion published in the *Discorsi*.[15] Their differences have revolved around the causes of that revolution, but they generally have accepted the same basic story.[16] The received narrative begins with the Copernican challenge to Aristotelian cosmology and Ptolemaic astronomy, continues with the discovery of Kepler's Laws in astronomy, Galileo's development of a new physics and the emergence of the mechanical view of

---

13 Ernst Mach, *The Science of Mechanics: A Critical and Historical Account of its Development*, trans. Thomas J. McCormack, 6th ed. with revisions through the 9th German ed. (Lasalle, Illinois: Open Court, 1960; originally published in German, 1883), pp. 39–45.

14 John McEvoy sees historians as having gone through three stages of analysis: "during the last fifty years, the discipline of the history of science has passed through three distinct historiographical stages, each characterized by a dominant, but not exclusive, interpretative style. The first stage was shaped by the positivist-Whig view of science as a teleologically structured corpus of experimental knowledge. This perspective was challenged, in the early 1960s, by the postpositivist identification of the history of science with the articulation and application of theoretical doctrines and research programs. The postpostivist hegemony was itself displaced, in the 1970s and '80s, by the postmodernist view of science as a sociological entity shaped by the contingent constraints of specific agents and local practices." John G. McEvoy, "Positivism, Whiggism, and the Chemical Revolution: A Study in the Historiography of Chemistry," *History of Science* 35 (1997): 1.

15 See, for example, Burtt, *The Metaphysical Foundations of Modern Science*; Edward W. Strong, *Procedures and Metaphysics* (Berkeley: University of California Press, 1936); Alexandre Koyré, *Études galiléennes* (Paris: Hermann, 1939); E. J. Dijksterhuis, *The Mechanization of the World Picture*, trans. C. Dikshoorn (Oxford: Clarendon Press, 1961; originally published in Amsterdam, 1950); and Westfall, *The Construction of Modern Science*.

16 Mach's influence explains the immense amount of detailed scholarship on Galileo's intellectual development. See for example Koyré, *Études galiléennes*; Stillman Drake, *Galileo at Work: His Scientific Biography* (Chicago: University of Chicago Press, 1978); Ernan McMullin, ed., *Galileo: Man of Science* (New York: Basic Books, 1967); Maurice Clavelin, *The Natural Philosophy of Galileo: Essay on the Origins and Formation of Classical Mechanics*, trans. A. J. Pomerans (Cambridge, Mass.: MIT Press, 1974); William R. Shea, *Galileo's Intellectual Revolution: Middle Period, 1610–1632* (New York: Neale Watson, 1974); Winifred Lovell Wisan, "Galileo's Scientific Method: A Reexamination," in *New Perspectives on Galileo*, ed. R. E. Butts and J. C. Pitt (Dordrecht: Reidel, 1978); and William A. Wallace, *Galileo and His Sources: The Heritage of the Collegio Romano in Galileo's Science* (Princeton: Princeton University Press, 1984).

nature that replaced Scholastic Aristotelianism, and reaches a triumphant climax with the Newtonian synthesis that bound these strands together into one coherent whole, thus heralding the triumph of modern science.[17] This story presumes that by 1700 there was a definitive rupture between the worn-out Scholasticism of pre-Copernican thought and the new science that emerged in the seventeenth century. The revolution in cosmology and metaphysics included the following radical transformations: the finite Aristotelian cosmos was replaced with an infinite Newtonian universe; nature was mathematized and mechanized; and experiment came to play an important role in the justification of scientific theories.[18] This is the story Westfall reiterates.

Without denying the reality of these changes in some areas of thought, historians have found it necessary to reexamine several assumptions that guided the traditional historiography.[19] One of these assumptions, drawn from nineteenth-century positivist classifications of the sciences, is that physics is the most fundamental science. Consequently, historians have highlighted the mathematization of physics in the seventeenth century, thereby marginalizing developments in other areas of natural philosophy.[20] They have concluded that the other sciences underwent their revolutions only when they became similarly mathematized. These assumptions lie at the root of the oxymoronic extension of the Scientific Revolution into a process that took many centuries to complete, since the revolutions in chemistry and biology did not occur until the eighteenth and nineteenth centuries, respectively.[21] A further assumption is that disciplinary boundaries

17 See, for example, Koyré, *Études galiléennes*; Alexandre Koyré "The Significance of the Newtonian Synthesis," in *Newtonian Studies*, ed. Alexandre Koyré (Cambridge, Mass.: Harvard University Press, 1965), pp. 3–24; Burtt, *The Metaphysical Foundations of Modern Science*. Similar ideas can be found in Hall, *The Revolution in Science, 1500–1750*; Dijksterhuis, *The Mechanization of the World Picture*; and Westfall, *The Construction of Modern Science*.
18 On the persistence of this account, see Westman, "Two Cultures or One?"
19 See Andrew Cunningham and Perry Williams, "De-centering the 'Big Picture': *The Origins of Modern Science* and the Modern Origins of Science," *British Journal for the History of Science* 26 (1993): 407–32. Cunningham and Williams delineate a number of the difficulties with the traditional understanding of the Scientific Revolution.
20 Comte's classification of the sciences, which embodies a reductionist and universalist criterion, is consonant with and perhaps historically connected to twentieth-century historians' of science privileging the mathematization of nature and the development of mathematical physics in their accounts of the Scientific Revolution. See Auguste Comte, *Cours de philosophie positive* (Paris, 1830), 1:47–95. Mach did not share Comte's reductionism. See *The Science of Mechanics*, p. 596.
21 Consider, for example, Butterfield's now notorious characterization of "The Postponed Scientific Revolution in Chemistry," not to mention his extension of the Scientific Revolution to a process that took five hundred years to complete. See Butterfield, *The Origins of Modern Science*, chap. 11.

have remained static throughout history. This assumption embodies an essentialism about science, according to which science is defined as unchanging and unambiguously identifiable in every historical era, and divisible into specific sciences that – as categories of intellectual activity – have always remained the same.[22] This essentialism creeps into the interpretation of the Scientific Revolution itself: having defined the nature of the Scientific Revolution, historians searched this event and explanations of it.[23]

A number of scholarly developments have challenged all of these assumptions. One of the early signs that the concept of the Scientific Revolution was coming under scrutiny was the publication of *Revolution in Science* (1985), in which I. B. Cohen traced the origin and development of the term "revolution" as applied to science.[24] The publication of this book was particularly significant because Cohen, one of the most eloquent advocates of the traditional understanding of the Scientific Revolution,[25] acknowledged – at least implicitly – that the concept itself was not static but had evolved from earlier usage. Questions about the outlines of the traditional historiography dominated several sessions at the meetings of the History of Science Society during the 1980s, culminating in one of the first attempts to survey new historiographical directions, *Reappraisals of the Scientific Revolution* (1990).[26] The chapters in that collection covered a wide range of topics, including the various traditional scientific disciplines, natural magic, natural history, metaphysics, and religion. Although differing in content and approach, the chapters in *Reappraisals* adopted a historicist analysis of their subject matters, attended to questions about context and audience, and reexamined the role of institutions. Despite the new approaches articulated in this volume, its editors remained committed to the historiographical utility of the concept of the Scientific Revolution, "which continues to provide organizing principles for the majority of practising histories of science, serving heuristic functions,

---

22 On the question of disciplinary boundaries, see Andrew Cunningham, "Getting the Game Right: Some Plain Words on the Identity and Invention of Science," *Studies in History and Philosophy of Science* 19 (1988): 365–89. See also Alan Gabbey, "The Case of Mechanics: One Revolution or Many?" in Lindberg and Westman, *Reappraisals of the Scientific Revolution*, pp. 493–528, and Donald R. Kelley, ed., *History and the Disciplines: The Reclassification of Knowledge in Early Modern Europe* (Rochester, N.Y.: University of Rochester Press, 1997).
23 Cohen's *The Scientific Revolution* is organized around just such an essentialist approach.
24 Cohen, *Revolution in Science*.
25 See, for example, I. Bernard Cohen, *The Birth of a New Physics*, rev. and updated (reprint, New York: Norton, 1985; first published 1960), and I. Bernard Cohen, *The Newtonian Revolution: With Illustrations of the Transformation of Scientific Ideas* (Cambridge: Cambridge University Press, 1980).
26 Lindberg and Westman, *Reappraisals of the Scientific Revolution*.

even if it no longer commands universal assent."[27] H. Floris Cohen, who undertook the daunting task of examining the entire historiography of the Scientific Revolution, remained committed to both the reality of the revolution and its historiographical utility. Surveying the historiography of the Scientific Revolution from Kant and Whewell through the 1970s and 1980s, Cohen focused on historians' answers to two questions: what was the Scientific Revolution, and why did it happen? Throughout his extensive discussion, Cohen never questioned the idea that the concept of the Scientific Revolution is essential to historical analysis of the period.

New lines of thinking that departed from the master narrative of the Scientific Revolution had started appearing during the 1960s, although historians of science have only recently become fully aware of the subversive nature of these developments. The publication of Frances Yates's *Giordano Bruno and the Hermetic Tradition* (1964) drew attention to the traditions of high magic, which had not played an important role in earlier accounts of the Scientific Revolution. Yates herself claimed to adhere to the traditional historiography. "With the history of genuine science leading up to Galileo's mechanics this book has had nothing whatever to do."[28] For that story, she was content to rely on the accounts of the traditional historians. Rather, she proposed to explain why the Scientific Revolution occurred when it did.

> It is here, as a historical study, and particularly as a historical study of motives, that the present book may have a contribution to make towards elucidating these problems. It is a movement of the will which really originates as an intellectual movement. A new centre of interest arises, surrounded by an emotional excitement; the mind turns whither the will has directed it, and new attitudes, new discoveries follow. Behind the emergence of modern science there was a new direction of the will, a new longing and determination to understand those workings and to operate with them.
>
> Whence and how had this new direction arisen? One answer to that question suggested by this book is "Hermes Trismegistus."[29]

To the recovery and development of the Hermetic philosophy Yates ascribed such developments as "the concentration on number as a road into nature's secrets," belief in the uniformity of nature, and the

---

27 David C. Lindberg, "Conceptions of the Scientific Revolution from Bacon to Butterfield: A Preliminary Sketch," in Lindberg and Westman, *Reappraisals of the Scientific Revolution*, p. 20.
28 Frances Yates, *Giordano Bruno and the Hermetic Tradition* (New York: Vintage, 1964), p. 447.
29 Ibid., p. 448.

acceptability of speculating about nature.[30] Although Yates introduced serious consideration of noncanonical subject matters and noncanonical historical figures, she retained the traditional account of the Scientific Revolution.[31]

Yates's work precipitated an increased interest in and controversies about the possible role of such noncanonical subjects as alchemy, natural magic, and Hermeticism in the "new science" of the seventeenth century.[32] The interest in alchemy contributed to a radical reinterpretation of the work of Isaac Newton, whose *Principia* and *Opticks* are almost universally considered to be the crowning achievements of the Scientific Revolution. Older accounts of Newton concentrated almost exclusively on his mathematics, physics, and optics, the contributions that were taken to embody modern science and its methods.[33] Although earlier scholars had noted the wealth of Newton's manuscripts, they were selective, concentrating on materials that seemed relevant to later scientific developments. As scholars began taking account of the mass of materials he left on theological and alchemical subjects and to situate him within the context of a European natural philosophical community in which such interests were widely shared, they recognized that Newton did not share the intellectual agenda of his twentieth-century commentators.[34]

It was only when historians were willing to relinquish their own presuppositions about the nature of science that they could begin to assimilate the documentary record and come to grips with Newton in

30 Ibid.
31 Yates's "thesis" spawned a controversy about the role of Hermeticism in the Scientific Revolution. See Robert S. Westman, "Magical Reform and Astronomical Reform: The Yates Thesis Reconsidered," in *Hermeticism and the Scientific Revolution*, ed. Robert S. Westman and J. E. McGuire (Berkeley: University of California Press, 1977).
32 An early locus for such studies was M. L. Righini Bonelli and William R. Shea, eds., *Reason, Experiment, and Mysticism in the Scientific Revolution* (New York: Science History Publications, 1975). The increasing interest and sophistication of scholarship in this area is evident in Brian Vickers, ed. *Occult and Scientific Mentalities in the Renaissance* (Cambridge: Cambridge University Press, 1984).
33 See, for example, A. Rupert Hall and Marie Boas Hall, *Unpublished Scientific Papers of Isaac Newton* (Cambridge: Cambridge University Press, 1962); Koyré, *Newtonian Studies*; Westfall, *Force in Newton's Physics*; and Cohen, *The Newtonian Revolution*.
34 This is not the first time that historians have emphasized the importance of these traditions for the development of early modern natural philosophy. Pioneering works include Yates, *Giordano Bruno and the Hermetic Tradition*; Paolo Rossi, *Francis Bacon: From Magic to Science*, trans. Sacha Rabinovitch (London: Routledge and Kegan Paul, 1968; originally published in Italian, 1957); Allen G. Debus, *The Chemical Philosophy: Paracelsian Science and Medicine in the Sixteenth and Seventeenth Centuries*, 2 vols. (New York: Neale Watson, 1977); and Charles Webster, *From Paracelsus to Newton: Magic and the Making of Modern Science* (Cambridge: Cambridge University Press, 1982).

his own terms.[35] P. M. Rattansi and J. E. McGuire took an early and important step in recognizing that Newton's intellectual agenda was markedly different from our own. Their seminal paper, "Newton and the 'Pipes of Pan,'"[36] was one of the first scholarly studies of the "other" Isaac Newton. They concluded their study of Newton's deep concern with the *prisca theologia* by remarking that,

> It is certainly difficult for us in the twentieth century to conceive one whose scientific achievements were so great, pursuing with equal interest and energy such other studies, especially when his efforts in those fields produced so little of enduring value. It is even more difficult for us to imagine the mechanics and cosmology of the *Principia* being influenced by Newton's theological views and his belief in a pristine knowledge. Sir Isaac Newton, however, was not a "scientist" but a Philosopher of Nature. In the intellectual environment of his century, it was a legitimate task to use a wide variety of material to reconstruct the unified wisdom of Creation.[37]

By highlighting the importance of the *prisca* tradition for Newton, Rattansi and McGuire situated him within the broader context of Renaissance humanism and its appeal to ancient sources.

Newton's interest in noncanonical subjects was extensive. He was preoccupied with alchemy for most of his life. At his death, he left a massive amount of manuscript material on alchemical subjects. Dobbs's groundbreaking book, *The Foundations of Newton's Alchemy or, "The Hunting of the Greene Lyon,"* was the first comprehensive, scholarly analysis of Newton's alchemical writings,[38] and Westfall fully incorporated this material into *Never at Rest*, his magisterial biography of Newton, a move that seems to indicate that he had altered the more austere approach evident in his earlier book, *Force in Newton's Physics*.[39] Although it is true that Westfall was willing to countenance Newton's alchemy, he wanted to understand it within the context of his "scientific" work, never considering that Newton's alchemical obsessions might provide a clue to the meaning of his

---

35 For an account of the history of Newton's alchemical and theological manuscripts, see B. J. T. Dobbs, *The Foundations of Newton's Alchemy or, "The Hunting of the Greene Lyon"* (Cambridge: Cambridge University Press, 1975), pp. 6–20. Similar considerations also apply to Boyle. See Lawrence M. Principe, *The Aspiring Adept: Robert Boyle's Alchemical Quest* (Princeton: Princeton University Press, 1998).

36 P. M. Rattansi and J. E. McGuire, "Newton and the 'Pipes of Pan,'" *Notes and Records of the Royal Society of London* 21 (1966): 108–43.

37 Ibid., p. 138.

38 Dobbs, *The Foundations of Newton's Alchemy*.

39 Richard S. Westfall, *Never at Rest: A Biography of Isaac Newton* (Cambridge: Cambridge University Press, 1980).

work in physics. "The significance of alchemy in his intellectual odyssey lay in the broader vistas it opened to him, additional categories to supplement and complete the narrow mechanistic ones. His enduring fame derived from his seizing the possibilities thus spread before him."[40]

Dobbs pushed the argument further, demonstrating that in contrast to the nineteenth- and twentieth-century view that places mathematical physics at the pinnacle of intellectual merit, Newton understood his life's work, which included tireless devotion to alchemy, "as a study of the modes of divine activity in the world."[41] His God, as Force argues, is "the Lord God of Dominion," and his voluntarist theology bound the diverse aspects of his thought into a coherent whole. In the present volume, McGuire underscores these connections in his exegetical tour de force in which he demonstrates the importance of theology to Newton's physics. His chapter reveals some of the implications of Dobbs's work even as it questions some of the details of her scholarship. As historians we cannot grasp the significance of Newton's physics or the meaning of his preoccupations with alchemy and theology until we acknowledge how deeply embedded what we call his scientific work was in his fundamentally theological quest.[42] If we insist, as Westfall does, on privileging Newton's mathematical physics and optics because those are the aspects of his thought that have prevailed in the end, we lose our hold on the concerns of the historical actor himself.

Since historians of science have interpreted Newton's work as the climax of the narrative they call the Scientific Revolution, this radical shift in our understanding of the meaning of his own work forces us, as Dobbs concludes, to reconsider many of the received opinions about the nature of the Scientific Revolution. Studies of Newton are no more an isolated phenomenon of scholarship than was he a man isolated from the main currents of European intellectual history. Recent studies, exemplified by the chapters in this volume, situate New-

---

40 Ibid., p. 301.
41 B. J. T. Dobbs, *The Janus Faces of Genius: The Role of Alchemy in Newton's Thought* (Cambridge: Cambridge University Press, 1991), p. 18.
42 "Newton's thought is a seamless unity of theology, metaphysics, and natural science. Newton's view of God's Dominion, i.e., the total supremacy of God's power and will over every aspect of creation, colors every aspect of his views about how matter (and the laws regulating the ordinary operation of matter) is created, preserved, reformed, and occasionally interdicted by a voluntary and direct act of God's sovereign will and power." James E. Force, "Newton's God of Dominion: The Unity of Newton's Theological, Scientific, and Political Thought," in *Essays on the Context, Nature, and Influence of Isaac Newton's Theology*, ed. James E. Force and Richard H. Popkin (Dordrecht: Kluwer, 1990), p. 84.

ton in the broad intellectual currents of his time. They also bring to light a number of unexamined presuppositions embedded in the canon: assumptions about the cast of individuals who played an active role in the Scientific Revolution and about the particular disciplines worthy of inclusion in the standard account.

Reading Newton in light of his own preoccupations rather than those of twentieth-century historians raises questions about the applicability of twentieth-century disciplinary boundaries to seventeenth-century intellectual history. Historians have privileged Newton's mathematical physics because, as Westfall asserts, that is the field of greatest prominence in their intellectual world. Newton, however, would have described himself as a natural philosopher. Indeed, he entitled his great masterpiece *Mathematical Principles of Natural Philosophy*. The early modern term "natural philosophy" had a different extension than does the modern term "science," encompassing God's creation of the world, his providential relationship with the creation, and the immortality of the soul, along with the chemistry, physics, anatomy, and physiology that we would expect. Natural philosophers did not establish criteria of demarcation between these issues and what we consider genuine scientific concerns, but regarded them as inseparable.[43]

In extending the range of topics to encompass all the concerns of early modern natural philosophers, historians of science have turned their attention to a number of areas that were traditionally ignored or at best categorized and dismissed as "occult," notably alchemy and astrology. While Newton's immersion in alchemy may be the most astonishing to traditional sensibilities, Kepler's Pythagorean numerology comes in as a close second. Alchemical concepts and Paracelsian matter theory appear in the writings of Gassendi the atomist as well as in those of Boyle the corpuscularian chemist.[44]

The inclusiveness of natural philosophy highlights another problem that the traditional historiography failed to solve, namely, the diffi-

---

43 This fact directly contradicts Brian Vickers's attempt at drawing demarcations. See his introduction in Vickers, *Occult and Scientific Mentalities in the Renaissance*. For evidence of the inseparability of what we would call "magic," "science," and "religion," see Amos Funkenstein, *Theology and the Scientific Imagination from the Middle Ages to the Seventeenth Century* (Princeton: Princeton University Press, 1986); Osler, *Divine Will and the Mechanical Philosophy*, and Dobbs, *Janus Faces*.

44 Pierre Gassendi, *Syntagma Philosophicum*, "Physics," sect. 3, pt. 1, bk. 3, chap. 6, "De Metallis, ac eorum Transmutatione," in *Opera omnia*, 6 vols. (Lyon, 1658; facsimile reprint, Stuttgart-Bad Cannstatt: Friedrich Frommann Verlag, 1964), 2: 135–43; Principe, *The Aspiring Adept*; and William R. Newman, "Boyle's Debt to Corpuscular Alchemy," in *Robert Boyle Reconsidered*, ed. Michael Hunter (Cambridge: Cambridge University Press, 1994), pp. 91–106, 107–16.

culty of incorporating various nonmathematical disciplines into the historical narrative that ran from Copernicus to Newton.[45] For example, although Boyle and others attempted to incorporate chemistry into the mechanical philosophy, the mechanization of chemistry did not produce the Chemical Revolution in the seventeenth century.[46] While William Harvey's famous work on circulation, *De motu cordis et sanguinis* (1628), seems consonant with the mechanizing and mathematizing themes of physics, his work on the generation of animals, *De generatione* (1651), retained a strongly Aristotelian cast.[47] Natural history, as practiced by John Ray and Francis Willughby, remained essentialist and descriptive, reflecting none of the dominant characteristics of the Scientific Revolution, except for an emphasis on direct observation.[48]

In addition to the inclusion of the noncanonical disciplines, studies of the social relations of science have also contributed to eroding the traditional concept of the Scientific Revolution by situating intellectual concerns within broader social and political contexts. Nevertheless, many studies of the social relations of early modern science, such as Robert K. Merton's classic *Science, Technology and Society in Seventeenth-Century England* and Charles Webster's *The Great Instauration: Science, Medicine, and Reform, 1626–1660*, implicitly adopted the received view of the Scientific Revolution and sought to explain the increased emphasis on science and technology in sociological, political, and reli-

---

45 Floris Cohen has noted this difficulty. See *The Scientific Revolution*. See also Thomas S. Kuhn, "Mathematical versus Experimental Traditions in the Development of Physical Science," *Journal of Interdisciplinary History* 7 (1976): 1–31; reprinted in Thomas S. Kuhn, *The Essential Tension: Selected Studies in Scientific Tradition and Change* (Chicago: University of Chicago Press, 1977), pp. 31–65.
46 See Thomas S. Kuhn, "Robert Boyle and Structural Chemistry in the Seventeenth Century," *Isis* 43 (1952): 12–36.
47 Westfall, *Construction of Modern Science*, pp. 97–8.
48 Hall, *The Revolution in Science*, chap. 13. Ray wrote, "Although I may have affirmed that universals do not exist in nature, but are merely figments of the human mind, I would not deny that they have a foundation in things truly agreeing or similar in special parts or properties. This agreement is so great, especially in living things, that individuals of the same species are seen as having been made according to the same exemplar or idea in the Divine Mind, just as in artificial things two machines of the same species have been made according to the same idea in the mind of the artificer. Whence it follows that species are essentially distinguished from one another and are intransmutable, and the forms or essences of these are either certain specific principles, that is, certain very small particles of matter, distinct from all others, and naturally indivisible, or certain specific seminal reasons enclosed by means of an appropriate vehicle." *De variis plantarum methodus dissertatio* (London: Smith and Walford, 1696), p. vi (translated by Philip Sloan). I am grateful to Paul Farber for alerting me to this passage.

gious terms.⁴⁹ Michael Hunter's *Science and Society in Restoration England* continued the story in this vein, situating the increasing interest in science and technology during this period within the context of "the growth of a fashionable, leisured culture focused on London."⁵⁰ Although studies such as these departed from the exclusive emphasis on intellectual issues found in the writings of historians writing in the tradition of Burtt and Koyré, they did not challenge the prevailing narrative about the Scientific Revolution. They placed the history of science within a social context, but they did not discuss the possible influence of that context on the content of the science.

Sociological studies of episodes in the Scientific Revolution promised to connect intellectual developments with their social and political contexts, but even the most innovative of these studies have not really questioned the accepted narrative of the Scientific Revolution. A case in point is *Leviathan and the Air-Pump*, a provocative and influential study in which Steven Shapin and Simon Schaffer linked Hobbes's debate with Boyle over the interpretation of the air-pump experiments with broader issues of social status, political philosophy, and social dimensions of experimental practice. Shapin pursued these themes in even greater detail in *A Social History of Truth*.⁵¹ Despite their attempt to explain many of the characteristics of Boyle's experimental philosophy by appealing to contemporary social practices, neither study questioned Boyle's status as a canonical figure in the traditional narrative or the traditional interpretation of Boyle as a herald of modern science.⁵²

---

49 Robert K. Merton, *Science, Technology and Society in Seventeenth-Century England* (New York: Harper, 1970; originally published in *Osiris* 4 [1938]), and Charles Webster, *The Great Instauration: Science, Medicine, and Reform, 1626–1660* (London: Duckworth, 1975). Debate about these issues, cast within an explicit acceptance of the received historiography of the Scientific Revolution, can be found in *The Intellectual Revolution of the Seventeenth Century*, ed. Charles Webster (London: Routledge and Kegan Paul, 1974). See also, Margaret C. Jacob, *The Newtonians and the English Revolution, 1689–1720* (Ithaca, N.Y.: Cornell University Press, 1976), and James R. Jacob and Margaret C. Jacob, "The Anglican Origins of Modern Science: The Metaphysical Foundations of the Whig Constitution," *Isis* 71 (1980): 251–67.

50 Michael Hunter, *Science and Society in Restoration England* (Cambridge: Cambridge University Press, 1981), p. 5.

51 Steven Shapin and Simon Schaffer, *Leviathan and the Air-Pump: Hobbes, Boyle, and the Experimental Life* (Princeton: Princeton University Press, 1985), and Steven Shapin, *A Social History of Truth: Civility and Science in Seventeenth-Century England* (Chicago: University of Chicago Press, 1994).

52 For example, Shapin does not discuss Boyle's interest in alchemy, one of the key factors leading to current revisions of our understanding of Boyle's natural philosophy. Shapin's most recent book, *The Scientific Revolution*, does nothing to displace

Westfall's prize-winning article, "Scientific Patronage: Galileo and the Telescope," marked the beginning of serious studies of the impact of patronage practices on early modern science.[53] Westfall argued for the importance of deploying actor's categories in order to explore the social relations of science during the Scientific Revolution, but he did not foresee that such studies would eventually undermine the historian's category itself. What became evident in the cultural histories of patronage – in Mario Biagioli's continued study of Galileo and patronage and even more so in Paula Findlen's study of collecting practices and natural history – is that the interests of patrons influenced the directions in which the content of science developed.[54] By considering natural history as an expression of the larger culture, Findlen discusses how natural history fit into the changing intellectual scene in early modern Europe without forcing it to conform to the model derived from astronomy and physics.[55]

What we would call the biological sciences – physiology, theories of generation, and natural history – never fit into the received historiography of the Scientific Revolution. For example, Butterfield tried to place William Harvey's discovery of the circulation of the blood within a linear tradition stemming from Paduan Aristotelianism, while Westfall attempted to incorporate Harvey's discovery into an account of the mechanical philosophy.[56] More recently, some historians have recognized that there was more than one tradition in the history of anatomy, physiology, and medicine and that the histories of these disciplines do not readily map onto the scheme used to describe the development of the mathematical and physical subjects. Accordingly, Andrew Cunningham has argued that there were at least three different traditions of Renaissance anatomy and that,

---

the standard historiography. In this book, he departs from the order in which topics are discussed, but his discussion focuses on the usual suspects and traditional topics, a case of old wine in new bottles.

53 Richard S. Westfall, "Scientific Patronage: Galileo and the Telescope," *Isis* 76 (1985): 11–30.
54 Biagioli, *Galileo Courtier*, and Findlen, *Possessing Nature*.
55 See William Ashworth Jr., "Emblematic Natural History of the Renaissance"; Andrew Cunningham, "The Culture of Gardens"; Paula Findlen, "Courting Nature"; Katie Whitaker, "The Culture of Curiosity"; Harold J. Cook, "Physicians and Natural History"; and Adrian Johns, "Nature History as Print Culture," all in Jardine, Secord, and Spary, *Cultures of Natural History*, pp. 17–37, 38–56, 57–74, 75–90, 91–105, and 106–24.
56 Butterfield, *The Origins of Modern Science*, chap. 3, and Westfall, *The Construction of Modern Science*, chap. 5. Robert G. Frank Jr. continues this linear story. See his book *Harvey and the Oxford Physiologists: A Study in Scientific Ideas* (Berkeley: University of California Press, 1980).

in general, their aim was to exhibit the handiwork of the Creator rather than to introduce observational and experimental methods to medical subjects as part of a broader revolution in science.[57] Thus Harvey's work can be understood within the context of natural philosophy. Accordingly, his work has come to be understood in the social, religious, and political contexts of his time, rather than as another contribution to a scientific revolution into which it does not easily fit.[58]

These complexities neither diminish the importance of mathematical physics as a part of natural philosophy in the seventeenth century nor negate the importance of the developments from Copernicus to Newton. Yes, there were thinkers concerned with mathematical physics – the names of Galileo, Huygens, Wallis, Wren, Leibniz, and others spring to mind. But mathematical physics was neither the primary concern of the much larger group of people who wrote about natural philosophy in the early modern period nor the only interest of those who did. One issue of central importance at this time was the formulation of a philosophy of nature to replace the Aristotelianism that they almost all rejected (although not in as thoroughgoing a fashion as we have been accustomed to believe).[59] While the mechanical philosophy was a serious contender as the replacement for Aristotelianism, so were the chemical philosophies that derived from the work of Paracelsus and van Helmont.[60] Principe shows the ways in which alchemical ideas were important to Boyle and Newton, who are frequently considered to be mechanical philosophers.[61] This claim runs counter to traditional history of science, which posited a great divide between alchemy and the mechanical philosophy and which denied that alchemy was of any real significance to the important contributions of either Boyle or Newton.[62] Discussions about the nature of

---

57 Andrew Cunningham, *The Anatomical Renaissance: The Resurrection of the Anatomical Projects of the Ancients* (Aldershot: Scolar Press; Brookfield, Vt.: Ashgate, 1997).
58 Roger French, *William Harvey's Natural Philosophy* (Cambridge: Cambridge University Press, 1994). See also Harold J. Cook, *The Decline of the Old Medical Regime in Stuart London* (Ithaca, N.Y.: Cornell University Press, 1986).
59 There is a growing literature on the persistence of Aristotelianism in seventeenth-century natural philosophy. See, for example, Charles B. Schmitt, *Aristotle in the Renaissance* (Cambridge, Mass.: Harvard University Press, 1983); Roger Ariew, "Theory of Comets at Paris during the Seventeenth Century," *Journal of the History of Ideas* 53 (1992): 355–372; and Peter Dear, *Discipline and Experience: The Mathematical Way in the Scientific Revolution* (Chicago: University of Chicago Press, 1995).
60 See Debus, *The Chemical Philosophy*.
61 See Marie Boas Hall, *Robert Boyle and Seventeenth-Century Chemistry* (Cambridge: Cambridge University Press, 1958), and Westfall, *Force in Newton's Physics*.
62 See for example Marie Boas Hall, "Newton's Voyage in the Strange Seas of Alchemy," in *Reason, Experiment, and Mysticism in the Scientific Revolution*, ed. M. L.

matter, the existence of the void, the relationship between body and soul, and the scope of human knowledge were all aspects of this search for a new philosophy of nature. These themes are central to some of the figures discussed in this volume. Janacek shows how Digby's ideas on matter and soul were intimately connected with each other and with his Catholic theology. Jenkins argues that Boyle's debate with Henry More on the interpretation of the air-pump experiments had more to do with their divergent understandings of God's relationship to the creation than with experimental results per se. Wojcik shows how Boyle and Newton disagreed about the nature and scope of human knowledge. For most natural philosophers, what we would call theological issues were as important for determining the acceptability of a philosophy as its success in explaining natural phenomena.[63]

Although most of these chapters lean toward the kind of historiography advocated by Dobbs, Westfall's voice continues to resonate throughout the volume. The essays may question and contextualize the canon, but they do not negate it. Indeed, the cast of characters receiving the most attention – Kepler, Boyle, and Newton – are major figures in traditional accounts of the Scientific Revolution. Reading the canon in context does not undermine the fact that the early modern period witnessed profound changes in the way natural philosophers viewed the world.

Listening carefully to the voices of the historical actors does not imply relinquishing the reality of historical change. Major changes in the way European thinkers understood the world are perceptible if we consider a sufficiently broad span of time. Not all disciplines underwent the same kinds of changes, and they did not all change at the same time. Individual thinkers appropriated ideas from a wide variety of sources and applied them to solve many different kinds of problems. They did not all share the same assumptions about how the world works or how we can find out about it. What we customarily refer to as science in the seventeenth century was not a single thing, and neither was the Scientific Revolution. By abandoning the search for essences, we can get on with the task of understanding changing ideas in context.

Righini Bonelli and William R. Shea (New York: Science History Publications, 1975), pp. 239–46.
63 See Osler, *Divine Will and the Mechanical Philosophy*, and Dobbs, *Janus Faces*.

PART I

The Canon in Question

2

Newton as Final Cause and First Mover

B. J. T. DOBBS

Friends and colleagues, it is an honor to be asked to speak to you this evening, and I thank you for the occasion. It is, however, entirely possible that you will regret granting me this forum, for I intend to undermine one of our most hallowed explanatory frameworks, that of the Scientific Revolution – almost always written with capital letters, of course.

I am well aware that something happened in the sixteenth and seventeenth centuries that human beings have since come to regard as revolutionary – revolutionary, that is, in the modern sense of that word: in I. B. Cohen's definition of political revolution, "a change that is sudden, radical, and complete." Or, in the words of Arthur Marwick, who has scant patience with overuse of the term, "a significant change in political structure carried through within a fairly short space of time."[1] Marwick's examples are the French revolutions of 1789, 1830, and 1848; the Russian one of 1917; the Mexican one of 1906. We must keep in mind that the modern meaning of revolution did develop in the political sphere. When we use it for scientific thought, we are in fact using a metaphor.

But as Cohen demonstrated so well in 1985, the word "revolution" hardly began to acquire its modern meaning until the eighteenth century; before then the term had carried the implication of a cyclical turning, a turning around, or even a turning back, a *re*turn to an original position, just as Copernicus used it in the title of his book, *On the Revolutions of the Celestial Spheres*. Cohen was unable to discover a single one of the canonical heroes of the sixteenth and seventeenth

---

B. J. T. Dobbs, who died on March 29, 1994, was Professor of History at the University of California, Davis. This lecture was delivered at the 1993 Annual Meeting of the History of Science Society.

1 I. Bernard Cohen, *Revolution in Science* (Cambridge, Mass.: Belknap Press, Harvard University Press, 1985), p. 51; and Arthur Marwick, *The Nature of History*, 3rd ed. (Chicago: Lyceum, 1989), pp. 269–70.

centuries who had appropriated the word *revolution* in its modern meaning for his own work. Only in the eighteenth century did Cohen find exemplars: Bernard Le Bovier de Fontenelle termed the invention of the calculus as a revolution in mathematics early in the eighteenth century; one W. Cockburn, M.D., in 1728 treated the work of Paracelsus as a medical revolution; Alexis Claude Clairaut in the 1750s cited Newton for a revolution in rational mechanics.² As historians know, or think they know now, it is often extremely problematic to impose later terminology upon an earlier period, a period in which the term in question had a different meaning. Modern usage arose in the context of the political sphere and carries with it the baggage of modern signification. There is a danger that we will thus impose the modern signification upon the past in ways that virtually ensure that we will misread the past.

Such retrospective judgments of the eighteenth century upon the work of the sixteenth and seventeenth centuries received a familiar progressive form in Jean Le Rond d'Alembert's "Preliminary Discourse" for the great Enlightenment encyclopedia project: Bacon, Descartes, Newton, Locke. D'Alembert used the metaphor of political revolution quite explicitly for Descartes: "He can be thought of as a leader of conspirators who, before any one else, had the courage to arise against a despotic and arbitrary power and who, in preparing a resounding revolution, laid the foundations of a more just and happier government, which he himself was not able to see established." Regretting the lack of space, d'Alembert mentioned in passing Kepler, Barrow, Galileo, Harvey, Huygens, Pascal, Malebranche, Boyle, Vesalius, Sydenham, and Boerhaave; he gave a bit more space to Leibniz, despite having little admiration for him apart from his work on the calculus.³ Except for the rather glaring omission of Copernicus, d'Alembert came close to our modern canonical list of participants in the construction of modern science during the sixteenth, seventeenth, and early eighteenth centuries. We thus see that already at mid-eighteenth century the narrative of the Scientific Revolution had taken shape in d'Alembert's work. Though it had not yet received its dramatic capitalized title, both the story line and the substance were there.

By the time d'Alembert wrote, Enlightenment optimism and belief in progress had achieved notable strength among the philosophes. Although d'Alembert mentioned the great battle between the ancients

---

2 Cohen, *Revolution in Science*, pp. 51–90.
3 Jean Le Rond d'Alembert, *Preliminary Discourse to the Encyclopedia of Diderot*, trans. Richard N. Schwab with Walter E. Rex, Library of Liberal Arts (Indianapolis: Bobbs-Merrill, 1968), pp. 74–85, on p. 80.

and the moderns, it is clear that he thought that the battle was essentially over and that the moderns had won.[4] "Progress" – in the sense of the improvement of human life on this earth – had largely supplanted millenarianism among the French philosophers, as Carl Becker so long ago explained, in a secularized version of Christian salvation history. And as Charles Paul demonstrated in 1980, the scientist was being carefully molded into a new type of cultural hero, as Fontenelle in his *éloges*, and later eulogists also, attributed the moral virtues of the idealized Stoic philosopher to recently deceased natural philosophers. Scientists came to be viewed as a superior breed of men engaged in a high moral venture, objective and selfless, dedicated to the mental and material improvement of humanity. Londa Schiebinger, with her fine sense of historical contingency, has argued that there was even a brief historical moment when women might have been included in the new elect, but negative judgments being forthcoming on that issue, the moment passed. Bringing women into the picture alerts us to the new science as an instrument of political and social power in the heavenly city of the eighteenth-century philosophers, and this point on the political and social significance of the new science is one to which I will return. Schiebinger's views must now be modified somewhat, however. Women did fully participate in the new science in some local contexts, as Paula Findlen shows in her recent study of Laura Bassi, the Minerva of Bologna, who lectured on Newtonian physics privately and at the university in the later eighteenth century.[5]

As science accumulated more and more social prestige in the later eighteenth, nineteenth, and early twentieth centuries, the image of Newton as principal cultural hero of the new science was handed on and further polished by succeeding generations of scientists and historians, the mathematical sciences taking the *Principia* as their pristine model, the experimental sciences relying on the *Opticks*. The fuller historical record was often reduced to a paragraph or two on Isaac Newton, the canonical "father of modern science." Thus Newton was seen more and more as the First Mover of modern science, the efficient cause in the Aristotelian sense, holding an unassailable primacy in

---

4 Ibid.; see esp. the discussion of the moderns in belles lettres (pp. 65–8).
5 Carl L. Becker, *The Heavenly City of the Eighteenth-Century Philosophers* (New Haven: Yale University Press, 1932); Charles B. Paul, *Science and Immortality: The Éloges of the Paris Academy of Sciences (1699–1791)* (Berkeley: University California Press, 1980); Londa Schiebinger, *The Mind Has No Sex? Women in the Origins of Modern Science* (Cambridge, Mass.: Harvard University Press, 1989), esp. pp. 214–77; and Paula Findlen, "Science as a Career in Enlightenment Italy: The Strategies of Laura Bassi," *Isis* 84 (1993): 441–69.

mathematics, enlightened rationality, and experimentation. A few romantics objected, it is true – claiming that Newton had destroyed the rainbow and the angels and that his theory of colors was all wrong; but even Einstein kept Newton's portrait on his wall, along with those of Michael Faraday and James Clerk Maxwell.[6] It is in part this massive apotheosis of Newton that has created problems with the concept of the Scientific Revolution, but of course that is not the only problem with the concept.

The concept of *the* Scientific Revolution of the sixteenth and seventeenth centuries achieved its own historiographic dominance in this century, primarily after World War II. Our own "founding father," George Sarton, was not much enamored of the concept of scientific revolutions, as Cohen observed, preferring to emphasize the cumulative character of "the inquiry concerning nature." But Martha Ornstein in 1913, Alfred North Whitehead in 1923, E. A. Burtt in 1925, John Herman Randall Jr. in 1926, Preserved Smith in 1930, and J. D. Bernal in 1939 all pressed forward with the concept of *the* Scientific Revolution as an event of major significance for the creation not only of modern science but of the modern world. I. B. Cohen has carefully examined the works of these authors in his fundamental study *Revolution in Science*, but he has gone on to demonstrate that the concept of *the* Scientific Revolution, as a major historiographic concept, was really fully developed in the late 1940s and the 1950s in works of Alexandre Koyré, Herbert Butterfield, and A. Rupert Hall.[7] To Cohen's list I would add *The Copernican Revolution*, published in 1957 by Thomas S. Kuhn. These works are all, I believe, still in print, some in revised editions, and they have probably been utilized in the training of virtually all the historians of science in this room. Their impact has been simply enormous, but, as Cohen observed, it was Kuhn's publication in 1962 of *The Structure of Scientific Revolutions* that gave the

---

6 Marjorie Hope Nicolson, *Newton Demands the Muse: Newton's Opticks and the Eighteenth-Century Poets* (1946; reprint, Princeton: Princeton University Press, 1966), pp. 1–2, 165–74; and Karl J. Fink, *Goethe's History of Science* (Cambridge: Cambridge University Press, 1991). On Newton's destruction of angels see William Blake, "Europe: A Prophecy," in *The Poetical Works of William Blake: Including the Unpublished "French Revolution" Together with the Minor Prophetic Books and Selections from "The Four Zoas," "Milton," and "Jerusalem,"* ed. John Sampson (London: Oxford University Press, Humphrey Milford, 1913), pp. 303–11, esp. p. 310, lines 145–9. On Einstein's wall decorations see Jeremy Bernstein, *Einstein*, Penguin Modern Masters (1973; reprint Harmondsworth: Penguin, 1976), p. 7.

7 Cohen, *Revolution in Science* pp. 22 (on Sarton), 389–404. "The inquiry concerning nature" is G. E. R. Lloyd's translation from the Greek in his *Early Greek Science: Thales to Aristotle*, Ancient Culture and Society (New York: Norton, 1970), unpaginated preface.

concept its widest currency, even as Kuhn broadened the concept to include other episodes in the development of science (Cohen himself has augmented the argument for many revolutions in science in considerable detail).

But however many scientific revolutions may be described, it is with the first one that I am principally concerned, for it is in the narrative of that first one that the narrative structure itself seems to demand Newton as Final Cause. If he was viewed as the First Mover or efficient cause in nineteenth- and early twentieth-century science, in our accounts of *the* Scientific Revolution we see Newton emerging as the Aristotelian Final Cause. Have you ever stopped to count the number of books that either begin or end with Isaac Newton?

No matter what one chooses to emphasize from the sixteenth and seventeenth centuries in telling the story, one must, it seems, bring the action to a dramatic climax with the work of Isaac Newton. The narrative has assumed all the characteristics of an inevitable progression; we have the sense that Newton must appear on the scene to pull the disparate strands of development into a grand synthesis. It is a teleological story we tell: Newton is the hidden end toward which the whole narrative is inexorably drawn, the Final Cause of the Scientific Revolution. Had he not really existed at that time and in that place, perhaps we would have had to invent him. Perhaps in some sense we did invent him, for the periodization and progressive ideology of the Scientific Revolution created such intractable interpretive problems associated with Newton that one may reasonably argue that our "Newton as Final Cause" is a historical construct bearing little resemblance to the historical record.

Of course, historians ask new questions of the past with every generation, every year, even every day. As François Furet has put it: "There is no such thing as 'innocent' historical explanation, and written history is itself located in history, indeed *is* history, the product of an inherently unstable relationship between the present and the past, a merging of the particular mind with the vast field of its potential topics of study in the past."[8]

"Written history is . . . the product of an inherently unstable relationship between the present and the past." Let us consider for a moment the publication of Butterfield's book, *The Origins of Modern Science*, in 1949. World War II had been over for only a few brief years, and, although Britain had been terribly battered, it had emerged on the winning side. The war itself had made the importance of science and science-based technology apparent as never before. Radar, Britain's

---

8 François Furet, quoted in Marwick, *Nature of History* pp. 389–90.

secret weapon, made it possible to survive Hitler's air attack in 1940, in the so-called Battle of Britain. From radar to the development and delivery of atomic bombs by the United States, the most advanced science of the period had generally been deployed by the winners. Small wonder, then, that the British historian Herbert Butterfield would say exuberantly in 1949 that the Scientific Revolution "outshines everything since the rise of Christianity and reduces the Renaissance and Reformation to the rank of mere episodes, mere internal displacements within the system of medieval Christendom." He claimed for his own generation an ability to see the importance of the Scientific Revolution "more clearly than the men who flourished fifty or even twenty years before us."[9] In that latter claim, Butterfield was entirely correct, because of his own precise location in time and space.

But was Butterfield's book good history? Butterfield, you will recall, was one of the persons who first mounted a serious critique of what he called the "Whig interpretation of history." That was a mode of historical interpretation common in nineteenth-century Britain that found in the past a steady progress toward liberal ideas and institutions. Butterfield pointed out that such views omitted vast sectors of human experience and in any case warped the evidence actually considered by imbuing the whole story with an aura of inevitability – it was somehow preordained that liberal ideas and institutions would prevail.[10] Yet Butterfield gave us the most Whiggish history of science imaginable. His book is pervaded with the conviction that the "winner" scientific ideas were right and good, and their triumph is made to seem inevitable.

I do not wish to belabor Butterfield's worthwhile book at too great length. It was and is a readable, nontechnical, and useful introduction to the origins of modern science. But I would like to point to a few problems with the concept of the Scientific Revolution it embodied.

One problem turns on the question of time in the use of the revolution metaphor. If a revolution is "a change that is sudden, radical, and complete," then there simply was no Copernican revolution, for by the end of the sixteenth century there were probably only about ten Copernicans in the whole world.[11] Perhaps partly to avoid that sort of

---

9 Herbert Butterfield, *The Origins of Modern Science, 1300–1800*, rev. ed. (New York: Free Press, 1957), pp. 7, 201. On science in wartime, see John M. Ziman, *The Force of Knowledge: The Scientific Dimension of Society* (Cambridge: Cambridge University Press, 1976), pp. 312–17.
10 Cf. Marwick, *Nature of History*, pp. 53–4, 56, 405; and Herbert Butterfield, *The Whig Interpretation of History* (1931; reprint, London: G. Bell, 1951).
11 Robert S. Westman, "The Copernicans and the Churches," in *God and Nature: Historical Essays on the Encounter between Christianity and Science*, ed. David C. Lind-

problem, Butterfield expanded the time frame to 500 years, half a millennium, 1300–1800. The time frame for the older narrative of the Scientific Revolution had already been expansive: from publication by Copernicus in 1543 to publication by Newton in 1687, some 144 years. Whether 144 or 500 years, the transfer of the term revolution from the political sphere to the scientific sphere has clearly detached the word from its root meaning in politics of a sudden change or one carried through in a fairly short space of time. One suspects, indeed, that we have forgotten we are using a metaphor at all.

In Butterfield's case, this expansiveness led him to develop the remarkable notion of "the postponed scientific revolution in chemistry." "It has often been a matter of surprise," he said, "that the emergence of modern chemistry should come at so late a stage in the story of scientific progress." That statement contains the hidden assumption that the forward motion of "scientific progress" is inevitable, and it also implies that modern chemistry would have come earlier but for some unfortunate obstruction in the path of "scientific progress." There are so many historiographic problems embedded in the idea of a "postponed" revolution in chemistry that one hardly knows where to start in disentangling them, but since Butterfield was primarily concerned to document the positive steps that led to modern science – that is, to us – he unhesitatingly accorded Lavoisier full honors for victory over prior "confusions" and the "strange mythical constitutions" previously invoked for various chemical substances. Lavoisier, he said, "belongs to the small group of giants who have the highest place in the story of the scientific revolution."[12]

One might suppose that I am here bedeviling a man and a concept that are not only made of straw but are also dead. But I have to tell you that the concept is not dead and that I am not so sure that it is made of straw, because we have rather recently, in 1990, been offered two perfectly fascinating essays on the chemical revolution by prominent historians of chemistry, Allen Debus and Maurice Crosland.

As Debus has pointed out, most accounts of the chemical revolution do indeed still reflect Butterfield's views, and Debus wants to correct Butterfield's problematic characterization by expanding the time period of the chemical revolution to some two and one-half centuries, from the early sixteenth century to the late eighteenth. Debus would thus relegate Lavoisier to latecomer status, whereas he would highlight the achievements of Paracelsus, van Helmont, the iatrochemists, Boyle, university professors of chemical pharmacy, Robert Fludd,

berg and Ronald L. Numbers (Berkeley: University California Press, 1986), pp. 76–113, esp. pp. 84–5; Westman's "count" here refers explicitly to astronomers.
12 Butterfield, *Origins of Modern Science*, pp. 203, 217.

Boerhaave, Becher, Stahl, and several medical doctors. Paracelsus was after all the contemporary of Copernicus, Vesalius, and the "German fathers of botany," Debus reminds us, and in certain ways he was much more radically revolutionary than they. The larger point Debus wants to make, however, is that including the medicochemical tradition would give us a much-needed corrective not only to current views of the chemical revolution but also to the general development of modern science.[13]

Crosland had been asked to speak about alchemy in the eighteenth century. Rather than offer such a short and negative paper, he said, he chose to present "the eclipse of alchemy in the 'age of enlightenment'" and discuss the chemistry that took its place. Crosland claims that alchemy was not so much attacked as ignored in the eighteenth century, a point to which I will return. He quite rightly emphasizes the epistemological and linguistic background to Lavoisier's new chemical nomenclature (which his own work has done so much to elucidate). Crosland's sketch of the chemical revolution is more traditional than that of Allen Debus, but he does implicitly challenge Butterfield's "postponed" revolution in chemistry by making eighteenth-century pneumatic chemistry a necessary precondition for Lavoisier's work. "Had [Lavoisier] been born a generation earlier he could not have propounded the oxygen theory," Crosland observes, nor could Dalton have done his work without Lavoisier. Crosland almost utilizes an evolutionary rather than a revolutionary model, even while retaining the phrase *chemical revolution*. But nevertheless he insists that nineteenth- and twentieth-century chemistry is the chemistry of Lavoisier and Dalton and not that of Paracelsus.[14]

Crosland and Debus, of course, are ever so much more sophisticated than Butterfield, even as they retain the terminology of revolution. So were Koyré and Kuhn, in the other most influential books from the 1950s and early 1960s, probing, as they both did, below the surface to reconstitute a broader context for scientific thought. But nonetheless they did think in terms of revolutions, and they have had many followers, including myself. Indeed probably all of us here absorbed elements of the idea that revolutions take place in science, and we use

---

13 Allen G. Debus, "Iatrochemistry and the Chemical Revolution" in *Alchemy Revisited: Proceedings of an International Congress at the University of Groningen, 17–19 April 1989*, ed. Z. R. W. M. von Martels, Collection de Travaux de l'Académie Internationale d'Histoire des Sciences 33 (Leiden: Brill, 1990), pp. 51–66.
14 Maurice Crosland, "The Chemical Revolution of the Eighteenth Century and the Eclipse of Alchemy in the 'Age of Enlightenment,'" in von Martels, *Alchemy Revisited*, pp. 67–77, on p. 71.

the idea for periodization as well as in reconstructing the origins of the thought patterns of modernity. The concept of *the* Scientific Revolution of the sixteenth and seventeenth centuries and the notion of an entire sequence of revolutions in science since that time have been enormously useful to us in creating the discipline of the history of science. But having said that, I must also say that I think perhaps those ideas are losing some of their utility. Recent trends offer many promising alternatives: local studies of natural philosophy in court, civic, and university cultures; studies like Steven Shapin and Simon Schaffer's *Leviathan and the Air-Pump* and Margaret Jacob's *The Cultural Meaning of the Scientific Revolution*, where the social and political significance and use of ideas are explicated; reconstruction of the contemporary context for the work of Bacon, Kepler, and Boyle.[15] But we are still encumbered with some of the baggage of the metaphor of revolution that obscures so much continuity in the midst of change and produces such improbable interpretations of historical actors, for in many ways we are still most intent upon explicating the changes that led to us. Here we are in the late twentieth century, still privileging the fragments of the past that we recognize as belonging to our own present. In some ways we have continued to act like nineteenth-century Whig historians.

It increasingly seems to me that these problems are not new ones. In the seventeenth century many participants in the new natural philosophies railed against Aristotle and against university intellectual culture as arid and barren – like thistles, Francis Bacon said. For three centuries we have accepted and repeated such maxims as fact rather than as part of a political struggle. Only recently, with the work of Charles Schmitt and Mordechai Feingold, for example, can we begin to appreciate the rich variety of university culture, where Aristotle was still at the forefront of the prescribed curriculum but where Bacon, Descartes, Locke, and Newton himself were appropriated and studied almost as soon as they published.[16] The universities reemerge

---

15 Steven Shapin and Simon Schaffer, *Leviathan and the Air-Pump: Hobbes, Boyle, and the Experimental Life* (Princeton: Princeton University Press, 1985); and Margaret C. Jacob, *The Cultural Meaning of the Scientific Revolution*, New Perspectives on European History (New York: Knopf, 1988).

16 Francis Bacon, *Novum organum*, bk. 1, aphorism 73, in *"Advancement of Learning" and "Novum organum,"* introd. James Edwin Creighton, rev. ed. (New York: Colonial Press, 1899), pp. 334–5; Charles B. Schmitt, *John Case and Aristotelianism in Renaissance England*, McGill-Queen's Studies in the History of Ideas, 5 (Montreal: McGill-Queen's University Press, 1983); Schmitt, *The Aristotelian Tradition and Renaissance Universities* (London: Variorum, 1984); Mordechai Feingold, *The Mathematicians' Apprenticeship: Science, Universities, and Society in England, 1560–1640* (Cambridge: Cambridge

from this new work as nurturing cradles of intellectual life, and it is now possible to reevaluate the major role of Aristotelian thought patterns – and other supposedly outmoded thought patterns – in the new ways of thinking. Similarly, we have carelessly accepted what now begins to look like Protestant propaganda. After Galileo's little problem with the Church of Rome, much was made of the dismal state of natural philosophy in Catholic countries; we only now begin to learn of complex and rigorous Jesuit scientific thought.[17]

But to my mind the issue of the proper interpretation of our scientific heroes has been the most pressing problem of all, a problem that was at least in part generated by the concept of the Scientific Revolution. I think the problem arises somewhat in this fashion: we choose for praise the thinkers that seem to us to have contributed to modernity, but we unconsciously assume that their thought patterns were fundamentally just like ours. Then we look at them a little more closely and discover to our astonishment that our intellectual ancestors are not like us at all: they do not see the full implications of their own work; they refuse to believe things that are now so obviously true; they have metaphysical and religious commitments that they should have known were unnecessary for a study of nature; horror of horrors, they take seriously such misbegotten ideas as astrology, alchemy, magic, the music of the spheres, divine providence, and salvation history. We become most uncomfortable and begin to talk about Copernicus as "conservative" or "timid," terms that hardly fit the commonsense concept of a revolutionary. Or we talk about Kepler as a "tortured mystic" or a "sleepwalker" or a "split personality."[18]

Already in the eighteenth century d'Alembert had trouble with his hero Francis Bacon. After calling the lord chancellor "immortal" and praising him to the skies for his bold new opinions and his attacks on

---

University Press, 1984); and Feingold, "The Oxford Curriculum in Seventeenth-Century Oxford," in *The History of the University of Oxford*, vol. 4, ed. Nicholas Tyacke (Oxford: Oxford University Press, 1997).

17 The interest in Jesuit science has accelerated in recent years; the latest effort on that topic is a collection of essays edited by Mordechai Feingold, *The Jesuits and the Scientific Revolution* (Princeton: Princeton University Press, 1997).

18 See Thomas S. Kuhn, *The Copernican Revolution: Planetary Astronomy in the Development of Western Thought* (Cambridge, Mass.: Harvard University Press, 1957), p. 294, where the index offers twelve page references for the "conservatism" of Copernicus. See also Alan G. R. Smith, *Science and Society in the Sixteenth and Seventeenth Centuries* (New York: Science History Publications, 1972), p. 91; I. Bernard Cohen, *The Birth of a New Physics*, rev. ed. (New York: Norton, 1985), p. 132; and Arthur Koestler, *The Sleepwalkers* (London: Hutchinson, 1959).

the scholastics (with which d'Alembert agreed, of course), d'Alembert waffled more than a little:

> If we did not know with what discretion, and with what superstition almost, one ought to judge a genius so sublime, we might even dare reproach Chancellor Bacon for having perhaps been too timid. He asserted that the scholastics had enervated science by their petty questions, and that the mind ought to sacrifice the study of general things for that of individual objects; nonetheless, he seems to have shown a little too much caution or deference to the dominant taste of his century in his frequent use of the terms of the scholastics, sometimes even of scholastic principles, and in the use of divisions and subdivisions fashionable in his time. After having burst so many irons, this great man was still held by certain chains which he could not, or dared not, break.[19]

D'Alembert "knew" – or thought he knew – in 1751 that scholasticism was all wrong. He knew that, or thought he knew that, partly on the word of Francis Bacon, lord chancellor of England, who had died some 125 years before. It seems never to have occurred to d'Alembert that what he himself perceived as rational and proper was not eternal, or that *his* rationality would have been less than obvious to previous generations. His hero was "still held by certain chains which he could not, or dared not, break."

Twentieth-century historians have produced equivalent wafflings about others in the canonical list. How could Copernicus have taken just that one small step of transposing Sun and Earth and not have gone further, retaining as he did the primacy of circular motion *and* the closed cosmos *and* Aristotelian physics? How could John Dee – at the forefront of scientific navigation in the sixteenth century – possibly have supposed he could communicate with angels? How could Kepler possibly have maintained his 1595 vision of the Platonic solids as God's architectural plan for the solar system even after he announced the three laws of planetary motion that were subsequently incorporated into *our* science? How could Galileo have ignored Kepler's ellipses even after he proclaimed the Book of Nature to be written in mathematical language? How could Descartes have persisted in believing deductive reasoning to be the way to true knowledge when all about him the experimental method was being pushed forward? How could Newton retain his belief in a Sun-centered cosmos after he himself had described an infinite universe? One cannot imagine that,

---

19 D'Alembert, *Preliminary Discourse*, p. 76.

mathematician as he was, he did not know that no center was possible for an infinite space.

But above all, how could Newton, the epitome of austere scientific, mathematical rationality, have pursued alchemy as he did, through exhaustive analysis of alchemical texts and through extensive experimentation? As you now realize, from the introduction this evening, I was assigned the topic of Newton's alchemy and chemistry by my illustrious mentors Michael McVaugh and Seymour Mauskopf for the great UNC-Duke Newton seminar of spring 1968, and, as I like to tell people, I have been working on that seminar paper ever since. The problem inexorably expanded into a dissertation, two books, and assorted articles, and it led me far afield from traditional history of science.

The problem, as I slowly – very slowly – came to perceive it, was indeed a historiographic one. During the same period that Newton came to be acclaimed as the "father of modern science," alchemy suffered a radical decline, or an eclipse, as Crosland has called it. But it was more than an eclipse or a decline, for alchemy was, by the nineteenth century, characterized as "delusion and superstition." I would like to quote here the words of Thomas Thomson in the introduction to his *History of Chemistry*, published in 1830. The year 1830 is exactly 103 years after Newton's death in 1727. Here are Thomson's words:

> Chemistry, unlike the other sciences, sprang originally from delusion and superstition, and was at its commencement exactly on a level with magic and astrology. Even after it began to be useful to man, by furnishing him with better and more powerful medicines than the ancient physicians were acquainted with, it was long before it could shake off the trammels of alchymy, which hung upon it like a nightmare, cramping and blunting all its energies, and exposing it to the scorn and contempt of the enlightened part of mankind. It was not till about the middle of the eighteenth century that it was able to free itself from these delusions, and to venture abroad in all the native dignity of a useful science.[20]

"The scorn and contempt of the enlightened part of mankind" – to that alchemy had been subjected, in Thomson's view, but rightly so, for alchemy was "delusion and superstition."

Consider, then, the consternation of Sir David Brewster, astronomer royal and the first major biographer of Newton, when he obtained

---

20 Thomas Thomson, *The History of Chemistry*, 2 vols. (London: Henry Colburn and Richard Bentley, 1830), 1:1.

access to Newton's private papers later in the nineteenth century. Brewster "knew" that Newton was correctly placed in "the enlightened part of mankind" – Newtonian celestial physics was Brewster's bread and butter – but Thomson's evaluation of alchemy echoed in his ears. With what horror, then, did he see Newton's alchemical papers: "[We] cannot understand how a mind of such power, and so nobly occupied with the abstractions of geometry, and the study of the material world, could stoop to be even the copyist of the most contemptible alchemical poetry, and the annotator of a work, the obvious production of a fool and a knave."[21] Brewster was shocked, and understandably so. Two totally divergent historical developments had created a chasm he could not span – the apotheosis of Newton as austere mathematician and experimenter who had set the better part of humanity on the right track, and the degradation of alchemy to "delusion and superstition."

Brewster's obvious first mistake was to assume that Newton's whole purpose was to study the material world, which was what he, Brewster, did, as a nineteenth-century scientist. His second mistake was to accept without hesitation the nineteenth-century evaluation of alchemy as delusion and superstition. But I do not know how he could have done otherwise, given those two very powerful historical trends plus several others of which he remained unaware. We in the twentieth century, with much, much more information to bring to bear upon the problem, can surely do better. I hope I have done so, but I regretfully take note that in Rupert Hall's new biography of Newton, the old problems remain.[22]

Hall argues that Newton wanted only "facts" from his alchemical and chemical experiments. Perhaps that is, in a sense, correct, but the word "fact" offers us another problem in historical semantics. Colleagues who know the languages tell me that the ancient Greeks had no word for fact, though in the seventeenth century the English word fact was acquiring its modern usage of socially validated knowledge of the natural world.[23] Yet alchemy for Newton had more than one ontological level: he supposed he could get from it not just knowledge

---

21 David Brewster, *Memoirs of the Life, Writings, and Discoveries of Sir Isaac Newton*, 2 vols. (Edinburgh: Thomas Constable; Boston: Little, Brown, 1855), 2: 374–5.
22 Rupert Hall, *Isaac Newton: Adventurer in Thought*, Blackwell Science Biographies (Oxford: Blackwell, 1992), p. 200.
23 G. Debrock, "Aristotle Wittgenstein, *alias* Isaac Newton between Fact and Substance," in *Newton's Scientific and Philosophical Legacy*, ed. P. B. Scheurer and G. Debrock, International Archives of the History of Ideas, 123 (Dordrecht: Kluwer, 1988), pp. 355–77; Barbara Shapiro, " 'Facts' in Early Modern Science and Law," History of Science Society Annual Meeting, 1993, Santa Fe, New Mexico; and Shapiro, "The Legal Origins of the Scientific 'Fact' " (unpublished manuscript).

of the natural world but knowledge of the supernatural world as well. If Newton was looking for "facts" in his alchemical work, I think the sort of facts he hoped to glean from alchemy have very little resonance with the Newton who is perceived only as a student of the natural world. What Newton hoped to gain from alchemy was a precise knowledge of the operations of the Deity in organizing and vivifying the inert particles of matter in the microcosm.[24]

Alchemy never was, and never was intended to be, a study of matter for its own sake or a study of the natural world for its own sake. Alchemists sought perfection or the knowledge thereof, not utility for this life, as chemistry later came to do. They sought the philosophers' stone, an agent of perfection for both nature and humanity – as they said, medicine for both men and metals. The alchemists' philosophers' stone was closely allied with the Christ of Christianity – both were, after all, agents of perfection and redemption. Newton saw an even broader significance: his Arian Christ, God's agent in the creation and governance of the created world, was closely allied with the philosophers' stone, the active principle of alchemy. If he, Newton, could but demonstrate the laws of divine activity in nature, the Christ operating in and governing the microcosm, then he could demonstrate in an irrefutable fashion the existence and providential care of the Deity – a grand goal, though hardly a modern one. Newton indeed hoped to restore the original true religion – to effect a "revolution," a return to a former state, in the earlier meaning of the word revolution.[25]

If Newton had succeeded in this aim, and had thereby stemmed the tides of mechanism, materialism, deism, and atheism, as he had hoped, we would of course live in a different world. But I do not much like to play the game of counterfactual history. Perhaps the world would have been better, perhaps not. But of course the tides of mechanism, materialism, deism, and atheism were already quite strong when Newton lived. He was well instructed in the dangers of those tendencies by the Cambridge Platonists, and he fought a valiant holding action against them, but he did not win. His system was very quickly co-opted by the very -isms he fought, and adjusted to suit them. He came down to us as co-opted, an Enlightenment figure without parallel who could not possibly have been concerned with alchemy or with establishing the existence and activity of a providential God. He did not win in the end, as Koyré knew already in 1957: "The mighty, energetic God of Newton who actually 'ran' the universe

24 B. J. T. Dobbs, *The Janus Faces of Genius: The Role of Alchemy in Newton's Thought* (Cambridge: Cambridge University Press, 1992).
25 Ibid.

... became, in quick succession, a conservative power, an *intelligentia supra-mundana*, a 'Dieu fainéant.' "[26] But surely we no longer need play down and make despicable Newton's valiant effort.

In conclusion, I would like to suggest that there may be some historical value in evaluating Newton in a different way: not as one of history's all-time winners, not as the First Mover of modern science, not as the Final Cause of *the* Scientific Revolution, but as one of history's great losers, a loser in a titanic battle between the forces of religion and the forces of irreligion. Perhaps he is no less a hero from that perspective, but a hero of a different sort, rather more like Roland at Ronceval, crushed by overwhelming odds in the rear guard, than like a peerless leader of the vanguard.[27] To reevaluate Newton in that way might also allow us to engage in some fresh and creative rethinking of the many changes taking place in the sixteenth and seventeenth centuries.

26 Alexandre Koyré, *From the Closed World to the Infinite Universe* (New York: Harper, 1958), p. 276.
27 *The Song of Roland*, trans. Frederick Bliss Luquiens (New York: Collier; London: Collier-Macmillan, 1952).

3

The Scientific Revolution Reasserted

RICHARD S. WESTFALL

I find myself in an ambiguous pose that I need to explain before I continue. I did more than count Betty Jo Dobbs as my friend. Rightly or wrongly, I regarded myself as her earliest admirer in the discipline – after her graduate professors, to be sure. We got into correspondence, first about Newton's handwriting, and then about his alchemical papers, while she was still a graduate student, and as a consequence I read her dissertation chapter by chapter as she completed each one. I wrote enthusiastically to Northwestern in support of her first appointment, and I continued to write whenever she needed a reference. I feel her loss both personally and professionally. For all that, I am going to devote this chapter, written in her memory, to taking issue with the last piece she published, her History of Science Society Lecture, "Newton as Final Cause and First Mover." Anyone who knew Professor Dobbs understood that you did not take a high tone with her. Neither do I here, even when there is no possibility of a retort. You also understood that she took the intellectual life seriously and thought that important issues required thorough discussion. I can do no better service to her memory than to take the issues she raised seriously and, in a low voice, to contribute what I can to their elucidation. The issues concern the Scientific Revolution, and I intend to make that topic, rather than details of Dobbs's essay, my focus.

In the introductory paragraph of her lecture, Dobbs announces her intention to undermine the concept of the Scientific Revolution. In contrast, I intend to defend it, and since, in undermining, she objects to capitalizing the words, I will register my sense of its importance by the use of capitals.

Let me begin with an issue that Dobbs makes prominent, Whig historiography. Is it Whiggish to assert, as Herbert Butterfield did, that the Scientific Revolution "outshines everything since the rise of Christianity and reduces the Renaissance and Reformation to the rank of mere episodes, mere internal displacements within the system of medieval Christendom"? I confess that I find the issue of Whig historiography perplexing. Consider another passage from Butterfield's book. It is difficult, he said, to arrive at a clear understanding of alchemy because "the historians who specialise in this field seem sometimes to be under the wrath of God themselves; for, like those who write on the Bacon-Shakespeare controversy or on Spanish politics, they seem to become tinctured with the kind of lunacy they set out to describe." Insofar as I understand what "Whiggish" means, I would apply it to this statement, and although I suspect I have made statements more provocative myself, I nevertheless think historians should avoid them. Butterfield's assertion about the Scientific Revolution appears to me to fall into a quite different category. Merely describing the past in its own terms does not constitute the historian's function in my notion of it. We are not antiquarians. We are called to help the present understand itself by understanding how it came to be. We strive to find a meaningful order in the multifarious events of the past and thus, explicitly or implicitly, we pass judgments on the relative importance of events. Before we dismiss Butterfield's statement on the Scientific Revolution by attaching the pejorative adjective "Whiggish" to it, let us pause to consider whether it may not be correct.

Recall the world about us. To me it appears that the existence of modern science is the precondition for most of the central features of our society. I think of such things as means of communication, from the mass media that bring the world to our homes each morning to individual devices such as the telephone and e-mail, which together have so expanded our lives in comparison with those of the people I know from the seventeenth century. Ease of transportation enabled scholars from all over the country and beyond to gather in New Mexico to hear Dobbs's lecture; we have incorporated the various dimensions of ready transportation into our lives to the extent that we have forgotten it was not always there. The level of material plenty has lifted the burden of poverty from the great majority. Modern medicine has more than doubled the average life span and driven pain and disease, once familiar members of every circle, to the margins of our existence. These features of our life are not evenly spread around the globe. In general, they prevail where modern science flourishes and are in shorter supply elsewhere.

Most people think of these characteristics as benefits. Almost no one considers other features of our world that are also derivative from science as benefits, though they are no less central. Scientifically based technology has accelerated the consumption of nonrenewable resources until we stand already face to face with their exhaustion. It has produced products that nature cannot degrade, so that we are well on the way to choking on our own refuse. It has conjured up weapons of mass destruction more hideous than earlier ages were able even to imagine. I do not think that I have compiled a partisan list. Every item on it appears incontrovertably true, and I am convinced that I could go on indefinitely listing similar ways in which science impinges, both positively and negatively, on our lives until I had more than satisfied everyone who finds my list wanting in some respect. The longer the list becomes, the less willing I am to dismiss Butterfield's judgment by tarring it with a pejorative adjective.

I note one other feature of the modern world that strikes us only when we stop to note its absence. When I open the morning newspaper, almost every story either involves science directly or involves factors that depend on science. Very little has to do with religion. On Saturdays, it is true, there is a spate of articles to keep the church-related population subscribing, and when the pope comes to visit, he does command the headlines. In America we hear about the religious right, but what we hear has to do with politics more than religion. That is close to the limit of the presence of religion in the news. It is worth recalling that the means by which the pope travels and the means which keep us informed about him are both direct products of modern science. Before the Scientific Revolution, theology was queen of all the sciences. As a result of the Scientific Revolution, we have redefined the word "science," and today other disciplines, which once took their lead from Christian doctrine, strive to expand their self-esteem by appropriating the word in its new meaning to themselves. Theology is not even allowed on the premises anymore. Here is the very heart of Butterfield's specific statement. A once Christian culture has become a scientific one. The focus of the change, the hinge on which it turned, was the Scientific Revolution of the sixteenth and seventeenth centuries. As you may gather, I am convinced that there has been no more fundamental change in the history of European civilization. Dispense with the concept of the Scientific Revolution? (The capital letters seem ever more important to me.) How can we dream of it?

I move on to the related issue of Scientific Revolution as metaphor. It is virtually impossible to discuss any complex issue without using metaphors. Take away metaphors, and you leave us mute. The issue as I see it is not whether the concept of the Scientific Revolution is a metaphor. Of course it is. The issue is whether it is a good metaphor that enhances understanding or a bad one that misleads. Let us take the two definitions of revolution that Dobbs offers. First, I. B. Cohen's definition of a political revolution: "a change that is sudden, radical, and complete"; and, second, that of Arthur Marwick: "a significant change in political structure carried through within a fairly short space of time." Obviously, in using "revolution" as a metaphor, we dispense with the political content. The two definitions are similar. Both stress a brief period of time and a thorough change. In my understanding of it, the concept of the Scientific Revolution rests primarily on the second of this pair. Those who framed it sought consciously to break with an earlier tradition that treated the history of science as a series of discrete discoveries. The authors of the concept compared the science that prevailed before the Scientific Revolution with that which prevailed after and found them disparate. It was not that scientists had learned that the sun is at the center of our system, that bodies fall with uniformly accelerated motion, that the product of the pressure and volume of a confined gas is constant. No; science had become a different enterprise, proceeding in a different direction down a different road. From this point of view, it is irrelevant to demonstrate, for example, that university Aristotelianism was not the dull exercise in futility that Bacon and Galileo pictured, or that Jesuit science was active during the seventeenth century. In no way does acceptance of the Scientific Revolution commit us to denigrating those activities or the scholarship that treats them. Many worthy activities went on during the seventeenth century. We omit them from this discussion if they do not bear on the Scientific Revolution. Whether academic Aristotelianism was brimming over with intellectual vigor or not, whether Jesuit science flourished similarly or not, the question is whether the enterprise of science as it was carried out after 1687 was radically different from that before 1543. Clearly I think that it was and that the transformation was a once and for all event that has never been reversed. Scientists of today can read and recognize works done after 1687. It takes a historian to comprehend those written before 1543.

To capture this discontinuity, those who framed the concept seized the metaphor of revolution. Go back and read Abraham Wolf's *History of Science, Technology, and Philosophy in the 16th and 17th Centuries* and William Dampier's *History of Science*, both substantial works much in use when I began to edge into the discipline, and decide whether the metaphor has enhanced understanding. We should not, of course, forever stand pat on a position defined in the 1940s and 1950s, but surrendering a valuable insight is not my idea of moving forward. Scholarship is not like art. We do not skip from style to style at random. Scholarship is cumulative. We expand understanding by incorporating what earlier scholars have established. In my view, it will not do to say, as I have heard it said, that the concept of the Scientific Revolution has ceased to be useful. If the concept ever had validity, if the scientific enterprise made a radical change of direction in the sixteenth and seventeenth centuries, the concept continues to have validity. Though we should not remain forever where historians of science arrived in 1950, neither should we wantonly abandon what they accomplished.

I think it is obvious that the period of time involved was not considered sufficiently important to invalidate the metaphor. There is more to be said on this issue, however. Although I do not recall arguments to this effect in Koyré, Burtt, et al., it is important, in regard to time, not to forget how relative the lengths of periods are to different areas of experience. If we accept Cohen's word "sudden," we should recall that political transformations are not like military ambushes, which again differ from explosions. The French Revolution did not occur in an instant. It did not occur in five minutes, or in a week, or in a year. Historians usually assign roughly ten years to it, and one might, if he chose, quarrel with the idea that anything taking ten years was sudden. That decade, in relation to the life-span of the French absolute monarchy, is not far different from the century and a half of the Scientific Revolution in relation to the life-span of Aristotelian natural philosophy. The word "revolution," applied to an intellectual transformation of this magnitude, does not make me particularly uncomfortable because it required a century and a half to run its course.

To sustain my argument I must sketch in, however briefly, the elements of discontinuity. In its most general terms, the Scientific Revolution was the replacement of Aristotelian natural philosophy, which aside from its earlier career had completely dominated thought about nature in western Europe during the previous four centuries. As soon as we put it this way, two further points present themselves. First, the concept of the Scientific Revolution does not involve disparagement of medieval science. Quite the contrary. There would have been no point in talking about revolution against ignorance or stupidity. Those who formulated the concept of the Scientific Revolution understood Aristotelianism as a serious system of natural philosophy and always treated it that way. They regarded its replacement, not as trivial recognition of the obvious, but as an intellectual labor of enormous dimensions. It is no accident that expanding appreciation of the accomplishments of medieval science has gone hand in hand with elaboration of the concept of the Scientific Revolution. It follows, second, that insisting on discontinuity does not entail denial of any and all continuity. To me, at least, it appears impossible to imagine the

Scientific Revolution without the background of rigorous thought about nature among the medieval Scholastics. The discontinuities were on another plane.

First, there was heliocentrism. It is almost universal to take Copernicus's *De revolutionibus* as the beginning of the Scientific Revolution, and the reordering of the heavens remained fundamental. As we all know, heliocentric systems had been proposed before. Only with the Scientific Revolution was the idea fully worked out to the extent that it could be taken seriously. If it is true, as Dobbs mentions, that there were no more than ten committed heliocentrists in 1600, it is equally true that by 1650 there was almost no one who counted and was free to follow his own intellect who had not accepted the new system. Already by then astronomers were having trouble imagining how anyone had ever thought otherwise. In 1995 it is impossible to imagine that this change is not final.

Heliocentrism was important on at least four levels. There was the picture of the universe itself, repudiating the accepted picture and recognizing the earth on which we live as a planet, one among six, circling a central sun. Second, with Kepler, heliocentrism marked the beginning of the mechanization of nature. Astronomy ceased to be a kinematic endeavor in which circles were added to circles to approximate observed planetary positions; Kepler treated celestial bodies as common matter moved by natural forces. Heliocentrism also rejected the anthropocentrism of medieval natural philosophy. Fourth, the new system marked the first step in that challenge to the commonsensical understanding of nature embedded in Aristotelian natural philosophy, which was one of the characteristics of the Scientific Revolution. The universe is not as it appears to be. At every level the new astronomy was a break, by no means trivial, with the prevailing system.

Closely allied to the new astronomy was the transformation of mechanics. Aristotle's conception of motion was no less central to his system than geocentrism. There was no way of dispensing with it and leaving the system intact, but if it were correct, it was quite impossible that the earth be in motion. Many of the common phenomena of motion could not occur as we do in fact observe them to occur. The early years of the seventeenth century witnessed a reconceptualization of motion, associated primarily with Galileo. The meaning of the word in the context of mechanics was radically restricted. Boys being educated ceased to count as examples of motion. Motion is simply the transfer of a body from the vicinity of one set of other bodies to the vicinity of another set. It is a state in which a body finds itself, and like other states it perseveres without the continuing action of a cause. Uniform motion in a straight line becomes indistinguishable from rest, perhaps even a synonym for rest. Only changes in the state of motion require a cause. Along with its rejection of an essential element in received Aristotelian philosophy, here was also another assault on common sense, for motions we observe do not appear to persevere in this manner. We know the outcome of a prolonged wrestling with the concept, which was not completed in a day, as the principle of inertia, and philosophers of science today consider it to be the central concept in the structure of modern science.

Galileo did not limit himself to the concept of motion. He proceeded on to define what had never existed before, a mathematical kinematics of terrestrial motions – of uniform motion on horizontal planes, of uniformly accelerated motion in vertical planes, of the combination of the two in projectile motion. The mathematization of nature, the search for geometrical simplicity and harmony, had formed an important strand in the new astronomy. It found confirmation now in the new mechanics. Together, the two set the mathematical character of modern science, one of its most striking contrasts with Aristotelian science. It is significant that Kepler and Galileo proposed the first of the mathematical laws of nature. The concept of natural law had long been around; a redefinition of what we mean by it was one further aspect of the Scientific Revolution.

For those who accepted the new astronomy and the new mechanics, Aristotelian natural philosophy had become untenable. If the transformation of scientific thought were to proceed, a new natural philosophy had become a necessity. What filled this need came to be called the mechanical philosophy. In its very formulation it embraced the new science. Every one of the different versions of mechanical philosophy accepted heliocentrism. In all its versions, it was unthinkable without the principle of inertia. The mechanical philosophy universalized the rejection of common sense. It is not just the heavens that are not as they seem to be. It is not just motion that is not as it seems to be. Nothing is as it seems to be. In the sixth and final day of his *Meditations*, Descartes stated the principle in what appears to me as the fundamental philosophic insight of the century. Having earlier, in his process of systematic doubt, temporarily cast away the existence of the external world, he now restored it to being, demonstrating to his own satisfaction that it must exist. But, he added, there is no corresponding necessity that it be in any way similar to the world our senses depict. We see about us a world filled with life and qualities. They are only illusions. There is nothing out there but particles of matter in motion. Some of the particles can join together in structures that produce the phenomena we call organic life. Particles can impinge on the nerve endings of sentient beings, causing sensations that philosophers have mistakenly projected onto the external world. They are only illusions, however. The world could hardly be more different from the way it appears.

Nevertheless, for all its agreement with the new science on one plane, the mechanical philosophy was in tension with it on another. It never hesitated to offer verbal pictures of particles of matter in motion producing the phenomena of nature. It was unable to deduce the phenomena in precise mathematical detail. Its vortices did not yield Kepler's three laws. Its explanations of heaviness did not yield either uniformly accelerated motion or the common acceleration of all bodies. Here was the role that Newton played in the Scientific Revolution. He resolved this tension by revising the mechanical philosophy. In addition to particles of matter in motion, there are forces between the particles, forces mathematically defined that do yield, in precise mathematical detail, those results that the verbal pictures of earlier mechanical philosophy could not.

To pick up on another theme in Dobbs's chapter, there was of course nothing inevitable about Newton's work. Like the rest of the Scientific Revolution, it was a free creation of the human spirit, meditating on the evidence that nature presents and forcing itself to conform to that evidence. When we look back from the distance of three centuries, however, we see that Newton was in fact familiar with the work of Kepler, Galileo, Descartes, and others and that he did in fact weave together themes of the Scientific Revolution, which perhaps as a consequence now appear to us as its principal themes, into what would be an enduring pattern. Possibly the same themes, or a different selection of themes, could have been woven together by another scientist into a different fabric. I confess that I have trouble imagining this, but perhaps only because it did not happen. However, what did happen had an internal logic, because Newton consciously built on the work of earlier men. Internal logic is not inevitability.

In any event, with Newton the new science and the new philosophy of nature found their definitive form in which they shaped the scientific tradition in the West for the coming two centuries. The very nature of the new enterprise that the Scientific Revolution inaugurated insured that the Newtonian system would be modified, as it has been, extensively, especially in our century. As long as we remember those modifications, it does appear to me that the system, in a broad sense that emphasizes its fundamental features in their contrast with earlier ones, continues to reign, and I see no prospect whatever that the reign will terminate.

Method was also an aspect of the Scientific Revolution, which increasingly built itself on experimental procedure. On this subject we must not overstate discontinuity, for experiment as such was not new with the seventeenth century. We do have to look carefully, however, to find experiments before the seventeenth century. Experiment had not yet been considered the distinctive procedure of natural philosophy; by the end of the century, it was so recognized. Among prominent examples, I think of the investigation, international in character and spread over roughly a quarter of a century, that took its beginning from Galileo's discussion of the observation that water cannot be syphoned over a height of more than about thirty feet. The investigation, experimental all the way and experimental in an increasingly sophisticated manner, passed by way of Torricelli in Italy and Pascal in France to Boyle in England, where it eventuated in another of the early mathematical laws of nature. In the science of optics, Newton's investigation of the heterogeneity of light and the production of colored phenomena assumed from the beginning that experimentation was its only possible mode. It too pursued a mathematical law, and although it did not find one, it was consistently quantitative. There had never before been an investigation like either of these or many others; in the future there would be little, and increasingly less, in science that was unlike them.

The elaboration and expansion of the set of available instruments was closely allied to experimentation. I have been collecting information on the scientists from this period that appear in the *Dictionary of Scientific Biography*, 631 in all. One hundred fifty-six of them, only a small decimal short of one-quarter, either made instruments or developed new ones. They are spread over every field of investigation. One hundred fifty-six may be a misleading number since it includes not only new instruments useful in scientific investigation, such as the telescope, barometer, and air pump, but instruments of practical use, in surveying and surgery, for example. Nevertheless, there had never been a similar proliferation of instruments before; there would never be a time without it thereafter.

At least one more dimension of the Scientific Revolution demands notice – a new relation between science and Christianity. Late in the sixteenth century, Galileo compiled a manuscript on natural philosophy that offers a base for comparison. As a young professor at the University of Pisa, Galileo apparently set out to improve his grasp of natural philosophy by studying lecture notes from the Jesuit professors at the Collegio Romano. What he set down summarized aspects of the long tradition of Scholastic philosophy, and what interests me here is the still unquestioned conviction, clearly if implicitly expressed in his manuscript, that nothing in contradiction to Christian doctrine could be admitted in natural philosophy. Compare that state of affairs with what obtained a century later. There is no suggestion here that we ignore the religious proclivities of scientists from the later period, especially English scientists, among whom Newton traveled. The net result, however, of their many treatises demonstrating the existence of God from natural phenomena appears to me to have been the separation of science from revealed theology. From the point of view of science, it does not seem excessive to speak of its liberation. Centuries before, as European civilization had taken form out of the chaos of the dark ages, Christianity had fostered, molded, and hence dominated every cultural and intellectual activity. By the end of the seventeenth century, science had asserted its autonomy.

Compare two incidents. Early in the seventeenth century, the Catholic Church, under the leadership in this respect of Cardinal Bellarmino, condemned Copernican astronomy because it conflicted with the overt meaning of certain passages of Scripture. Sixty-five years later Newton engaged in a correspondence with Thomas Burnet about Burnet's *Sacred Theory of the Earth*. Burnet had convinced himself that the Scriptural account of the creation was a fiction, composed by Moses for political purposes, which could not possibly be true in a philosophical sense. In the correspondence, Newton defended the truth of Genesis, arguing that it stated what science (chemistry in this case) would lead us to expect. Where Bellarmino had employed Scripture to judge a scientific opinion, both Burnet and Newton used science to judge the validity of Scripture. To speak merely of the autonomy of science does not seem enough; we need to speak rather of its authority, to which theology had now become subordinate. The positions of the two had been reversed. That change also has never been reversed anew.

The whole meaning of the Enlightenment of the eighteenth century was the authority of science over the intellectual life of Europe. Its authority was spreading also beyond the intellectual into the material realm as it demonstrated its capacity to transform technology. As far as I am aware, no one has seriously addressed the question of what it was about the new science that made it adaptable to technological use. Its quantitative character probably played a role. Its experimental method appears to have supplanted blind trial and error. Perhaps – dare one offend sensibilities by expressing such a hazardous opinion? – it even offered a more satisfactory account of nature. In any event, never before, not only in Europe but anywhere on the earth, had natural philosophy provided a grip on technology. Everyone agrees that increasingly European technology acquired a scientific foundation. Some of us even suspect that the immense disparity that has come to separate the Western world from the rest of the globe stems directly from the technological command that modern science has delivered into the hands of the civilization that brought it into being. Dispense with the concept of the Scientific Revolution? I am convinced that it is the key not only to the history of science but to modern history as well.

I cannot refrain from remarking, on the basis of Dobb's chapter, that the concept appears to offer insights not so easily shucked off, so that even her challenge to it bears the mark of its continuing influence. In her penultimate paragraph, she speaks of "the tides of mechanism, materialism, deism, and atheism" against which Newton fought a valiant but vain holding action. We might note in passing how readily the language of inevitability flows even from a reluctant pen when one discusses these matters. I think that Dobbs would agree with me that nothing contributed as much as the new science to raising the tides of mechanism, materialism, deism, and atheism. They bear witness to the new relation between science and Christianity that I see as one dimension of the new order that the Scientific Revolution ushered into being. They are one more reason why I will not part with the concept.

In the final section of her chapter, after she has challenged the concept of the Scientific Revolution in general, Dobbs turns to two aspects of Newton's thought that appear particularly in tension with the concept, his alchemy and his theology. They are subjects on which I claim to have earned the right to comment. I agree that they bear directly on the concept of the Scientific Revolution.

Alchemy is a complex subject that does not reduce to the common denominator of other topics that come up with the Scientific Revolution. Earlier I quoted with disapproval Butterfield's forthright opinion on alchemy, but something within me keeps wanting insistently to agree with it. I recall an occasion when a wild-eyed student, who must have heard that I was working on Newton's alchemy, came to me for help in getting Maydew. The gulf between us was at least a thousand times as wide as the desk that separated us. Obviously he was trying to practice alchemy. I do not even know what practicing alchemy might mean, nor, to be truthful, do I want to know. To me, alchemy is solely a historic phenomenon. I am convinced that Dobbs also shared this opinion, for I remember an occasion in which she expressed her disapproval of a student of alchemy who appeared to be personally involved in the art. It seems to me relevant as well that Newton kept his own interest in alchemy rather secret, so that only in our age have we learned about its full extent. If Crosland is correct, as I think he is, in saying that alchemy was not so much attacked in the eighteenth century as ignored, surely that means people had by then ceased to take it seriously – perhaps to be seen as a step beyond keeping one's interest secret, but a step further down the same path. Since then, very few have taken it seriously as anything but a historic phenomenon. All of this seems to me relevant to a discussion of alchemy and the Scientific Revolution.

What is equally relevant, alchemy was not part of the Aristotelian system that the Scientific Revolution overturned. There had been alchemy during the Middle Ages, and inevitably it had expressed itself in Aristotelian terms, though it was never part of Scholastic philosophy. The great age of alchemy was the late sixteenth and the seventeenth centuries. The majority of the classics in the art come from that period, and they expressed themselves in the extravagant language of postmedieval, Neoplatonic natural philosophies. For those who have not done so, I recommend that they read in the work of the alchemist who appears most frequently in Newton's alchemical manuscripts, Irenaeus Philalethes (the pseudonym of George Starkey, a somewhat older contemporary of Newton), and try to imagine how Saint Thomas would have reacted to it. However they interpret Newton's interest in alchemy, they are unlikely after that exercise to consider it as a lingering echo of the old order.

I have a third preliminary comment. If Newton devoted extensive attention to alchemy, he was also a mathematician of the most distinguished order and the author of the *Principia* and the *Opticks*. Although he abandoned alchemy about the time he moved to London, so that we find almost no alchemical manuscripts that he composed after 1696, he continued to regard his achievements in mathematics and physics as badges of honor to be defended with passion. He engaged in a ferocious struggle with Leibniz to vindicate his priority in the invention of the calculus and willingly extended the struggle into a debate on their respective natural philosophies. He published his *Opticks* plus second and third editions, arranged for its translation into Latin to insure its circulation beyond Britain, and cooperated in a French translation. He participated in two additional editions of the *Principia*, devoting immense energy to the first of these, which introduced a number of substantial emendations into the text. At almost the last moment in this same edition, when he learned that Bernoulli had found a mathematical error in proposition X, book II, he interrupted the progress of the edition while he located the source of the error and corrected it, lest the proposition appear to reveal a deficiency in his command of mathematics. In my opinion, when we discuss Newton's alchemy, we must do so in the context of his whole career in science in order that we not slip unconsciously into regarding alchemy as the primary key to his thought.

The only satisfactory solution to the question of alchemy that I see is to accept it as part of the rejection of Aristotelian natural philosophy and to incorporate it into our account of the Scientific Revolution. I know very well that prevailing accounts of the Scientific Revolution do not do this. As I indicated earlier, I have fought on this field, even before Dobbs appeared on it, but on what came to be her side, and I know how the battle goes. For all that, or more accurately because of all that, I am convinced that we are most apt to understand seventeenth-century alchemy when we learn to see it, not as foreign to the Scientific Revolution, but as a part of it, a part moreover that was assimilated into its final product. I myself have advanced the argument, to which I continue to subscribe with undiminished conviction, that alchemy helped Newton transcend the limitations of conventional mechanical philosophy, and that he abandoned alchemy only after he had incorporated transmuted alchemical concepts into his notion of force. Dobbs held similar ideas about the origin of Newton's conception of force, and I believe that had she continued to reflect on the issue, she would have come to agree that the role of alchemy in Newton's career reinforces rather than undermines the concept of the Scientific Revolution.

In regard to religion, I want likewise to insist that we not associate Newton's prolonged endeavors in theology with the old order that the Scientific Revolution replaced. By the time, toward the end of his undergraduate years, when Newton began his career in science, and the time, roughly eight years later, when his serious study of theology commenced, the transformation in the relation of science with Christianity was already well advanced. Newton participated in the continuing process. He quickly became one of the most radical religious thinkers of his age. At a time when almost no one else went so far, he embraced Arianism, or a position so close as to be almost indistinguishable from it. I have described his most radical theological manuscript, the fragmentary treatise *Theologiae gentilis origines philosophicae*, as the first of the deist tracts, and I continue to believe that is correct. Whether the adjective "scientific" applies to this effort, the word "revolution" does not seem out of place.

We should not lightly dismiss the implications of adopting Arianism. The oaths imposed by Cambridge University on those taking degrees included affirmation of their belief in the Thirty-nine Articles of the Church of England, and the terms of the Lucasian chair explicitly included religious orthodoxy. In 1669, the university summarily ejected Daniel Scargill, a fellow of Caius, for holding opinions that sound similar to those Newton embraced not long thereafter. Later, after Newton moved to the Royal Mint in London, the law of the land excluded anyone with views like his from public office. Whatever we may think of his personal choice, Newton never considered mounting the barricades, and the need to keep his heretical opinions secret imposed a constraint on his entire adult life. Only within a restricted circle of confidants did he reveal himself. One of those he appears to have instructed, William Whiston, his successor in the Lucasian chair, chose to go public. Before he could blink an eye, Whiston was driven from the university; he went on to become the best-known religious pariah of Augustan England and further convinced Newton that silence was the only option. Another confidant, Samuel Clarke, although he was one of the most highly regarded theologians in the Anglican Church at the time, escaped Whiston's fate only by drawing back as Convocation prepared to discipline him. Clearly Newton's own age did not regard his religious stance as a defense of the existing order.

Dobbs speaks correctly of the "tides of mechanism, materialism, deism, and atheism" that swept over eighteenth-century Europe. The question to be settled is Newton's relation to them. On one issue there can be no debate. Because of his silence, we cannot see Newton as a moving force that raised the tides. In view of his documented position, however, I do not see how to avoid the conclusion that he too felt the force that raised them and that he moved more in harmony with the tides than in opposition. I think that talk of a valiant holding action gets his role in the religious drama wrong. Newton's "salvation history," and his belief in the "providential care of the Deity" that Dobbs mentions were not, in the way Newton understood them, notions that traditional Christianity would have embraced, for Newton set them squarely in an Arian context. He saw himself as one of the remnant chosen to preserve and restore pristine Christianity, that is, Arian Christianity. He looked upon trinitarianism, the religion of the established order, not just as a mistake, but as a fraud deceitfully imposed in the fourth century by evil men and sustained since by greed. Had Newton not chosen silence and isolation, he might well be recognized today as one of those whose religious thought helped to generate the tides of mechanism, materialism, deism, and atheism. To be sure, as Dobbs argues, the tides mounted far beyond anything that Newton wanted. It was, however, not the first time in history that a movement once set going had careened on past the goal its originators intended, nor was it the last.

I wish Professor Dobbs were still here. She did not put forth ideas casually, and although I have attempted to address her with the respect she deserves, she would assuredly have defended herself with spirit. The issues she has raised deserve the attention of all historians of science because the concept of the Scientific Revolution has been our central organizing idea. I have defended it as best I could, being convinced that without it our discipline will lose its coherence and, what is more, the cause of historical understanding take a significant step backward.

PART II

Canonical Disciplines Re-Formed

4

The Role of Religion in the Lutheran Response to Copernicus

PETER BARKER

Betty Jo Teeter Dobbs not only shifted the boundaries of Newton scholarship, she changed its center. Her first book *The Foundations of Newton's Alchemy* shifted the boundaries of inquiry by insisting on the importance of disciplines outside the canonical sciences in understanding the scientific tradition. In *The Janus Faces of Genius* she took the further step of arguing that religion, not science, was the real center of Newton's thought. Although Dobbs's main work focused on Newton, it would be a great loss to historical scholarship if we failed to draw the wider historiographical consequences of her work – consequences she herself made plain in her 1993 History of Science Society Lecture "Newton as Final Cause and First Mover." In the present chapter I support Dobbs's contention that these historiographical shifts and recenterings are important in understanding the history of science well before Newton and his time. The period I consider begins with the Reformation and concludes with the career of Johann Kepler. My main concern involves Lutheran natural philosophers and astronomers, and especially those trained at Wittenberg after its reform by Philip Melanchthon. After briefly reviewing earlier opinions on the relation between science and religion in this period,[1] I go on to docu-

This chapter presents preliminary results from a long-term collaborative project with Bernard R. Goldstein on Kepler's unification of physics and astronomy. The author gratefully acknowledges his help, and absolves him of blame for the defects of the current presentation. The author also gratefully acknowledges the support of the National Science Foundation and the National Endowment for the Humanities for portions of this research. Earlier versions of this chapter were presented under the title "Science and Religion in the Sixteenth Century – Enemies or Friends?" at the University of Oklahoma, December 1994, and "Religion and Natural Philosophy in the Lutheran Response to Copernicus," at Tel Aviv University, May 1995. I would also like to thank Roger Ariew, Eve Bannet, Mordechai Feingold, Margaret Osler, Ronald Schleifer, and the organizers of the conference on Jewish Responses to the Early Modern Science, Tel Aviv and Jerusalem, May 1995.

1 There is much more to be said about the relation between science and religion than can be considered here. Quite apart from the importance of Catholic thinkers, and

ment the role of Lutherans in spreading Copernicus's ideas, and the pervasive positive effect of their religion on Lutheran scientific thought. I suggest that Kepler's early work, in particular, should be read as centrally religious.

Science and Religion

A number of false presuppositions have impeded the historical understanding of sixteenth-century science. The first of these is that science and religion are natural enemies.[2] Second has been a tendency to expect sixteenth-century scientists to be the kind of people, and to have the kind of intellectual concerns, that emerged after the Scientific Revolution and became especially clear in the twentieth century. Astronomy is a special problem. It existed as a separate subject with its modern name even before the Scientific Revolution. And some practitioners, Brahe and Kepler, for example, devoted their full time to the subject (more or less) in the manner of their twentieth-century successors. This gives a false impression of exclusivity in subject matter and study. Lutheran astronomers like those from Wittenberg were not trained as specialists in astronomy. They are better understood as

Protestant thinkers who were not Lutheran, any general discussion of the sixteenth century should certainly mention the importance of Islamic and Jewish contributions.

The Islamic contribution to science has been consistently undervalued, perhaps because some founders of the history of science as an academic field were actively hostile to the idea of a real Islamic contribution. See, for example, the remarks of Pierre Duhem in "History of Physics" (1911), in *Essays in the History and Philosophy of Science*, trans. Roger Ariew and Peter Barker (Indianapolis: Hackett, 1996), chap. 8, and the rebuttal by F. Jamil Ragep, "Duhem, the Arabs and the History of Cosmology," *Synthese* 83 (1990): 201–14. The persistence of the Islamic contribution is documented in G. A Russell, ed., *The "Arabick" Interest of the Natural Philosophers in Seventeenth Century England* (Leiden: Brill, 1994).

On the Jewish contribution, also insufficiently appreciated until recently, see in particular Bernard R. Goldstein, *The Astronomy of Levi ben Gerson (1288–1344)* (New York: Springer, 1985), and "Scientific Traditions in Late Medieval Jewish Communities," in *Les juifs au regard de l'histoire*, ed. G. Dahan (Paris: Picard, 1985), pp. 235–47, and many other papers by the same author; and David B. Ruderman, *Jewish Thought and Scientific Discovery in Early Modern Europe* (New Haven: Yale University Press, 1995).

2 The centrality of religion to scientific work in the sixteenth and seventeenth centuries has been argued most forcefully by B. J. T. Dobbs, *The Janus Faces of Genius: The Role of Alchemy in Newton's Thought* (Cambridge: Cambridge University Press, 1991), in the case of Isaac Newton. Andrew Cunningham has made an important general case for the role of religion in the enterprise of natural philosophy. See, for example, "How the *Principia* Got Its Name; Or, Taking Natural Philosophy Seriously," *History of Science* 29 (1991): 377–92, and R. K. French and A. Cunningham, *Before Science: The Invention of the Friars' Natural Philosophy* (Aldershot: Scholar Press, 1996).

humanists with high-level astronomical competencies. Such competence in astronomy was not a prominent feature of southern European humanism, and its appearance in northern Europe should be seen as the consequence of uniquely Lutheran concerns with God's providential plan for the world. Although members of other confessions accepted various forms of this key idea, only the Lutherans, in their doctrine of the Real Presence, diffused their deity throughout the cosmos, so that study of any of its parts was coming to know more about God.[3]

A third difficulty in historical understanding has been the tendency to read some remarks by Lutherans as skeptical about the attainability of scientific knowledge in general, often reinforced by an erroneous instrumentalistic reading of sixteenth-century astronomy.[4] But a Lutheran *must* be able to achieve knowledge of God's providential plan.[5] Skepticism here would risk atheism. To the extent that science – or, more properly, natural philosophy – reveals the providential plan, there can be no question that knowledge is attainable. On the details of the attainment of knowledge in natural philosophy and astronomy, Lutherans are orthodox late Renaissance Aristotelians, as we will see.

The doctrine of the Real Presence, the doctrine that the Savior was really present in the host during Communion, is a pillar of sixteenth-century Lutheranism. The doctrine as it applies to the communion

---

[3] The connection between the Lutheran doctrine of the Real Presence and the investigation of the natural world is forcefully argued, with supporting quotations from Luther, by John W. Montgomery, "Cross, Constellation and Crucible: Lutheran Astrology and Alchemy in the Age of the Reformation," *Ambix* 11 (1963): 65–86, esp. pp. 67ff. A recent book-length treatment of similar issues is Sachiko Kusukawa's admirable *The Transformation of Natural Philosophy: The Case of Philip Melanchthon* (Cambridge: Cambridge University Press, 1995). Additional useful information may be found in Bruce Moran, "The Universe of Philip Melanchthon: Criticism and Use of the Copernican Theory," *Comitatus* 4 (1973): 1–23.

[4] The allegation that sixteenth-century astronomy is instrumentalistic begins with Pierre Duhem, ΣΩZEIN TA ΦAINOMENA, Essai sur la notion de théorie physique de Platon à Galilée (Paris: Hermann et Fils, 1908), translated as *To Save the Phenomena: An Essay on the Idea of Physical Theory from Plato to Galileo*, trans. E. Dolan and C. Maschler (Chicago: University of Chicago Press, 1969). On the issue of skepticism, see N. Jardine, "The Forging of Modern Realism: Clavius and Kepler against the Skeptics," *Studies in History and Philosophy of Science* 10 (1979): 41–73. Both themes are considered in N. Jardine, "Epistemology of the Sciences," in *The Cambridge History of Renaissance Philosophy*, ed. C. B. Schmitt and Quentin Skinner (Cambridge: Cambridge University Press, 1988), pp. 685–711. On the status of astronomy, see Peter Barker and Bernard R. Goldstein, "Realism and Instrumentalism in Sixteenth Century Astronomy: A Reappraisal," *Perspectives on Science* (forthcoming).

[5] On the use of the verb *agnitio* for both knowledge of God and knowledge of nature, see the perceptive remarks by Kusukawa, *Melanchthon*, pp. 95–7, 179 (Rheticus's use of the term in connection with astronomy), and 184.

host is a particular case of a more general doctrine, that Christ is ubiquitous. This doctrine caused a bitter dispute between Lutherans and Calvinists, who denied ubiquity and hence the Real Presence. It equally reinforced Lutheran separation from the Catholic Church, which taught that a miraculous transformation (transubstantiation) created a real presence in the host only during Communion.

Important as it was in sixteenth-century polemics, the doctrine that the *host* embodied the Real Presence was only a special case of the more fundamental doctrine of ubiquity. The host was only one object among many. For Luther and his followers, the Real Presence was distributed throughout all objects. As Luther put it, the Savior is "substantially present everywhere, in and through all creatures, in all their parts and places."[6] The specifically Lutheran doctrines of ubiquity and the Real Presence of Christ in the host are the basis for the Lutheran belief in the universal presence of a providential deity, whose design or plan may be known through the study of nature. These specifically Lutheran doctrines informed their attitude to science in general. They motivated the special Lutheran interest in astronomy that spread Copernicus's mathematical models and his fame and provided Kepler with the resources to give the strongest and most lasting defense of Copernican cosmology.

The Wittenberg Astronomers

Presenting Lutheranism as a positive force in the spread of the new science runs counter to many things that have been said by earlier historians about the relation between science and religion. In 1896 Andrew Dickson White – then president of Cornell University – set the tone for much twentieth-century discussion with a book entitled *A History of the Warfare of Science with Theology in Christendom*.[7] Taking the confrontation between Galileo and the Catholic Church as an emblem, many people have seen the relations between science and religion during the sixteenth and seventeenth centuries as irrelevance at best, and active antagonism at worst.[8] Chief among the causes of

6 Montgomery, "Cross, Constellation and Crucible," 70. The quotation is from Luther, *D. Martin Luthers Werke: Kritische Gesamtausgabe* (Weimar: H. Bohlau, 1883–1919), 23: 134.
7 Andrew Dickson White, *A History of the Warfare of Science with Theology in Christendom*, 2 vols. (New York: D. Appleton, 1896).
8 For more recent views of Galileo, see Mario Biagioli, *Galileo Courtier* (Chicago: University of Chicago Press, 1993); Richard J. Blackwell, *Galileo, Bellarmine and the Bible* (Notre Dame, Ind.: University of Notre Dame Press, 1991); and Rivkah Feldhay, *Galileo and the Church* (Cambridge: Cambridge University Press, 1995).

religious antagonism to the new science was supposed to be Copernicus's idea that the Sun, and not the Earth, was the center of planetary motions. This was the central issue in Galileo's difficulties with the Catholic Church, but the same antagonism was supposed to be equally well established among the Protestants.[9]

Thomas Kuhn's *Copernican Revolution* attracted great interest to the history of science in general and the sixteenth and seventeenth centuries in particular.[10] Unfortunately, Kuhn trusted White as his source on the Lutheran response to the new science. He not only made Protestants into early and vigorous opponents of Copernicanism, but quotes Luther himself, and his most important follower Melanchthon, in condemning Copernicus:

> Citation of Scripture against Copernicus began even before the publication of *De Revolutionibus*. In one of his "Table Talks," held in 1539, Martin Luther is quoted as saying: "People gave ear to an upstart astrologer who strove to show that the earth revolves, not the heavens or the firmament, the Sun or the Moon.... This fool wishes to reverse the entire science of astronomy: but sacred scripture tells us that Joshua commanded the Sun to stand still and not the Earth." Luther's principal lieutenant, Melanchthon, soon joined in the increasing clamor against Copernicus. Six years after Copernicus' death he wrote: "The eyes are witness that the heavens revolve in the space of 24 hours. But certain men, either from love of novelty, or to make a display of ingenuity, have concluded that the earth moves; and they maintain that neither the eighth sphere nor the Sun revolves.... Now it is want of honesty and decency to assert such notions publicly, and the example is pernicious. It is part of a good mind to accept the truth as revealed by God and to acquiesce to it."[11]

Both of Kuhn's quotations are straight from Andrew Dickson White.

There were immediate indications that all was not well with these ideas. As early as 1928 Charles Singer – one of the founders of history of science as a serious academic field – described White's book as

---

9 White, *Warfare*, 1:126–7. Similar positions, in some cases based directly on the evidence adduced by White, are taken by Dorothy Stimson (1917), Ernst Zinner (1943), Thomas S. Kuhn (1957), and Marie Boas Hall (1962). See the reply by John R. Christianson, "Copernicus and the Lutherans," *Sixteenth Century Journal* 4 (1973): 1–10, and the discussion of Kuhn (cited in note 10).

10 Thomas S. Kuhn, *The Copernican Revolution* (Cambridge, Mass: Harvard University Press, 1957). On the *fortuna* of Kuhn's book, see Robert S. Westman, "Two Cultures or One? A Second Look at Kuhn's *The Copernican Revolution*," *Isis* 85 (1994): 79–115.

11 Kuhn, *Copernican Revolution*, 191; White, *Warfare*, 1:126–7.

"lacking in depth despite its learning."[12] More recently, historians have questioned the dismissal of Copernicanism by Luther. The damning quotation offered by White, and repeated by Kuhn, is not something written by Luther himself. It is a note supposedly made by a dinner companion, and not published until after his death. Luther, therefore, neither authorized nor corrected this quotation. If we look in Luther's works for criticisms of Copernicus in the period after the publication of *De revolutionibus* in 1543, they are conspicuous by their absence. So, it may be doubted whether Luther said these words, and even if he did, it may be doubted that Copernicus is the "upstart astrologer" in question.[13] But supposing we concede both these points. If we suspend judgment on the antagonism between religion and science for a moment, then the quotation from Luther becomes evidence for something else. It shows that Luther knew about Copernicus's work, not only four years before its publication, but a year before Rheticus published the first nontechnical description of it. Taken at face value the quotation is evidence for Lutheran *interest* in Copernicus.

Even stronger reservations apply in the case of the quotation from Melanchthon. White quotes the first edition of Melanchthon's introductory textbook on natural philosophy, *Initia doctrinae physicae* (Wittenberg: Lufft, 1549), which admittedly does have some negative things to say about Copernicus. White does not tell you that in the same year Melanchthon, in his own words, "[began] to love and admire Copernicus more,"[14] and that in the later editions of the book (Wittenberg: Lufft, 1562; Wittenberg: Prato, 1567) he removed or revised his criticisms of Copernicus, used Copernican figures for planetary parameters in preference to the Ptolemaic ones, and reserved special praise for Copernicus's theory of the moon.[15] More important still, Melanchthon fostered a strong astronomical tradition at Lutheran universities, and this tradition was favorably inclined toward at least

---

12 Charles Singer, *Science and Religion Considered in Historical Relation* (London: E. Benn, 1928), p. 79.
13 Christianson, "Copernicus and the Lutherans," pp. 1–10.
14 "His et similibus observationibus moti, magis Copernicum admirari et amari coepimus." Philip Melanchthon, *Corpus reformatorum, Philippi Melanchthonis opera, quae supersdunt omnia*, ed. C. G. Bretschneider and H. E. Bindsell, 28 vols. (Halle: Schwetschke and Son, 1834–60; reprint, New York: Johnson Reprint, 1963), 11:839. Subsequently referred to as *CR*.
15 Robert S. Westman, "The Melanchthon Circle, Rheticus and the Wittenberg Interpretation of the Copernican Theory," *Isis* 66 (1975): 165–93. For Melanchthon's remarks on Copernicus, see *CR* 13:244 (Moon model praised), 13:262 (Copernicus's apogees of superior planets preferable to Ptolemy's).

the technical aspects of Copernicus's work. All this is quite out of keeping with any "want of honesty and decency to assert such notions publicly." The evidence shows that Lutherans did just that.

For our purposes it is important to remember that the Reformation began in universities. Two assume particular importance in our story: Wittenberg and Tübingen. Melanchthon began his career at Tübingen, and was appointed to Wittenberg just in time to provide Luther with the benefit of his superlative training in rhetoric and languages. The early part of the Lutheran reformation was very much a double act. While Luther's firm – perhaps dogged – defense of his central ideas anchored the new confession, it was Melanchthon who provided the administrative and diplomatic skill that ensured the survival of Lutheranism. He was also the main author of the first important statement of Lutheran principles: the Augsburg Confession (1529).

As universities were the primary training ground for clergy, reform of the church entailed reform of the universities. Set in motion by Luther, this reform was carried through by Melanchthon, who became the educational leader of the German-speaking world. After reforming the curriculum at Wittenberg itself, Melanchthon was called in to write a new constitution for his old university of Tübingen, and played a role in the reforms at Frankfurt an der Oder, Leipzig, Rostok, and Heidelberg. New Lutheran universities were also founded at Marburg, and Königsberg, but the demand for education in the new confession so outstripped supply that in many places "academies," "gymnasia," or "high schools" were founded that offered courses at the same level as the early years of university. In some cases these academies matured into universities – at Jena and Altdorf, for example.

Melanchthon became a Lutheran, but even before he was a Lutheran he was a humanist. His humanist training provides one reason for his attention to astronomy, as part of his general concern to rescue the curriculum from Scholasticism, as well as to make it safe for Lutherans. The quadrivium, the four basic subjects studied by all university students, had always contained astronomy as one part, often with a strong admixture of astrology. Astrology itself was a second reason for Melanchthon's interest. Although he denied that the influence of the heavens could bring about specific events, Melanchthon and his followers believed this influence could predispose events to move in a certain direction, and they believed that the influence of the heavens had a profound effect on people's personalities, which again predisposed, but did not strictly determine, how things turned out on earth. Melanchthon and his successors also believed that signs in the heavens were a means for God to communicate with the human race.[16]

Astronomy achieved a new importance in the Lutheran curriculum, the old and new universities founded new chairs to support its study, and Melanchthon filled them with his students. It is this group that has come to be called the "Wittenberg astronomers."[17] Given the Lutheran interest in astrology, the boundaries of this group have perhaps been drawn too narrowly. Also, Melanchthon himself has not been counted as a Wittenberg astronomer, although he has been acknowledged as the intellectual center of the group. Reinhold, the author of the *Prutenic Tables*, has been the defining figure, with his achievements in astronomy the model for judging others.[18] A brief survey of some important Wittenberg astronomers shows that their careers accommodate both science and religion as serious intellectual pursuits within a broader humanist framework.

From the viewpoint of historical evidence, there are relatively few genres of written work that can be used to classify an author as a Wittenberg astronomer. The most important genres are the two main parts of the astronomy curriculum, *sphaericae*, or commentaries on Sacrobosco, and *theoricae*, by this time most likely in the form of commentaries on Puerbach. In addition there are astronomical tables themselves, and books explaining their use. However, we might also consider two other genres. The first is new translations of classical works on astronomy and astrology, in line with the wider humanist program of the recovery of ancient learning. This program was strongly supported by Melanchthon and is also apparent in the work of his students. The second additional genre comprises works on astrology and prognostication. As already noted, a central motivation for Melanchthon's interest in and support for astronomy was his belief in the importance of astrology.

[16] For Melanchthon's views on astrology, see "Oratio de dignitate astrologiae," CR 11: 261–6; the thesis defended by Rheticus, mentioned below, CR 10:712–15, and the opening sections of *Initia doctrina physicae*, CR 13:179–95, which employs a definition of physics that explicitly includes the influence of the heavens on the Earth. For a later statement, see C. Peucer, *Commentarius de praecipuis divinationum generibus* (Wittenberg: Crato, 1553). In general see D. Bellucci, "Mélanchthon et la défense de l'astrologie," *Bibliothèque d'Humanisme et Renaissance* 50 (1988): 587–622. Lutheran concerns with astrology may also have been a natural consequence of the early Protestants' belief that they were living in the last days, that the pope was perhaps the Anti-Christ, and that spectacular celestial events like the comets of 1532-3, might be signs that the end of the world was at hand (Kusukawa, *Melanchthon*, pp. 167–73).

[17] L. Thorndike, *A History of Magic and Experimental Science* (New York: Columbia University Press, 1929–58), vol. 5, chap. 27. Westman, "The Melanchthon Circle."

[18] Erasmus Reinhold, *Prutenicae Tabulae coelestium motuum* (Tübingen: 1551).

Given these categories, Melanchthon himself deserves to be counted as a Wittenberg astronomer. While still at Tübingen he made a partial Latin translation of Aratus's *Phaenomena*, and published a Greek edition in 1521.[19] While at Wittenberg, according to Thorndike, he assisted Schöner in editing the works of al-Battānī and al-Farghānī, and personally made calculations on the motion of the Sun.[20]

Another member of the group, though not entirely typical, is Georg Joachim Rheticus, who graduated from Wittenberg in 1536 after defending a thesis showing that Roman law did not prohibit astrological predictions, as long as they were based on physical causes.[21] For the next two years he taught arithmetic and geometry at the university. In 1538 he began an academic tour spending time with leading astronomers including Schöner in (Lutheran) Nuremberg, Peter Apian in (Catholic) Ingolstadt, and, in the summer of 1539, Nicholas Copernicus in (Catholic) Frauenburg. With Copernicus's encouragement, Rheticus then wrote a nonmathematical description of Copernicus's new astronomy, published in 1540 as the *Narratio prima*. Rheticus returned to Wittenberg in 1541, and was elected dean of liberal arts. In 1542 he was called to a chair in mathematics at Leipzig. On the way to his new appointment he stopped in Nuremberg to arrange for the publication of Copernicus's main book, *De revolutionibus*.

By the time Copernicus's detailed presentation of his new astronomical models became public, Lutheran mathematicians had a good idea what to expect. Another Lutheran, Andreas Osiander, wrote a preface for the book that expressed what came to be the orthodox Lutheran view of Copernicus: it was all right to use the models for purposes of calculation – indeed they had many advantages – but there was no need to accept the corresponding physical assertions, for example that the Earth was a planet or that the Sun was (nearly) the center of the planetary motions.

Erasmus Reinhold, who had been apppointed to a chair of mathematics at Wittenberg in the same year as Rheticus, began to produce a new set of tables for the position of the Sun, Moon, and planets, almost immediately after the publication of *De revolutionibus*. Although he used the mathematical models introduced by Copernicus in preference to the older Ptolemaic ones, he continued to assume that

---

19 William Hammer, "Melanchthon, Inspirer of the Study of Astronomy, with a Translation of his Oration in Praise of Astronomy (De Orione, 1553)," *Popular Astronomy* 59 (1951): 309.
20 Thorndike, *A History of Magic and Experimental Science*, 5:360, and esp. p. 393 (work with Schöner); p. 385 and n. 25 (calculations).
21 Melanchthon, CR 10:712–15.

the Earth was the center of the planetary system, as did all his Wittenberg-trained colleagues except Rheticus. This was the sense in which Rheticus was unusual: he saw the value of Copernicus's new mathematical techniques, but he also accepted the cosmic plan that went with them. Reinhold's new astronomical tables, however, were enormously popular, and people trained in his Copernican but geocentric mathematical techniques spread throughout the German-speaking part of Europe.

Let us briefly consider the career of a previously unrecognized member of the group, a career that displays the full range of humanist skills taught at Wittenberg. This student of Melanchthon also has unusually strong credentials in astronomy proper (he wrote a book on the mathematical techniques introduced by Reinhold) and has things to say about Copernicus that show something new and important about Wittenberg astronomy. Hilderich von Varel (Edo Hildericus) was born in 1533 at Jever. He went from the evangelical school there to Wittenberg, where he studied theology under Melanchthon himself and mathematics under Caspar Peucer (Reinhold's successor). After Wittenberg, he served for three years as professor of mathematics at Jena (1564–7), but his main appointments began with the rectorship of the Protestant school at Magdeburg (from 1573). He taught oriental languages and history at Frankfurt an der Oder for a year in 1577, until he became professor of oriental languages and theology at the Heidelberg Protestant high school in 1578. But in 1580, when all teachers were obliged to subscribe to a statement of Lutheran orthodoxy called the Formula of Concord, he left rather than sign.[22] In 1584 he finally achieved a permanent position, becoming professor of theology at the important Protestant academy at Altdorf, an academy that leaned toward Calvinism, and did not require its faculty to sign the Formula of Concord. There he married Suzanne Seidelmeier on March 14, 1592. Von Varel taught moral philosophy and Greek literature at Altdorf, until his death in 1599. One example can give some

22 The unfortunately named Formula of Concord was an attempt to heal the breach between strict Lutherans and less-strict Phillippists and others sympathetic to Calvinism. Issued on June 25, 1580 (the fiftieth anniversary of the Augsburg confession), it was accepted by Saxony, Württemberg, and Baden, but rejected by Hess, Nassau, and Holstein. The free cities of Hamburg and Lübeck adopted it. Bremen and Frankfurt rejected it. Wherever it was accepted, state employees – for example, university professors – were obliged to subscribe to the formula, on pain of dismissal. The employment of the formula thus deepened the divisions it was intended to heal, by obliging people like von Varel, and indeed Kepler, to choose between a new orthodox Lutheranism that opposed Calvinism, and the older Phillipist version that tolerated it. Kepler's response was to follow a career that avoided the Formula. Von Varel moved to a Calvinist institution.

indication of what his students were learning: on December 9, 1595, von Varel presided at a public thesis defense where the respondent was a student called Michael Piccartus from Nuremberg. The subject of the thesis was angels. At first glance this is not the career of an astronomer.

Von Varel's writings, however, tell a slightly different story. Sprinkled among the books on languages and theology are books that are unquestionably on astronomy. His first two important publication were orations on, respectively, the life of Demosthenes (1562) and on the introduction of the biblical languages (Greek, Latin, and Hebrew) in the curriculum of Christian academies (1567). But immediately after that he published a text on astronomical calculations (1568). Next we find an oration on the political and theological organization of the Jewish people (1570). This is followed by a set of propositions on the dimensions of the terrestrial globe (1576), a book of poems about Melanchthon (1580), a defense of the truth and certainty of the Christian religion (1582), a book on the second coming of Christ (1586), and the first Latin translation of a Greek work on astronomy by Geminus of Rhodes (1590).[23]

The first of these astronomical works is nothing less than a public presentation of the calculational techniques used in the Wittenberg interpretation of Copernicus – a "how to" book about Wittenberg astronomy. Erasmus Reinhold, as the author of the *Prutenic Tables*, is the hero of the book. The calculations, adapted from Copernicus, are the center of attention and not the cosmic scheme that Copernicus had supplied with them.

But in his second important astronomical work, matters are different. On September 8, 1576, at the Frankfurt Academy, Hilderich von Varel publicly defended a set of "cosmographical propositions on the dimensions of the terrestrial sphere."[24] Proposition VI is rather surprising:

> VI. Nicolaus Copernicus, another Ptolemy of our days, has rejected the ninth and tenth orbs, and retains as many of the first eight celestial orbs as our eyes show to us. And he retains the

---

23 Chronologically, von Varel's known publications are as follows: *Oratio de vita Demosthenis* (Wittenberg, 1562); *Oratio de instituenda in academis Christianis publica praelectione trium biblicarum linguarum* (Jena: Richtzenhan, 1567); *Logistice astronomica* (Wittenberg: Schwertel, 1568); *Oratio de politia et hierarchia populi Judaici* (Wittenberg, 1570); *Propositiones cosmographicae de globi terreni dimensione* (Frankfurt: n.p., 1576); *Carmen de Philippo Melanchthonis* (Basel: Perna, 1580); *Propositiones de veritate et certitudine religionis Christianae* (Altdorf, 1582); *On the Second Coming of Christ* (Altdorf, 1586); *Gemini . . . elementa astronomiae* (Altdorf: Lochner, 1590).

24 *Propositiones cosmographicae de globi terreni dimensione ab Edone HILDERICO in inclyta Academica Francofordiana ad orderam propositae . . . 8 Sept. 1576* (Frankfurt: n.p., 1576).

order of that part of the world, except that, leaving the Moon in its orb, he transfers the orb of the Moon enclosing the globe of the elements from the middle of the world to the orb of the Sun, and, in turn, the body of the Sun itself out of its orb, to the middle of the world. And this would be the order of the parts of the world according to Copernicus. The Sun is placed in the middle of the world, the orb of Mercury follows, [and] next the orb of Venus. In fourth place [is] the Great Orb, as he calls it himself, now the vehicle of the globe of the elements, and the orb of the Moon. Above [these] four orbs, from there in order follow the orbs of Mars, Jupiter, Saturn, and last [the orb] of the fixed stars that for Copernicus is unmoving. This order is wonderful [*mirificus*], [but] I would say no more about it now, following the customary order of the parts of the world in these propositions.[25]

Note the elegant way von Varel presents Copernicus's system to a sixteenth-century audience. He makes them imagine the world pattern they already know and then switches two established parts, leaving as much in place and as much familiar as possible.

Von Varel calls this order of the world *mirificus* – a term that ranges in meaning from "wonderful," through "a cause for wonder," to "astonishing." Of course, if we take the last meaning, then the comment can be read negatively, but there is really no reason to attach a negative weight to the word. Nothing negative is said before or after. Notice also what is left *unsaid* – he does not immediately tell us that this is contrary to Scripture and hence must be rejected. There are at least two reasons for this.

First, the general Lutheran attitude to the relations between the Bible and the results of natural philosophy was the one called Accommodationism: the Bible was not intended as a book on natural philosophy, which could only be discussed by learned people who used a special technical vocabulary. The Bible was supposed to be under-

---

25 Hilderich von Varel (Edo Hildericus), *Propositiones cosmographicae*, A2V: "VI. Nicolaus Copernicus nostrae aetatis alter Ptolomaeus, orbes nonum et decimum reijcit, et tantum priores octo orbes coelestes, quot oculi nobis monstrant, retinet, retinet et eundem ordinem partium mundi, nisi quod orbem Lunae cum globo elementari incluso, Luna in suo orbe relicta, transfert ex medio mundi ad orbem Solis, et vicissim ipsum Solis corpus ex suo orbe ad medium mundi. Ut hic sit ordo partium mundi iuxta Copernicum. In medio mundi ponitur Sol, sequitur orbis Mercurii, inde orbis Veneris, Quarto loco orbis magnus, ut ipse vocat, iam vehiculum globi elementaris, et orbis Lunae. Supra quartum orbem, deinde ordine sequuntur orbes Martis, Iovis, Saturni, et ultimus stellarum fixarum Copernico immobilis. Mirificus hic est ordo, de quo iam plura non non dicam, sequens in his propositionibus usitatum partium mundi ordinem."

stood by everybody, hence it was written in the common speech, and neither used the technical language nor attempted to express the technical ideas of natural philosophy. There might very well be seeming disagreements between the Bible and natural philosophy, and these disagreements were a matter for reasonable discussion among reasonable people. The "theses" from which von Varel's remarks are extracted *are* a technical discussion, and so von Varel would not have felt it was necessary to reconcile the section on Copernicus with the Bible, although if you were a member of his audience you might be expected to be able to defend an opinion of your own on whether these matters were ruled out by Scripture. However, von Varel did begin the whole set of theses with a Lutheran commonplace that all the matters to be treated fell under a consideration of *how* God created the world, with what parts and in what order. We have already noted that Lutherans saw the study of nature as a means of coming to know God through his providential plan. Von Varel's opening remarks are an instance of this general attitude, showing how Lutheran religious ideas inform the general study of nature without determining specific details. Lutheran views of God's construction plans assumed a new and specific importance with Kepler.

A second reason for von Varel's silence on the scriptural status of Copernican cosmology is that, in the conventional scheme of knowledge for von Varel and his contemporaries, astronomy is subordinated to natural philosophy and natural philosophy is subordinated in turn to theology.[26] This meant that the ultimate explanatory principles of each discipline had to be found in the next highest discipline. However, it was perfectly acceptable to discuss astronomy or natural philosophy without raising connections to other disciplines – although the audience would clearly be aware that there were apparent contradictions between Scripture and the Copernican cosmic scheme. Von Varel's audience, then, may have thought that discussion of Copernicus's cosmology was unlikely to undermine truths they regarded as well established in theology and physics, which were superordinate disciplines to astronomy and cosmology. The nature of the sixteenth-century hierarchy of knowledge leads to a complex of issues surrounding sixteenth-century views on demonstration

---

26 J. A. Weisheipl, "Classification of Sciences in Medieval Thought," in J. A. Weisheipl, *Nature and Motion in the Middle Ages*, ed. W. E. Carroll (Washington, D.C.: Catholic University of America Press, 1985), pp. 203–37; Roger Ariew, "Descartes and the Tree of Knowledge," *Synthese* 92 (1992): 101–16. That Copernicus subscribed to conventional ideas on the order of the sciences is shown by the canonical list of middle sciences he gives in the first paragraph of the manuscripts of *De revolutionibus*. The passage is deleted from the first printed version.

in science. These views made the case of astronomy special and legitimated the consideration of models that were clearly at variance with physics.

Astronomy as a posteriori Science

A fundamental question is why Wittenberg astronomers felt entitled to appropriate Copernicus's mathematical models, based as they were on a cosmology that was radically incompatible with accepted physics and theology. The answer lies in the peculiar status of astronomy during the sixteenth century according to the accepted theory of demonstration, or as we might say, scientific explanation. At the time that Melanchthon's initiatives created a new impetus for the study of astronomy and mathematics in German-speaking Europe, practitioners of the mathematical sciences accepted a characteristic version of (what they took to be) Aristotle's account of demonstration in science.[27] In the *Posterior Analytics* Aristotle had distinguished two kinds of demonstration: ὅτι and διότι. In Latin the equivalent labels were *demonstratio quia* and *demonstratio propter quid*, for what are today called "demonstration of the fact" and "demonstration of the reasoned fact," respectively.[28] During the sixteenth and seventeenth centuries it was also common to distinguish these different kinds of demonstration as reasoning from effects to causes and reasoning from causes to effects. Consequently they were also known as demonstration a posteriori and demonstration a priori.[29]

Although both kinds of demonstration were syllogistic patterns, the first specified a *possible* cause in explaining an effect; the second specified the *actual and unique* cause of that effect. Clearly it was possible to specify many demonstrations a posteriori for a single effect, but only one demonstration a priori. To move from the former to the latter one began with a wide field of a posteriori demonstrations, and by a process variously known as *negotatio intellectus, consideratio mentalis*, or, more simply, *meditatio*, established that only one among the original candidates was worthy of continued consid-

---

27 On the conceptual differences between Aristotle's account and sixteenth-century views, see Jardine, "Epistemology of the Sciences," pp. 686–93.
28 Aristotle, *Posterior Analytics* I.2.71b17–72b4, I.13.78a22ff. See also W. A. Wallace, *Causality and Scientific Explanation* (Ann Arbor: University of Michigan Press, 1972), 1:10–18.
29 Melanchthon, *Initia doctrina physicae*, in CR 13: 194; B. Keckermannn, *Systema logicae . . . editio ultima, prioribus correctior* (Hanoviae: P. Antonius, 1620), pp. 531, 540; Stephan Chauvin, *Lexicon philosophicum* (Rotterdam: P. Vander Slaart, 1692), V3r, V3v. For text and discussion, see Barker and Goldstein, "Realism and Instrumentalism in Sixteenth Century Astronomy: A Reappraisal."

eration.³⁰ This remaining demonstration gave a causal explanation of the original effect that satisfied Aristotle's strictest principles for scientific knowledge – its principles were true either on the grounds that they were necessary or that they were certified by derivation from already certified principles of a higher science (physics, mathematics, or in some cases theology). In this way causal knowledge of the original effect would be secured, and the resulting demonstration could be neatly located in the established hierarchy of knowledge, its premises or principle linking the higher science that certified them with the subject matter of the lower science that investigated the original effect.

As an example, consider a reconstruction of the reasoning used by Ioannes Pena to establish the nature of comets.³¹ Peter Apian had discovered in the 1530s that the tail of a comet always points away from the Sun. The effect here is the direction of the tail, which Apian found to lie on a great circle from the Sun, through the head of the comet, on the side away from the Sun.³² What could cause this effect? Aristotle had taught that comets were long-lived fires in the upper air. During the fifteenth century Regiomontanus had proposed that the tail of a comet was smoke rising from this fire. But according to the physical consensus of the day, smoke and other fiery matter should move directly upward from the Earth, which stood at the center of the universe.³³ So the comet's tail should point radially away from the center of the Earth, not radially away from the Sun. Only one phenomenon linked the Sun with the head of the comet in the correct orientation: the rays of light from the Sun.³⁴ According to accepted optical

30 On this vocabulary, see H. Mikkeli, *An Aristotelian Response the Humanism: Jacopo Zabarella on the Nature of Arts and Sciences*, Studia Historica 41, (Helsinki: SHS, 1992), p. 99, and Jardine, "Epistemology of the Sciences," p. 687, reading "negotatio" for "negotiatio" in the latter.

31 Ioannes Pena, "De usu optices," preface to *Euclidis optica et catoptrica* (Paris: Andreas Wechulus, 1557); Peter Barker, "Jean Pena (1528–58) and Stoic Physics in the Sixteenth Century," in *Recovering the Stoics: Spindel Conference 1984*, ed. R. H. Epp, supplement to the *Southern Journal of Philosophy* 13 (1985): 93–107.

32 Peter Barker, "The Optical Theory of Comets from Apian to Kepler," *Physis* 30 (1993): 1–25; Peter Barker and Bernard R. Goldstein, "The Role of Comets in the Copernican revolution," *Studies in History and Philosophy of Science* 19 (1988): 299–319.

33 Regiomontanus, *Ioannis de Monteregio Germani, viri undecunque doctissimi, de cometae magnitudine, longitudineque ac de loco eius vero, problemata XVI*, ed. Johannes Schöner (Nuremberg: apud Fridericum Peypus, 1531). A facsimile is reproduced in J. Jervis, *Cometary Theory in Fifteenth Century Europe*, Studia Copernicana 26 (Warsaw: Polish Academy of Sciences, 1985).

34 It is important to note that although sixteenth-century optics made use of the concept of a light ray, it did not necessarily endorse the view that light was material: David

theory, these rays should cast a shadow behind the head of the comet and radially away from the Sun. But the tail of a comet is bright, unlike a shadow. Pena suggested that the tail must be solar rays passing through the comet's head, made visible because they were not dispersed but focused by their passage. He concluded that the only possible explanation for the appearance of the tail was that the head of the comet was a spherical lens.[35] Recognizing the requirement that the head focuses the rays of the Sun provided an additional argument against Aristotle's account of comets (and one Pena makes explicitly): fires do not have the property of focusing light rays, so the comet's head cannot be a fire.

In Pena's discussion we have two possible causes for the appearance of the comet's tail: Aristotle's account of the comet's composition and an alternative optical account. A variety of arguments exclude one possible explanation and favor the other. If we accept the lenslike nature of the comet's head as the only possible explanation for the direction and appearance of its tail, then we can explain the tail's direction and appearance a priori by assuming that the head has this nature. Of course, the crucial step in these considerations is to rule out any other possible causes than the lenslike comet's head. This may strike a modern reader as unreasonable, but apparently the identification of the tail as a focused set of light rays was so compelling that almost all the figures we now consider important in sixteenth-century astronomy accepted the view that the head of a comet was a lens.[36] And even to the modern reader the central argument remains compelling ("If the tail is focused rays, the head *must* be a lens").

As the example of comets shows, it was supposed to be possible to establish demonstrations of fact based on true causes in higher sciences like physics, and even some subordinate sciences like optics. Matters were less satisfactory in astronomy, the science of predicting planetary positions. It had been known since antiquity that models

---

C. Lindberg, *Theories of Vision from al-Kindi to Kepler* (Chicago: University of Chicago Press, 1976); "The Science of Optics," in *Science in the Middle Ages*, ed. David C. Lindberg (Chicago: University of Chicago Press, 1978), pp. 338–68; "The Genesis of Kepler's Theory of Light: Light Metaphysics from Plotinus to Kepler," *Osiris*, 2nd ser., 2 (1986): 5–42.

35 Pena, "De usu optices," p. bb ii r, lines 18–20: "Quia enim cauda in partem a Sole aversam tendit, est igitur pyramis lucida procreata a concursu radiorum solarium, sese ad perpendicularem frangentium occursu corporis perspicui aere densioris. Necesse ergo est Cometen esse corpus perspicuum et diaphanum vitri instar perlucidum: ab iis enim tantum corporibus fieri refractionum pyramidas docet Optice."

36 In addition to Pena, Cardano, and, probably, Gemma Frisius, we may definitely count Rothmann, Brahe, Maestlin, and the young Kepler. Kepler later became an important critic of Pena's account. See Barker, "Optical Theory of Comets."

using a single eccentric circle on which the planet moved would yield identical predictions to models using an Earth-centered circle carrying an epicycle, which in turn carried a planet. The angular position of the planet viewed from the Earth was demonstrably the same in both constructions, a result usually attributed to Apollonius.

Eccentrics and epicycles gave no *causal* explanation of planetary motions. It was generally accepted that the planets were moved by the otherwise invisible substance of the heavens, although each was confined within a zone bounded by spherical surfaces concentric to the center of the universe. The zones for different planets were non-overlapping. In the *Planetary Hypotheses* Ptolemy elaborated this picture by suggesting a series of devices that fitted inside the concentric shells for each planet. These devices replaced the eccentric circles and epicycles of his earlier *Almagest* with three-dimensional constructions in the aether, which were assimilated and improved by Arab astronomers during the Western Middle Ages, and became a standard feature of the basic Latin texts on planetary astronomy, the *theoricae*, during the fifteenth and sixteenth centuries.[37]

In the *theoricae*, the eccentric circle used to represent the motion of the Sun was replaced by three interlocking shells or "orbs" (Figure 4.1). The innermost surface of the inner orb (C), and the outermost surface of the outer (E), were concentric with the Earth, which was the center of the universe (B). The outer surface of the inner orb, and the slightly larger inner surface of the outer orb were centered on a single eccentric point some distance from the Earth (A). These latter two surfaces defined an eccentric orb of uniform thickness (D) in which the body of the Sun was embedded. As the orbs rotated about diameters that passed through the center of the universe, the path followed by the Sun was the same as the eccentric circle in the original two dimensional model. But now the motions of the orbs could be imagined as the causes creating this motion.

Ptolemy had not been able to produce a model for the motion of the Moon or planets using only an eccentric circle. He had been obliged to add an epicycle carried by the eccentric (and additional complications in the cases of the Moon and Mercury). These epicycles could also be incorporated into a causal account by replacing the epicycle with a sphere of the same diameter, with the planet embedded in it

---

37 For Ptolemy's original models see Bernard R. Goldstein, *The Arabic Version of Ptolemy's Planetary Hypotheses, Transactions of the American Philosophical Society* 57 (1967), part 4. On the *theorica* tradition, see O. Pedersen, "The *Theorica Planetarum* Literature of the Middle Ages," *Classica et Mediaevalia* 23 (1962): 225–32; E. J. Aiton, "Peurbach's *Theoricae novae planetarum*: a translation with commentary," *Osiris*, 2nd ser., 3 (1987): 5–44; and Barker and Goldstein, "Realism and Instrumentalism."

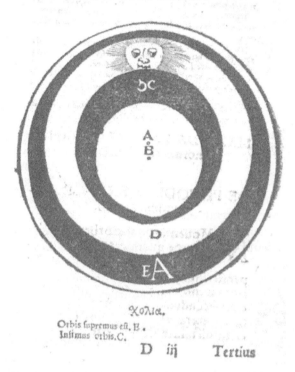

Figure 4.1. The Sun's three orbs. From Erasmus Reinhold, *Theoricae Novae Planetarum Georgii Purbachii Germani ab Erasmo Reinholdo Salveldensi auctae...* (Wittenberg: Lufft, 1542), f. Diii r.

(Figure 4.2). Just as the body of the Sun was supposed to move in its model, this little epicycle sphere (or *sphaerula*, F) would then move in the eccentric orb M defined by the inner and outer orbs (here shown in the model for the Moon as G and K). The motion corresponding to the rotation of the epicycle was now the rolling of this sphere. Again, if the heavens were imagined to contain such devices, Ptolemy's extremely successful planetary models could be understood causally, and connected with higher sciences such as physics and cosmology.

However, a special difficulty prevented causal explanation in astronomy: Apollonius's theorem made it easy to specify alternative possible causes for planetary motion, and the vantage point of an

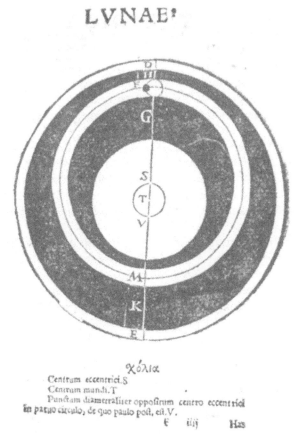

Figure 4.2. The Moon's orbs, including its epicycle. From Erasmus Reinhold, *Theoricae Novae Planetarum Georgii Purbachii Germani ab Erasmo Reinholdo Salveldensi auctae*... (Wittenberg: Lufft, 1542), f. Fiiii r.

observer confined to the Earth gave no way of choosing between them. Although sixteenth-century *theoricae* used an eccentric model for the motion of the Sun, by Apollonius's theorem the same results could have been achieved using a circle concentric to the Earth carrying an epicycle.[38] The two models would correspond to completely different

[38] In the *Almagest* (iii.3) Ptolemy presents both the eccentric and the epicyclic models for the Sun and shows that they are equivalent (G. J. Toomer, *Ptolemy's Almagest* [New York: Springer, 1984], p. 143). However, in the *Planetary Hypotheses*, the later work giving three-dimensional constructions, he clearly prefers the eccentric model (O. Neugebauer, *History of Ancient Mathematical Astronomy* [New York: Springer,

patterns of orbs. In the lunar and planetary models matters were even worse. These models used both eccentric circles and epicycles. Hence they could be replaced by equivalent models with concentrics and double epicycles, or an eccentric carrying an eccentric with no epicycles, and various other permutations. All these models would lead to different sets of orbs and different causal accounts. An observer on the earth had no grounds for eliminating any of these a posteriori possibilities to achieve a priori demonstration in astronomy. Astronomy was therefore confined to a posteriori proofs.[39] As Kepler's teacher Michael Maestlin put it, in a set of theses defended at Heidelberg in 1582, "We judge that the investigation of the circles and celestial spheres must be begun a posteriori, that is from the particular appearances of the motions, . . . but, not a priori, since no one is able to ascend into the aethereal region, where he would see everything in person."[40]

When Rheticus arranged for the publication of Copernicus's book, Nicolas Osiander added a preface that insisted on reading it in the standard astronomical mode – that is, a posteriori. In the preface Osiander reminded readers that epicyclic and eccentric models were equivalent; that astronomers had no way to choose between them and that although the physicist aspired to causal explanation the astronomer had to settle for what was useful or convenient.[41] This had the

1975], p. 924, and fig. 91, p. 1403). My thanks to Bernard R. Goldstein for these references.

39 Some astronomers speak of ὅτι demonstrations that establish models from observations of planetary position, and oppose them to διότι proofs that take the *theorica* models for granted and derive planetary positions. However, they do this without connecting the models to physics or offering causal explanations, as indeed they cannot, given the problem raised by Apollonius's theorem. Examples are Erasmus Reinhold, *Theoricae novae planetarum* (Wittenberg: Lufft, 1542), B6v–B7r (unpaginated), and C. Wursteisen, *Quaestiones vero in theoricas planetarum Purbachii* (Basel: Henricpetri, 1569), pp. 6–7.

40 *De astronomiae hypothesibus sive de circulis sphaericis et orbibus theoricis, disputatio ad discutiendum proposita, a' M. Michaele Maestlino Goeppingensis . . . respondente Matthia Menero Mulhusano* (Heidelberg: Mylius, 1582): "A2r Quaestio. . . . Anne tales circuli et orbes revera in coelo existant, an vero mera figmenta Mathematica sint. Thesis I. Investigationem Circulorum et orbium coelestium censemus a posteriori inchoandam esse, hoc est, ab apparentiis motuum particularibus, quae universali experientiae aequalitatis reclamare videntur: non autem a priori, siquidem in aetheream regionem nemo ascendere potest, qui omnia coram spectet." On Maestlin's Copernicanism, see Charlotte Methuen, "Maestlin's Teaching of Copernicus: The Evidence of His University Textbook and Disputations," *Isis* 87 (1996): 230–47.

41 The unsigned preface appears in N. Copernicus, *De revolutionibus orbium coelestium*, (Nuremberg: Petreius, 1543), iV–iiR (Facsimile reprint, New York: Johnson Reprint Corporation, 1965).

advantage of deflating in advance objections to Copernicus's astronomy based on its incompatibility with accepted physics and theology, but also denied Copernicus's real aspirations to provide causal explanations where none had appeared before. Wittenberg astronomers like Reinhold, Peucer, and Hilderich von Varel continued to use Copernicus in this way. Confining astronomy to a posteriori demonstrations, it was therefore legitimate to use Copernicus's mathematical models. On this view, there was no way *in astronomy* to determine whether the Earth or the Sun was the true center of the planetary motions. The issue was quite simply irrelevant for positional astronomy.

It is important to emphasize that the Wittenberg a posteriori reading of Copernicus was not equivalent to instrumentalism or fictionalism, as it has sometimes been called by more recent writers. Instrumentalists or fictionalists do not regard the investigation of the real state of the world corresponding to their theories as a legitimate part of science. Wittenberg astronomers and their contemporaries *did* continue to regard such investigations as a legitimate part of science. If one were ever to take the instrumentalist position, surely it would be in the case of an account, like Copernicus's, that had the additional liabilities of contradicting the accepted higher sciences of physics and theology. Someone who took this position should simply never mention Copernicus's cosmology. Hilderich von Varel's detailed description of Copernicus is therefore clearly incompatible with an instrumentalist agenda applied to astronomy. Further evidence is provided by a frequent trope of sixteenth-century astronomers, that they would be in a position to complete a priori proofs in astronomy if only they were transported to the heavens, or had reliable testimony from spirits that had been there.[42] We should remind ourselves that, for all these astronomers and natural philosophers, the ascent of the soul was a theological certainty. Remarks like that of Maestlin, quoted earlier, indicate only a temporary incapacity. After death, the soul of the virtuous astronomer would pass through regions of the heavens that would supply direct information on the real configuration of the spheres and orbs. Passages in which sixteenth-century astronomers compare their present situation unfavorably with observers in the heavens emphasize their belief that there was something to be found there. A priori causal knowledge of the world remained the scientific

---

42 Spirits from the heavens would know: Wojciech de Brudzewo, *Commentariolum super theoricas novas planetarum Georgii Purbachii*, ed. L. A. Birkenmajer (Cracow: Jagellonian University Press, 1900), pp. 26–7. Those who have visited the heavens would know: Wursteisen, *Quaestiones*, b6r (unpaginated). For text and translations, see Barker and Goldstein, "Realism and Instrumentalism."

ideal, even for scientists who, in astronomy, were limited to a posteriori proofs.[43]

The continuing importance of the ideal of causal proof, even to astronomers, is further supported by the speed with which Copernicus's followers developed a set of arguments for his new system specifying matters that Copernicus had explained a priori, where conventional astronomy had offered only a posteriori proofs. Rheticus was the first to contrast Copernicus's approach with the usual manner of proceeding in astronomy.[44] In a letter to Stadius published in 1556 Gemma Frisius writes: "Ptolemy assumes that the three superior planets (for example) rising achronychally, that is diametrically opposite the Sun, would always be in the perigee of their epicycles; and this is τὸ ὅτι [a posteriori]. The hypotheses of Copernicus entail this same truth by necessity, and demonstrate [it] δι ὅτι [a priori]."[45]

Similarly, writing to Tycho Brahe in 1588, Christopher Rothmann highlights the conversion of a posteriori demonstrations to a priori in the new system. Ptolemy and Brahe himself explain only *how* (*quomodo*) stations and retrogressions occur. By contrast: "The Copernican [account] shows not only that [the planets'] stations and retrogressions occur, but also *why* [*quare*] they occur as they do.... these retrogressions appear in this way to us on account of the movement of the Earth" (emphasis added).[46] Maestlin describes Kepler's project in just

---

43 The case made briefly here is made at greater length in Barker and Goldstein, "Realism and Instrumentalism."

44 For example, Georg J. Rhäticus, *Narratio prima*, ed., trans., and comm. Henri Hugonnard-Roche and Jean-Pierre Verdet with Michel-Pierre Lerner and Alain Segonds. (Wrocław: Ossolineum, 1982), sec. ix, p. 110: "Cum autem tum in physicis, tum in astronomicis, ab effectibus et observationibus ut plurimum ad principia sit processus."

45 Gemma Frisius, "Epistola ad I Stadius" in *Ephemerides novae et exactae ab a. 1554 ad a. 1570*, ed. I. Stadius (Coloniae Agrippinae: A. Birkmann, 1556), A2r: "Nam quod tres superiores (ut demus exemplum) ἀκρονυχοι, sive e diametro Solis positi, semper sint in perigeo suo epicycli, assumit Ptolomaeus, atque hoc est τὸ ὅτι. Verum Copernici hypotheses idem illud necessario inferunt, ac demonstrant δι ὅτι." On Gemma Frisius's Copernicanism, see Bernard R. Goldstein, "Remarks on Gemma Frisius's *De radio astronomico et geometrico*," in *From Ancient Omens to Statistical Mechanics*, ed. J. L. Berggren and B. R. Goldstein (Copenhagen: University Library, 1987), pp. 167–79.

46 Rothmann to Brahe, 13 Cal. Oct. 1588, *Tychonis Brahe opera omnia*, ed. J. L. E. Dreyer (Copenhagen: Libraria Gyldendaliana, 1919), 6:158, line 124 "Etsi enim ostendit, quomodo..."; lines 128–32: "At Copernicea, non tantum, quod fiant regressiones et stationes illae, ostendit, sed etiam quare sic fiant... sed quod regressiones illae ita nobis propter motuum Terrae appareant." The passage is also noted by Bruce Moran, "Christopher Rothmann, the Copernican Theory, and Institutional and Technical Influences on the Criticism of Aristotelian Cosmology," *Sixteenth Century Journal* 13 (1982): 100. On Rothmann, see also Bernard R. Goldstein and Peter Barker,

these terms, writing to the duke of Württemberg in 1595: "Up to now everything that has been disputed and written about the height and size of all celestial spheres has its basis only from astronomical observations, that is a posteriori. But no one had even understood that astronomy might have some help a priori."[47]

By the time of Kepler's education at Tübingen in the 1580s two distinct positions on Copernicus's work had emerged in northern Europe. The majority, and conservative, position, represented by Reinhold and his followers, read Copernicus in the light of accepted doctrines of demonstration that limited astronomy – but not other sciences – to a posteriori demonstrations. The minority, and radical, position, represented by Rheticus, Gemma Frisius, and Rothmann, suggested that Copernicus could supply a priori proofs in astronomy. These positions were not entirely contradictory. The conservative position regarded positional data and nothing else as the "effect" to be demonstrated or explained, and denied the possibility of reasoning to a unique model for the mechanism generating it. Copernicus had not discussed orbs explicitly, and his two-dimensional models could be converted into many different configurations by applying Apollonius's theorem. The original difficulty that limited astronomy to a posteriori demonstration therefore remained even in Copernicus's system.

Radical Copernicans like Gemma Frisius and Rothmann saw the possibility of giving a priori demonstrations for matters that were undeniably astronomical but had not been demonstrated, or explained, in earlier accounts. An important example, as we saw earlier, was the phenomenon of retrogression. In astronomy as understood by the writers of *theoricae*, the reasoning proceeded a posteriori from a set of observations for each planet that shows a retrogression has taken place, to a model using an epicycle that accounts for these observations and predicts future retrogressions of that planet. By contrast, once the relative sizes of the planets' spheres are given together with their heliocentric periods, the Copernican system entails the *existence* of this phenomenon for all planets.[48] This is a large part of what makes

"The Role of Rothmann in the Dissolution of the Celestial Spheres," *British Journal for the History of Science* 28 (1995): 385–403.

47 Maestlin to Duke Friedrich of Württemberg, March 12, 1596, in *Johannes Kepler gesammelte Werke*, ed. Walther von Dyck and Max Casper (Munich: C. H. Beck, 1937), 13: no. 31: lines 20 ff. (hereafter *KGW*). Translated by James R. Voelkel, in "Development and Reception of Kepler's Physical Astronomy, 1593–1609" (Ph.D. diss., Indiana University, 1994), UMI no. 9518529, p. 42.

48 Radical Copernicans claimed as a virtue that their account explained the existence of phenomena like retrogression. But earlier accounts had not *failed* to explain retrogression, and, of course, to be predictively accurate they had to *incorporate* the

Copernicus's account a system (a term he did not apply himself) in contrast to the set of Ptolemaic models found, for example, in the *theorica* literature.

Astronomy as a priori Science

The idea of providential design lay in the background of any Lutheran discussion of Copernicus, seeming to reinforce the conservative position. In the usual hierarchy of the sciences, theology subsumed physics, which in turn subsumed astronomy. Physics seemed to demonstrate the centrality of the earth and to establish a pattern of the world that doomed Earth-bound astronomers to a posteriori demonstrations in astronomy. These patterns were reinforced by the argument from design – an argument linking natural philosophy and theology. But theology also offered resources that Kepler would use to counterbalance the conservative influence of physics – ideas that concerned the nature of God and the nature of the human soul, although not central to theology proper.

During the fifteenth and sixteenth centuries the Neoplatonist idea that God was a geometer had been frequently repeated. A related theme treated God as an architect, geometry being one of the prerequisites for architectural success. Although prominent in pre-Lutheran sources (for instance, Nicholas of Cusa), the idea of the geometer God assumed special importance with the Lutheran emphasis on the providential plan. A geometer god would have a geometrical plan for his providentially ordered universe.

A second useful doctrine, again not original with Lutherans but assuming a special importance for them, concerned the faculty of the human soul known as the natural light. This faculty gave access to knowledge innate in the soul since its creation by God. Melanchthon was clearly familiar with this doctrine before he became a Lutheran. He mentions it in an address given before he moved to Wittenberg.[49] At Wittenberg, however, Melanchthon developed a specifically Lutheran application of the doctrine. Its most important instance was the

> phenomenon. Earlier accounts had simply not aspired to explain the existence of the phenomenon. Not surprisingly, the radical Copernican arguments sometimes sound circular and self-congratulatory. A cynic might accuse them of first inventing a requirement that no one else accepts and then congratulating themselves for meeting it. Thomas Kuhn deserves special praise for recognizing that situations like this were a recurrent feature of scientific change, as well as for drawing the attention of historians to its noncumulative aspects.

49 *De artibus liberalibus oratio Philippo Melanchthone Tubingae anno 1517*, in J. R. Schneider, *Philip Melanchthon's Rhetorical Construal of Biblical Authority* (Lewiston, N.Y.: Edwin Mellen, 1990), pp. 38–43.

theory of "natural law" – the set of basic moral principles that all human beings were supposed to be born with:

> The law of nature, therefore, is a common judgement to which all men give the same consent. This law which God has engraved on the mind of each is suitable for the shaping of morals. For just as there are certain common principles in the theoretical branches of learning, in mathematics for instance (they might be called "common thoughts" or "*a priori* principles," such as "The whole is greater than its parts"), so there are certain common axioms and *a priori* principles in the realm of morals: these constitute the ground rules for human activity. (We must use these terms for pedagogical reasons.) These rules for human activity are rightly called "laws of nature."[50]

These fundamental principles were employed by Melanchthon as the underpinning for human institutions such as marriage, private property, the rule of law generally, and submission to legally constituted secular authority in particular.[51] But Melanchthon makes it clear that the natural light was not limited to moral principles. It also gave access to fundamental truths of arithmetic and geometry that since antiquity had been paradigms of certainty. Kepler recognized this aspect of the doctrine:

> [T]here are, as I began to say above not a few more special principles of mathematics, which are understood by means of the common light of nature, do not need demonstration and are at first associated with quantities, and then are applied to other things, insofar as they share something with quantities. Of those principles there are more in mathematics than in other contemplative sciences, because of the very nature of human understanding itself, which seems to be such, by the law of creation, that it cannot know anything perfectly except quantities or by means of quantities. This is why it is the case that the conclusions of mathematics are the most certain and unquestioned.[52]

And again, near the beginning of the *Apologia pro Tychone contra Ursum*, Kepler tells us that geometers relied on the natural light to

---

50 P. Melanchthon, *Loci communes* (1521), trans. L. J. Satre and W. Pauck, Library of Christian Classics, vol. 19 (Philadelphia: Westminster Press, 1969), p. 50.

51 Kusukawa, *Melanchthon*, pp. 62–74, and esp. pp. 70ff. Melanchthon's students contributed books on natural law enumerating these moral principles and their corresponding virtues.

52 Kepler, *De Quantitatibus*, in chap. 1, *Johannis Kepleri astronomi opera omnia* (1871), ed. C. Frisch, 8:148; trans. G. Cifoletti, "Kepler's *De Quantitatibus*," *Annals of Science* 43 (1986): 224. I have argued elsewhere that the *De Quantitatibus* was most likely composed between 1596 and 1599, and that it shows us Kepler's views on the origin of mathematical certainty, at least early in his career; see "Kepler's Epistemology,"

guarantee the certainty of their conclusions before the historical development of logic:

> The first use of the word [hypothesis] was by geometers. Before the birth of logic as a part of philosophy, when they wanted to expound their demonstrations by the natural light of the mind, they used to start their demonstration from some established beginning. . . . Those things that were certain and acknowledged by all, they used, therefore, to call by the special name "axioms," that is to say opinions that had authority with all.[53]

The association of (perhaps debatable) moral principles with apparently certain mathematical truths no doubt added plausibility to their assertion by Melanchthon. But accepting the certainty of geometrical truths through the natural light offered Kepler a route to circumvent the obstacles to a priori proofs in astronomy. On the received view, the detailed structure of the heavens should be revealed by a three-step process. First, a posteriori demonstrations showing possible models for the observed positions of a planet needed to be constructed. As a second step, one among these models had to be shown to be the real cause of the initial observations, before proceeding to the third step, the a priori demonstration of the observations from their actual causes. But if God was a geometer, then the geometrical part of his providential plan for the world would be accessible to human beings through the natural light, without need for previous steps of a posteriori reasoning or *negotatio*. Astronomical proofs could begin a priori, with the knowledge of principles supplied by the natural light. This, in effect, was the opportunity Kepler seized in his first important book, *Mysterium cosmographicum* or *Secret of the Universe*.

The *Mysterium* has not generally been read as an exposition of God's providential design of the world, although Kepler states the point quite clearly at the outset. In the "Greetings to the Reader," preceding the dedication, Kepler tells us that the book is about "The nature of the universe, God's motive and plan for creating it."[54] The preface begins by repeating this motif, and again in chapter 4 of the *Mysterium*, we are told: "I think that from the love of God for man a great many of the causes of the features in the universe can be deduced. Certainly at least nobody will deny that in fitting out the dwelling

---

in *Method and Order in Renaissance Philosophy of Nature*, ed. D. DiLiscia, E. Kessler, and C. Methuen (Aldershot: Ashgate Publishing, 1997), pp. 355–68.

53 Kepler, *Apologia*, chap. 1, trans. N. Jardine, *The Birth of History and Philosophy of Science: Kepler's A Defense of Tycho against Ursus with Essays on Its Provenance and Significance* (Cambridge: Cambridge University Press, 1984), p. 137.

54 Kepler, *Mysterium Cosmographicum*, trans. A. M. Duncan (New York: Abaris Books, 1981), p. 49.

place of the universe God considered its future inhabitant [i.e., Man] again and again. For the end both of the whole creation and of the universe is Man."[55]

The theme of providential design appears again when Kepler lists the central questions he will answer: "There were three [questions] in particular about which I persistently sought the reasons why they were such and not otherwise: the number, size and motion of the orbs."[56] Kepler's ambition to answer all of these questions shows that the *Mysterium* cannot be read exclusively as a contribution to astronomy. Astronomy had indeed addressed the motion of the orbs. But the question of their sizes ventures beyond astronomy into natural philosophy, although it was customary to treat the sizes and distances of celestial objects in the same books that presented *theorica* models. One simple reason for this failure to keep the disciplines separate was that the method of calculating distances introduced by Ptolemy depended on the dimensions derived from the models of the planets' orbs, although additional principles from natural philosophy had to be added to this astronomical information.[57] The very first question takes us outside even natural philosophy. The reason for the *number* of the orbs – why there are six planets and not some other number – can only be answered in terms of God's providential plan for the world.[58]

Kepler's use of the providence doctrine is unusual in several respects. Unlike Mersenne and other contemporaries who use arguments from design to establish the existence of God, Kepler assumes the existence of a geometer God in order to argue for the existence of a certain kind of plan – a geometrical plan. As is well known, Kepler suggests that the geometer God constructed the world by using each Platonic regular solid just once to define the spacing of the planets, whose orbs are circumscribed and inscribed between the solids. The

---

55 Ibid., p. 107.
56 Ibid., p. 63.
57 On the principles from natural philosophy needed to make these calculations, see Goldstein and Barker, "Role of Rothmann," pp. 386–9. In general, see A. Van Helden, *Measuring the Universe* (Chicago: Chicago University Press, 1985).
58 Kepler's contemporaries also connected the questions of the number and arrangement of the planets with divine providence. In 1624, for example, Marin Mersenne wrote: "It is impossible that there would be the *number* of planets and stars such as there is, and that the heavens could keep the *distance* between them, if there were not someone who has given them these proportions and who has made them in this number, rather than another." M. Mersenne, *The Impiety of the Deists, Atheists, and Libertines of This Time Fought against and Defeated Point by Point Using Reasons Derived from Philosophy and Theology*, in *Texts in Context: Descartes' Meditations*, ed. Roger Ariew, John Cottingham, and Tom Sorell (Cambridge: Cambridge University Press, 1998), p. 145. See also Barker, "Kepler's Epistemology," pp. 365ff.

geometrical truth that there are only five Platonic solids explains the cosmological truth that there are only six planets. Through the doctrine of the natural light, these last results achieve complete certainty and may serve as principles for a priori demonstrations providing causal knowledge in astronomy.

Now the causes of planetary positions, the basic data in astronomy, are the positions of the planetary orbs. The cause of the positions of the planetary orbs is God's geometrical plan. There are two possible schemes for ordering planetary positions and planetary orbs, Ptolemy's and Copernicus's. Only one of these accords with God's geometrical plan for the universe. This happens to be Copernicus's. In discovering that the Copernican planetary distances and only the Copernican distances accord with God's providential plan, Kepler has provided the strongest possible defense of Copernicus's system available to a sixteenth-century Christian. The stated mystery that God constructed the universe using each regular solid only once therefore becomes a defense of an even more important unstated mystery.

Conclusion

Lutheranism was not antagonistic to the new astronomy but appropriated it for its own purposes. Wittenberg astronomers were well versed in Copernicus's ideas, and complimentary about them. It might almost be said that spreading Copernicus's ideas was a special Lutheran project. Lutherans expressed early and strong interest in Copernicus's work, even arranging for its publication. Although *De revolutionibus* received attention from Catholics as well, there is no Catholic effort anything like the systematic spreading of Copernican mathematical methods undertaken by astronomers trained at Wittenberg and by their students. By the end of the sixteenth century if you were a Protestant studying almost anywhere in German-speaking Europe, you would be taught the ideas of Reinhold, and taught that they originated with Copernicus. This was Copernicus as the Lutherans knew him, without equants but also without heliocentrism. It is also striking that Lutherans were responsible for the first two book-length treatments of Copernicus as we know him, the author of the cosmic scheme that places the Sun in the center of the planets' motions. These were Rheticus's *Narratio prima* in 1540, followed in 1596 by Kepler's *Mysterium cosmographicum*.

The earliest Lutheran humanist astronomers were Lutherans first, humanists second, and astronomers after that. In Hilderich von Varel we have a scholar whose main interests were theological, but who did important work in astronomy. His attitude to Copernicus is representative of first-generation Lutheran astronomers, while his career contrasts with those of two other Lutherans, Michael Maestlin and Johann Kepler. Like von Varel, Maestlin and Kepler both studied theology and mathematics. Maestlin served as a Lutheran deacon before becoming professor of mathematics at Tübingen, a post he held for the rest of his life. Kepler came to Tübingen to study theology under Haffenreffer, a student of Melanchthon, but spent a great deal of his time studying astronomy with Maestlin. Kepler was obliged to abandon any real thought of a career in the Lutheran church when he developed religious views that would have made it impossible for him to sign the Formula of Concord,[59] but he had ambitions to support himself in another way, as the client of a wealthy patron who could support a resident mathematician or advisor in astronomy and astrology.

A fairly usual method of attracting a patron's attention, at this moment in history, was to dedicate a book to him or her. The books that Maestlin and Kepler used for these purposes are another interesting contrast. Although he was a convinced Copernican, Maestlin wrote an *Epitome of Astronomy* that is completely conventional – and Earth-centered.[60] Also a convinced Copernican, Maestlin's student Kepler wrote a book, boldly entitled *Secret of the Universe*, that is an ingenious application of Lutheran theological ideas to natural philosophy. In Kepler's early work, then, we see religion as not merely supportive of the new science, but religion contributing to the details of its development.

For Kepler to defend Copernican cosmology in the *Mysterium* he needed to draw on a variety of resources. He required a detailed knowledge of Copernican mathematical techniques. This was provided by *De revolutionibus* itself and by his teacher Maestlin, but also by the books of Melanchthon's followers like Reinhold, Peucer, and Hilderich von Varel. But to make a case for the cosmic scheme – to place the Sun in the center of the planetary motions – Kepler needed to circumvent the long-established arguments that denied causal knowledge in astronomy. To do this Kepler drew on two additional Lutheran ideas, both directly stemming from Melanchthon: on the one hand, the idea that all natural philosophy, *including* astronomy, revealed God's providential plan, and on the other hand, the doctrine of the natural light that made knowledge of geometry certain. Kepler brought these ideas together by taking more seriously than his con-

---

59 Max Caspar, *Kepler* (New York: Collier, 1962) pp. 49, 217.
60 *Epitome Astronomiae... M. Michaelem Maestlinum Goeppingensem, Matheseos in Academia Heidelbergensi Professorem* (Heidelberg: Jacobus Mylius, 1585; Tübingen: Gruppenbachius, 1588). See also Methuen, "Maestlin's Teaching of Copernicus."

temporaries the commonplace that God was a geometer. This combination enabled Kepler to fulfill the Copernican aspiration to produce a priori proofs where before there were only a posteriori ones. Maestlin acknowledged Kepler's success when he wrote:

> Whenever I think about your demonstrations, in which you most skillfully prove Copernicus's hypotheses *a priori*, I am exceedingly glad that some learned man is found, who dares to assail the utterances of that throng of ignorant mathematicians speaking against these hypotheses of Copernicus. And the more so, that he even lays open the plan of the creator in the creation of the world, and with the utmost ingenuity the measure of the spheres, which even for Copernicus himself had to be tracked down . . . , only *a posteriori* by means of observations. It seems now to have an open access *a priori*.[61]

Few of the ideas that Kepler built on were the exclusive property of Lutherans,[62] but all were especially prominent in Lutheran education during the sixteenth century. This combination of commitments (to the doctrines of demonstration a posteriori and a priori, the doctrine of the natural light, and the Lutheran principle of divine ubiquity resulting in a providential design for the universe) allowed Kepler to understand not only the stated mystery of the universe – that God constructed the universe by inscribing and circumscribing the spheres around the Platonic solids – but the even more important unstated mystery, that God was a Copernican.

61 Maestlin to Kepler, February 27, 1596, *KGW* 13: no. 29, lines 1–10, trans. Voelkel, "Kepler's Physical Astronomy," p. 40.
62 As Funkenstein and Osler show, in different ways, for "providence": A. Funkenstein, *Theology and the Scientific Imagination* (Princeton: Princeton University Press, 1986), esp. chap. 4.; M. J. Osler, *Divine Will and the Mechanical Philosophy* (Cambridge: Cambridge University Press, 1994).

5

Catholic Natural Philosophy: Alchemy and the Revivification of Sir Kenelm Digby

BRUCE JANACEK

On May 1, 1633, Lady Venetia Digby, the wife of Sir Kenelm Digby, died. Before her body was prepared for burial, however, Digby called upon his good friend, Anthony Van Dyck, the renowned Dutch painter of the Caroline court, and asked him to come to his home immediately and paint a portrait of Venetia as she lay in her deathbed. Van Dyck had already painted her portrait twice before, once sitting alone and again in a family portrait with Digby and their two children. Van Dyck agreed immediately to undertake a final painting of Digby's beloved wife.

Serenity dominates the result of his efforts, titled *Venetia Stanley, Lady Digby, on Her Deathbed* (Figure 5.1). In the portrait we view Venetia through parted bed curtains and, if we did not know better, we might think she was merely falling asleep. Her head is propped up and resting delicately on the open palm of her hand right hand. She is dressed in a white gown and cap, a pearl necklace gracing her neck, reclining in luxurious comfort, supported by numerous pillows and enveloped in sumptuous, velvety bedding. Yet it is the moment that captivates us: her left eye is almost but not quite closed, as if we are forever witnessing Venetia's last moments on earth. Drifting into an eternal slumber, her portrait conveys to us not only serenity but the immediacy of the moment as well. The only liberties that Van Dyck took were to add to the painting the pearl necklace, symbolic of Digby's eternal and perfect love, and a scattering of withered rose petals across her lap, representing the transience of earthly beauty. The painting hung in Digby's room for the rest of his life and he often remarked on the great comfort it brought him.[1]

I am grateful to Margaret J. Osler and Paula Findlen for their careful readings of earlier versions of this chapter.

1 Clare Gittings, "Venetia's Death and Kenelm's Mourning," in *Death, Passion and Politics: Van Dyck's Portraits of Venetia Stanley and George Digby*, ed. Ann Sumner (London: Dulwich Picture Gallery, 1995), pp. 54–68.

Figure 5.1. *Venetia Stanley, Lady Digby, on Her Deathbed*, by Anthony Van Dyck. By permission of the Trustees of Dulwich Picture Gallery, London.

However, Venetia's partially opened eye conveyed more than simply a poignant moment in time. For the remaining thirty years of his own life, the portrait reminded Digby of eternal life. After her passing, Digby slipped into a two-year period of mourning. Contemporaries whispered that Digby himself had been responsible for his young wife's death, brought on by his insistence that she drink "viper wine," a drink thought to stave off the aging process. An autopsy was performed, and Digby was cleared of any charges. Nevertheless Digby's grief was intense. Whether his sorrow was based on guilt or simply the longing of lost love, his letters written between 1633 and 1635 indicate that he suffered a wide range of emotions, from mental pain, guilt, anger, an idealization of Venetia, and even hallucinations – all revealing the depth of his bereavement.[2]

2 Ibid., pp. 61–7. Gittings notes that Digby's symptoms of bereavement are common in our time as well.

Eventually Digby found solace by returning to two traditions from which he had drifted away since a young man: Catholicism and the occult tradition of alchemy. The reasons why Digby returned to the Catholic fold are complex. For one, Digby had been born into a notorious Catholic family: his father, Sir Everard Digby, had been executed in 1606 for his involvement in the foiled Catholic rebellion known as the Gunpowder Plot in the previous year. Digby, born in 1603, was an infant at this time and was raised Catholic by his mother. As a young man, he joined the Church of England, concerned more about his rising political star in a society increasingly dominated by Puritan values and norms than he was about family tradition and devotion. So at a time of intense grief and sorrow, surely his family's long ties to the Roman Catholic Church must have had some effect on his reconversion.

Other factors led to his decision as well. The Catholic Church placed great emphasis on the material elements of religion: the veneration of relics, prayers accompanied with rosary beads, the lighting of candles for departed souls, all of which spoke to the deep tactile connection the Catholic Church had with the temporal world. England in the 1630s, however, despite the efforts of Charles I and Archbishop William Laud to narrow the perceived if not the real distance between the Churches of England and Rome, was experiencing a rising tide of radical Protestant sects, most notably Puritanism.[3] Digby found the Protestant rejection of formal liturgy, transubstantiation, and free will to be distasteful if not offensive. For him, Protestantism represented division and discord while the Catholic Church stood for perfection, inclusiveness, and unity. Catholic doctrine comforted Digby: the miracle of transubstantiation that occurred at every mass, providing God's grace for all communicants and not just the preordained elect as the Protestant denominations proclaimed, the perfection and symmetry of the Trinity, the antiquity and origins of the church were all spiritually and intellectually compelling, making Catholicism a powerful force in Digby's intellectual life.[4]

Indeed, Digby's return to Catholicism was even related to his alchemical studies. Just as B. J. T. Dobbs argues that Newton's Arianism

---

3 Kevin Sharpe argues that the efforts of Charles I and Laud to instill a more uniform liturgy into the Church of England were an outgrowth of their policy to achieve order and consolidate monarchical power, not to bring the Church of England closer to the Roman Church. See Kevin Sharpe, *The Personal Rule of Charles I* (New Haven: Yale University Press, 1992), pp. 275–402.

4 See, for example, Sir Kenelm Digby, *A Discourse Concerning Infallibility in Religion. Written by a Person of Quality, to an Eminent Lord* (Amsterdam, 1652); *A Conference with a Lady about Choice of Religion* (London, 1638); *Letters between Lord George Digby and Sir Kenelm Digby Concerning Religion* (London, 1651).

permeated his alchemy, so had Catholicism saturated not only Digby's natural philosophy but his alchemical studies as well.[5] Digby's interest in the Catholic understanding of the resurrection of the dead, the revivification of matter, and his desire for a unified and inclusive Christendom, one that would be reborn and renewed after the disastrously divisive effects of the Reformation, led him to study alchemy.

Digby believed that the transmutation of metals was only possible with techniques that could be replicated – a common requirement of numerous practitioners of the day. In her study of Digby, Dobbs argued that he explained his weapon salve or "powder of sympathy" to accord with his mechanical universe, thereby eliminating the "astral and spiritual influences involved in Paracelsian theory."[6]

Digby approached alchemy in much the same way: he attempted to explain the process of transmutation as nearly as he could according to the principles of the mechanical philosophy. At the same time Digby integrated his religious belief into the mechanical philosophy and alchemy. Like the miracle of transubstantiation, alchemical transmutation had to be assured when all the components were correctly in place. In effect, an alchemical process had to be as reliable as the Catholic mass. For Digby, alchemy combined faith and reason more perfectly than any other tradition.

Digby's interest in alchemy, however, was also driven by less lofty and far more personal reasons. Although his alchemical studies probably began before 1620, while he was a student at Oxford, and continued sporadically until the 1660s, Digby did not begin to study alchemy in earnest until about 1635, about the same time as he emerged from his mourning for Venetia. At that time he set up a laboratory in Gresham College, London, focusing particularly on the process known as palingenesis, the revivification of plants from their calcined ashes. Given that he returned to study alchemy at this time with a particular focus on palingenesis suggests that there was a profound psychological aspect to his alchemical studies.[7]

Digby is known to historians of seventeenth-century England as a public figure who intermittently and deftly served as a diplomat and counselor to two Stuart monarchs and the Cromwellian Protectorate,

---

5 Betty Jo Dobbs, "Studies in the Natural Philosophy of Sir Kenelm Digby," *Ambix* 18 (1971): 1–25, esp. 13.

6 The weapon salve or "powder of sympathy" was a substance that that treated the cause of the injury rather than the injury itself. For example, a knife that caused a stab wound would be treated with the powder of sympathy rather than the wound itself. See B. J. T. Dobbs, *The Janus Faces of Genius: The Role of Alchemy in Newton's Thought* (Cambridge: Cambridge University Press, 1991), pp. 81–4.

7 Betty Jo Dobbs, "Studies in the Natural Philosophy of Sir Kenelm Digby: Part II, Digby and Alchemy," *Ambix* 20 (1973): 143–63, esp. 148.

all the while brandishing his Catholicism. After his reconversion, Digby became an unapologetic advocate of his church, even when it was politically imprudent to be so. Digby's personal beliefs and public persona ran strikingly parallel courses.

To historians of science, Digby's studies in the mechanical philosophy especially, but also his work in botany and alchemy, have made him a figure of moderate scholarly interest.[8] The purpose of this essay is to suggest that to understand Digby's natural philosophy one must begin by understanding his religious belief. John Henry has written on Digby's interpretation of certain aspects of Catholic doctrine and his role in the so-called Blackloists, a group of English Catholics who in the late 1640s tried to establish an English Catholic Church that was separate from Rome.[9] Henry's study is significant not least because it identifies clearly the role of Catholicism in Digby's thought; it was *the* single defining element in his life, dominating not only his natural philosophy but also his public life. Catholic dogma provoked troubling implications of the mechanical philosophy for Digby. However, Digby did not resort to the traditional religious sources to defend his faith but rather natural philosophy. Alchemical principles allowed him to resolve the theological problems the new matter theory presented. Digby's intellectual philosophy was an integration of his deepseated faith in the teachings of the Roman Catholic Church and principles of natural philosophy. Clarifying the roles of faith and reason in Digby's life will help us to obtain a clearer understanding of his contribution to the Scientific Revolution.

Catholicism and the Mechanical Philosophy

In 1635 Digby left England and traveled to Paris to escape the increasingly intolerant religious atmosphere in England. He resided there throughout most of the next twenty-five years. France was particularly appealing to Digby because the religious toleration provided by the Edict of Nantes was certainly a model that he admired while he was in his self-imposed exile.[10] Digby always hoped that England would

8 Between 1971 and 1974 Dobbs published a series of articles in *Ambix* that examined Digby's natural philosophy and alchemy and they remain thus far the most sophisticated treatment of Digby's natural philosophy. In addition to the articles cited in nn. 5 and 7, see her third and concluding study of Digby: "Digby's Experimental Alchemy – The Book of Secrets," *Ambix* 21 (1974): 1–28.
9 John Henry, "Atomism and Eschatology: Catholicism and Natural Philosophy in the Interregnum," *British Journal for the History of Science* 15 (1982): 211–39.
10 A proclamation by Henry IV in 1598, the Edict of Nantes provided limited toleration, liberties, and security to Huguenots. Thought to be too generous by Catholics and too restrictive by Huguenots, the edict suffered some erosion in the early decades of

eventually arrive at the same solution for Catholics as France had for Protestants.[11]

Shortly after he first arrived in Paris, Digby composed his first theological treatise, *A Conference with a Lady*. The woman in question was Lady Purbeck, Frances Coke Villiers, daughter of one of England's most luminous jurists, Sir Edward Coke. Lady Purbeck had fled London to avoid an arranged marriage and Digby, who had only recently reconverted himself, used this opportunity to convert her to Catholicism.[12] He published the manuscript the following year, in 1636.

In his treatise, Digby argued that the Roman Catholic Church's ability to identify its lineage to the time of Christ was evidence that the church was unable to teach a false doctrine. However, while Catholic doctrine was the most infallible expression of God's will on earth, it was not the only source of spiritual edification. Natural philosophy also taught theological truths: "fayth thus delivered [by the Catholic Church], is absolutely more certaine and infallible then any naturall science whatsoever. And yet sciences are so certaine (I meane such as depend of experience and demonstration) as he were not a rationall man that should refuse his assent unto them."[13] Catholicism was to Digby a science, or more accurately, a science in the classical Latin sense, a *scientia* – knowledge that could be perceived and understood through rational means. Catholic doctrine was infallible and natural philosophy was nearly so. Therefore, for Digby, natural philosophy could be used to frame and reinforce Catholic positions, even the most controversial ones.

One of the most contentious doctrines was the church's position that grace was not predestined but that individuals had the free will to receive God's grace. The doctrine of predestination was perceived by many in the Church of England to be *the* defining element that set them apart from Roman Catholicism, the doctrine that Reformers used to define themselves as the "true church."[14] Distinguishing the Church

   the seventeenth century but continued to allow religious liberty until Louis XIV finally revoked that last vestige of the edict in 1685.
11 MS 215 Coddrington Library, All Souls College, Oxford.
12 Lady Purbeck did convert to Catholicism but whether because of Digby's argument remains unclear. Given the difficulties she had had in England, French Catholicism may have simply have seemed to her both a reprieve and a revenge. Lacking money to support herself, Lady Purbeck was sent briefly to a convent but she could not abide the strict rules of behavior. She eventually returned to England and was able to, as Petersson observed, "live out her life relatively unmolested." See R. T. Petersson, *Sir Kenelm Digby, the Ornament of England, 1603–1665* (London: Jonathan Cape, 1956), pp. 139–42, esp. p. 141.
13 Digby, *Conference with a Lady*, pp. 55–6.
14 See Dewey D. Wallace Jr., *Puritans and Predestination: Grace in English Protestant Theology, 1525–1695* (Chapel Hill: University of North Carolina Press, 1982), p. ix.

of England from the Roman Catholic Church became especially critical in the seventeenth century because both James I and Charles I wanted their church to be recognized for its universal and inclusive appeal. In his first speech to Parliament, James I acknowledged "the Roman church to be our mother church although defiled with some infirmities and corruptions."[15] When Charles I communicated with papal agents in the 1630s, he maintained that he was a good Catholic and not a schismatic.[16]

For Digby, predestination was the antithesis of universality precisely because of its exclusivity. He remarked: "The Religion taught by the reformers, hath yet a greater restriction then that: from even in its owne nature, it is not for all sorts of persons and for all capacities."[17] Digby's hostility to predestination was unyielding: in 1643 he later referred to "the abysse of Predestination."[18] Digby's unity – and the unity of Charles I's monarchy – was the kind of unity that the English associated so closely with tyranny: the liturgy and prayers were prescribed, papal authority crossed all sovereign borders, and religious dissent, while always present in the Catholic Church, would also always remain subservient to papal will. It was precisely for these reasons that virtually all English Protestants found Catholicism so abhorrent. Digby was advocating a position that sectarians and indeed most members of the Church of England would never accept and was doing so in a political atmosphere that was becoming increasingly volatile and intolerant of such dissent with each passing year. Yet it was precisely this issue of dissent that made Catholicism so compelling for Digby. For beyond the question of authority lay his larger goal of a unified Christendom.

Digby argued that the Catholic Church espoused a faith "more cer-

---

15 Erica Veevers, *Images of Love and Religion: Queen Henrietta Maria and Court Entertainments* (Cambridge: Cambridge University Press, 1989), p. 183. For more on Charles I, politics, and Catholicism, see Gordon Alvion, *Charles I and the Court of Rome: A Study in Seventeenth-Century Diplomacy* (London: Burns, Oates and Washbourne, 1935); John Bossy, *The English Catholic Community, 1570–1850* (New York: Oxford University Press, 1976); Martin J. Havran, *The Catholics in Caroline England* (Stanford, Calif.: Stanford University Press, 1962); Brian Magee, *The English Recusants: A Study of the Post-Reformation Catholic Survival and the Operation of the Recusancy Laws* (London: Burns, Oates and Washbourne, 1938); Daniel Massa, "Giordano Bruno's Ideas in Seventeenth-Century England," *Journal of the History of Ideas* 38 (1977): 227–42; David Mathew, *Catholicism in England, 1535–1935: Portrait of a Minority, Its Culture and Tradition* (London: Longmans, 1936); Graham Parry, *The Golden Age Restor'd: The Culture of the Stuart Court, 1603–42* (Manchester: Manchester University Press, 1981).
16 Veevers, *Images of Love and Religion*, p. 183.
17 Digby, *Conference with a Lady*, p. 83.
18 Sir Kenelm Digby, *Observations upon Religio Medici. Occasionally Written by Sir Kenelme Digby, Knight* (London, 1643), pp. 19–20.

taine and more infallible then any naturall science whatsoever." Then turning to natural philosophy, he proceeded to argue that it was devoted to the study of matter, which was corrupt and imperfect. However, he said, "faith depending upon the indefectibility of humane nature, which is infinitly more noble then they, and whose forme is elevated beyond the reach of matter.... It followeth of consequence that faith must be lesse subject to contingency, and lesse lyable to error then naturall sciences are. And they being in universall infallible and certaine; faith must likewise be so too; and more if more may be."[19]

Apparently, individuals' ability to record the intent of God were, for Digby, a prelude to their ability to read the Book of Nature accurately. The perfection of faith expressed in Catholic doctrine provided his entrée into the study of the perfect and regular order of the natural world. In making this argument Digby was not simply equating Catholicism with unity but demonstrating the triumph of Catholicism. For the Reformers to be right, the church fathers and Christian tradition must be wrong. However, such a consideration was not possible because, he reasoned, God would not allow an imperfect Church to exist for a millennium and a half.[20] Further, his church was far more widely dispersed and universal than any Protestant confession. The perfection of faith expressed by individuals and codified in the doctrine of the Catholic Church was the standard that all laws and principles in natural philosophy must meet.

Digby's Catholic apologies were always framed in terms of toleration. In a letter to Charles I that was copied and circulated, Digby appealed to the monarch's sense of justice to justify religious toleration: "I cannot make it too smooth and plaine. It is evident in nature, Reason, Justice and conscience, that punishments ought not to be inflicted where no crime hath bin committed."[21] Digby used principles of natural philosophy as tools to demonstrate the irenic nature of his position. Digby believed that the church was more perfectly aligned with the natural world than Protestantism. By adhering consistently to the fundamentally *natural* aspects of toleration, Digby could advocate his religious belief – even in the eyes of his opponents – with at least a measure of legitimacy.

However, Digby's religious beliefs caused more than political and practical difficulties; they could also prove to be philosophically inconvenient as well. Even as Digby was proving the universal truths of Catholic doctrine through principles of natural philosophy, his studies in natural philosophy, particularly the mechanical philosophy, were

19 Digby, *Conference with a Lady*, pp. 70–1.
20 Ibid., pp. 78–81.
21 MS 215 Coddrington Library, All Souls College, Oxford. This letter is an undated copy of the original and was not written in Digby's hand.

proving themselves to be theologically troubling. This problem required Digby to focus much of his energy toward its complex – and ultimately alchemical – resolution.

While he lived in Paris, Digby was associated with the so-called Newcastle Circle, a group of English mechanical philosophers turned émigrés also centered in Paris, composed of Thomas Hobbes, Charles Cavendish, William Cavendish, then marquis of Newcastle, and John Pell who alone resided in the Netherlands. Digby, through Hobbes it seems, joined the group upon its arrival in Paris from Newcastle. During the nearly twenty-five years that Digby lived in Paris, he lived among enclaves of both French and English Catholics, all the while keeping abreast of events in England. These groups believed that universality was, to a great extent, the raison d'être of the Catholic Church. Digby and his English Catholic colleagues hoped that by using widely accepted principles of natural philosophy, some universal theological truths could be clarified.

During this same time, Digby, along with other English exiles, met and discussed the subtleties of the mechanical philosophy with virtually all of its most illustrious proponents of the day: Isaac Beeckman (1588–1637), René Descartes (1596–1650), Pierre Gassendi (1592–1655), Thomas Hobbes (1588–1679), and Marin Mersenne (1588–1648). They discussed and responded to each others' work and Digby was a part of this circle throughout the 1640s and 1650s.[22]

The mechanical philosophy was, from its inception, a philosophy that began with one single assumption: matter is passive. Change in matter did not occur because of an inherent tendency in matter to change, but rather as a result of external forces that compelled inert matter to change. Change, therefore, according to mechanism, was a result of motion.[23] Digby believed that the fundamental properties of bodies were quantity, density, and rarity, and these qualities accounted for the nature of matter.[24] Digby's mechanical philosophy certainly began with these premises but he also incorporated Stoic, Catholic, and alchemical ideas. Nevertheless, despite Digby's inclusive approach to the mechanical philosophy, he was surrounded by the foremost mechanical philosophers of the day and was deeply influenced by Descartes, Hobbes, and Mersenne.[25] Digby's considerations

---

22 See Robert Hugh Kargon, *Atomism in England from Hariot to Newton* (Oxford: Clarendon, 1966), pp. 63–76.
23 Gary Deason, "Reformation Theology and the Mechanistic Conception of Nature," in *God and Nature: Historical Essays on the Encounter between Christianity and Science*, ed. David C. Lindberg and Ronald L. Numbers (Berkeley: University of California Press, 1986), pp. 167–91, esp. p. 168.
24 Marie Boas Hall, "Digby, Kenelm," in *Dictionary of Scientific Biography*, ed. Charles Coulston Gillespie, 18 vols. (New York: Charles Scribner's Sons, 1970–80), 4:95–6.
25 Dobbs, "Studies in the Natural Philosophy of Sir Kenelm Digby," pp. 13–25.

on the mechanical philosophy may have been uniquely his but they were not wholly unorthodox either.

Like Mersenne and his colleague, Pierre Gassendi, Digby considered the theological implications of the mechanical philosophy very carefully. Digby always maintained that notions of natural philosophy were inherently inferior to theology. Earlier we noted that Digby recognized that even though natural philosophy could help untangle a theological knot, it was nevertheless inferior to pure theological reasoning. Like many natural philosophers of the seventeenth century, Digby considered the Book of Nature to be theologically complementary but inferior to the Book of God. Digby blurred the distinction between the two books, cleaving to the natural laws demonstrated through his studies and reason but keeping the sacred goals and nature of his work intact. The role of mechanical philosophy was prominent in his discussions of the soul. Although Digby outlined his position on the resurrection of the dead in his text, *Observations upon Religio Medici*, he examined it in much greater detail when he published his largest and most ambitious work, *Two Treatises: The Nature of Bodies*; [and] *The Nature of Mans Soule*.

The *Two Treatises* is a vast volume, first printed in a folio edition in Paris in 1644. The first treatise, devoted to physical bodies, comprises fully three-quarters of the entire text. Still, Digby intended to make the first part the foundation for what he believed was the more important second part, examination of the soul. In the preface he remarked:

> This writing was designed to have seene the light under the name of one treatise. But after it was drawne in paper; as I cast a view over it, I found the proemiall part (. . . which treateth of Bodies) so ample in respect of the other . . . that I readily apprehended my reader would thinke I had gone much astray from my text, when proposing to speake of the immortality of mans Soule, three parts of foure of the whole discourse, should not so much as in one word mention that soule, whose nature and proprieties I aymed at the discovery of. To avoyde this incongruity, occasioned mee to change the name and unity of the worke; and to make the survay of bodies, a body by it selfe: though subordinate to the treatise of the soule.[26]

He later wrote in the preface dedicated to his son, "Lett them draw the entire thridde through their fingers, and lett them examine the consequentnese of the whole body of the doctrine I deliver."[27] The

26 Sir Kenelme Digby, *Two Treatises: In the One of Which, The Nature of Bodies; In the Other, The Nature of Mans Soule; Is Looked Into: In Way of Discovery, of the Immortality of Reasonable Soules* (Paris, 1644), pref., n.p.
27 Ibid.

treatise devoted to the soul could not be understood without understanding the physical principles of the mechanical philosophy.

Digby's religious beliefs were prominent in his argument. Digby believed that the immortal soul could and indeed must be understood by natural philosophical principles if individuals, particularly non-Catholics, would accept his arguments.[28] Although Digby's concerns were not uniquely Catholic – later we will see how Protestant natural philosophers certainly could and did share them – Digby had to frame his arguments such that they would be acceptable to his mostly Protestant reading audience. This strategy did not mean that Digby distanced himself from his faith when it became inconvenient but it begins to explain why Digby turned to natural philosophy in order to integrate his religious argument with principles of natural philosophy.

The issues that Digby and his Catholic community discussed had already been resolved officially by the Fifth Lateran Council in 1513, which established the immortality of the soul as official dogma. In an almost haunting anticipation of Luther a few years hence, Pope Leo X's council asked that Christian natural philosophers "use all of their powers" to demonstrate that the immortality of the soul can be understood by natural reason and not by faith alone.[29] Digby's circle was certainly engaged in these issues on their own. Pierre Gassendi in particular, sought to resolve the troubling theological implications of the mechanical philosophy.[30]

Therefore it is crucial that we see the *Two Treatises* for what Digby and his colleagues saw it: as an answer to the theological problem that confronted Digby and his colleagues in Paris in the 1630s and 1640s.

28 In their preface to the *Two Treatises*, Holden and Tyrrel wrote that Digby was "a large & lofty soule, who not satisfyed with unexamined words & ambiguous termes, longing to know dyves deeply into the bowells of all corporeall & compounded things: and then *devinely speculats the nature of immateriall & subsistent formes*. Nor this by wrangling in aerie names with chimericall imaginations & fained suppositions of unknowne qualities, but *strongly stryving to disclose [t]he reall & connaturall truth of each thing in it self, and of one constant & continued thridde*, weaves his whole worke into one webbe." While Holden and Tyrrel's interests were in Digby's advocacy of English toleration of Catholics, they were also aware of Digby's strategy to use natural philosophy to express their shared theological positions more precisely.

29 Emily Michael and Fred S. Michael, "Two Early Modern Concepts of Mind: Reflecting Substance vs. Thinking Substance," *Journal of the History of Philosophy* 27 (1989): 29–48, esp. 31. Michael and Michael note that for the next 150 years, "personal immortality was viewed as a critical and pressing philosophical problem" (32). See also Margaret J. Osler, *Divine Will and the Mechanical Philosophy: Gassendi and Descartes on Contingency and Necessity in the Created World* (Cambridge: Cambridge University Press, 1994), p. 62.

30 See Osler, *Divine Will and the Mechanical Philosophy*, pp. 84–101.

Aquinas's four-hundred-year-old synthesis of Aristotle and Christianity had largely – though never wholly as the Condemnations of 1270 and 1277 had demonstrated – identified the role of God in Aristotelian physics.[31] Therefore, if Aristotelianism was to be replaced by mechanism, then the role of God had to be clarified once more.

Digby's philosophy of the soul at the time in which he composed the *Two Treatises* was firmly grounded in the soil of Aristotelianism. Aristotle's *De anima* opens with the statement that "the soul is, so to speak, the first principle of living things. We seek to contemplate and know its nature and substance and then the things that are accidental to it."[32] Like Aristotle, Digby believed that the soul was "the first principle of living things." Digby argued that the soul was an "orderer," which could both direct itself through its own volition and also communicate its dicta to order other things. Like matter, the soul was subject to motion but, in addition, the soul could "communicate it unto such thinges, as are to be ordered."[33] Digby was satisfied that he had demonstrated that the soul had intelligence and, especially, free will:

> But where experience falleth short, reason supplyeth, and sheweth us that of her owne nature she [the soul] is communicative of order; . . . it is manifest, that the action must by nature and in the universall consideration of it, beginne from the orderer (in whom order hath its life and subsistence) and not from that which is to receive it: then, sithence ordering is motion, if followeth evidently, that the soule is a moover and a beginner of motion.[34]

---

31 The Condemnations of 1270 and 1277 were conservative backlashes led by the bishop of Paris, Etienne Tempier. Tempier was concerned about the liberal and radical effort by some academics to extend autonomy to philosophy, especially Aristotelianism. The Condemnations, which included as many as twenty teachings from Aquinas as well as the Aristotelian propositions such as the eternity of the world, denial of personal immortality, determinism, denial of divine providence, and denial of free will, represent the thirteenth-century declaration of the subordinacy of philosophy to theology. A succinct discussion of the Condemnations appears in David C. Lindberg's *The Beginnings of Western Science: The European Scientific Tradition in Philosophical, Religious, and Institutional Context, 600 B.C. to A.D. 1450* (Chicago: University of Chicago Press, 1992), pp. 236–9. For more on the Condemnations, see Van Steenberghen, *Aristotle in the West*, trans. Leonard Johnston (Louvain: Nauwelaerts, 1955), chap. 9; John F. Wippel, "The Condemnations of 1270 and 1277 at Paris," *Journal of Medieval and Renaissance Studies* 7 (1977): 169–201; and Edward Grant, "The Condemnation of 1277, God's Absolute Power, and Physical Thought in the Late Middle Ages," *Viator* 10 (1979): 211–44.
32 Aristotle, *De anima (On the Soul)*, trans. Hugh Lawson-Tancred (New York: Penguin Classics, 1986), p. 126.
33 Digby, *Two Treatises*, pp. 411–12.
34 Ibid.

Digby's belief that the soul could communicate its will to material bodies was an implicit statement that individuals had free will. If the soul could not communicate its will to move matter or bodies, then free will was not possible – which was precisely what Digby rejected.

It may not be possible to exaggerate the incendiary nature of Digby's argument when it was read in England. When the *Two Treatises* was published in 1644, England was in the midst of its Civil War, a war ignited in part by this very issue of free will. The predestinarian theology of grace was the fundamental divide between the Church of England and the Catholic Church that began to form in the Elizabethan period and continued through the first half of the seventeenth century. James I's acceptance of the Synod of Dort's 1618 statement reaffirming the role of predestination of the soul as a tenet of the Church of England was a ringing endorsement of unconditional predestination, limited atonement for the elect, and irresistible grace.[35] This official statement and position was, however, merely a desperate sandbag trying to hold back the rising waters of Arminianism in England. During the mid- to late 1620s the English gentry was nearly hysterical in its fear of Catholics, and this fear was intimately related to the Arminian and predestination controversies.[36]

It was in this tumultuous setting that Digby approached his study of the soul according to the mechanical philosophy. Digby defined

---

35 Wallace, *Puritans and Predestination*, pp. 79–111, esp. pp. 81–2.
36 The Tyacke thesis, that the rise of Arminianism – the doctrine that reintroduced the role of free will in the Protestant interpretation of grace – in the 1620s was one of the crucial causes of the English Civil War, needs to be taken into account in any discussion of Arminianism in seventeenth-century England. The discussion, both for and against it, is vigorous. While it has not been entirely accepted, a vanguard of scholarship largely supports the notion that the English Civil War was caused by religious controversies, Arminianism being one of the central causes. For more on this debate, see Nicholas Tyacke, "Puritanism, Arminianism and Counter-Revolution," in *The Origins of the English Civil War*, ed. Conrad Russell (New York: Barnes and Noble, 1973), pp. 119–43; Peter White, "The Rise of Arminianism Reconsidered," *Past and Present*, no. 101 (1983): 34–54; William M. Lamont, "The Rise of Arminianism Reconsidered," *Past and Present*, no. 107 (1985): 227–31; Nicholas Tyacke and Peter White, "Debate: The Rise of Arminianism Reconsidered," *Past and Present*, no. 115 (1987): 201–29; Peter Lake, "Calvinism and the English Church, 1570–1635," *Past and Present*, no. 114 (1987): 32–76; Kenneth Fincham and Peter Lake, "The Ecclesiastical Policy of King James I," *Journal of British Studies* 24 (1985): 169–207; Jonathan M. Atkins, "Calvinist Bishops, Church of Unity and Arminianism," *Albion* 28 (1986): 411–28; Nicholas Tyacke, *Anti-Calvinists: The Rise of English Arminianism, c. 1590–1640* (Oxford: Oxford University Press, 1987); review of *Anti-Calvinists* by Kevin Sharpe and subsequent correspondence with Conrad Russell, Ian Greeen, and Thomas Cogswell, *Times Literary Supplement*, August 14, 21, 28, September 4, 18, 1987, pp. 884, 899, 925, 955, 1017. Sharpe's *The Personal Rule of Charles I* has become the leading moderating force to Tyacke's thesis.

the nature of a "separated soul" – that is, a soul that existed outside a body, in terms of place, time, and activity. He said that a spiritual substance did not reside in a single place but in all places. Similarly, just as a soul was not bound to a single place, neither was a soul bound to the constraints of time. Finally, Digby said, "A third property we may conceive to be in a separated soule, by apprehending her to an Activity; which that we may rightly understand, lett us compare her, in regard of working, with a body: reflecting then upon the nature of bodies, we shall find, that not any of them will do the functions they are framed for, unlesse some other thing do stirre them up, and cause them so to do."[37] Later Digby observed in a similar vein that "the soule, ... by its nature, motion may proceed from it, without any mutation in it, or without its receiving any order, direction, or impulse, from an extrinsecall cause."[38] In other words, unlike matter, the soul was not inert and could move itself without an outside force.

The mechanical philosophy served as his point of comparison to the soul. When Digby wrote, "a seperated [sic] soule, is of a nature to have, and to know, and to governe all thinges," Digby was deliberately distinguishing between spiritual matters and physical matter. This distinction was essential to Digby in order for him to distinguish his thought from the taint of predestination. Indeed, it was that issue that gave Digby's colleague, René Descartes, such notoriety.[39]

Free will and predestination, however, were not the only theological distinctions and clarifications that Digby had to make. Another equally crucial issue for Digby was locating the doctrine of the Trinity in the mechanical philosophy. Descartes's system was essentially the-

---

37 Digby, *Two Treatises*, p. 425.
38 Ibid. pp. 424–6.
39 Descartes's mechanism was seen by some to be a denial of God's power on earth but Margaret Osler has argued that Descartes's apparent denial of miracles ought not to be seen necessarily as a diminishment of God's power, but rather as Descartes's affirmation of the omnipresent stability of God. In contrast to the mechanical philosophers, such as Gassendi, Charleton, Boyle, and Newton – who explicitly allowed for the possibility of miracles – Descartes seems to have believed that God freely created necessary laws and having once accomplished his Creation, applied the laws without exception or qualification. Universal laws did not exist apart from God but once created, their stable existence provided certainty and demonstrative knowledge. See Margaret J. Osler, "Eternal Truths and the Laws of Nature: The Theological Foundations of Descartes' Philosophy of Nature," *Journal of the History of Ideas* 46 (1985): 349–62. The question of whether Boyle was an atomist is an especially subtle one examined by J. J. MacIntosh's "Boyle on Epicurean Atheism and Atomism," in *Atoms, Pneuma and Tranquillity: Epicurean and Stoic Themes in European Thought*, ed. Margaret J. Osler (Cambridge: Cambridge University Press, 1991), pp. 197–219.

istic: he had reserved a place for God the Creator but not one for God the Redeemer, much less the Holy Spirit. Thus, contemporary theologians' accusations that Descartes's system was essentially Pelagian – that individuals could enter the Kingdom of Heaven without divine grace – had some cause.[40] Although not all mechanical philosophers cared about this issue, Digby was deeply concerned about the doctrinal implications of the philosophy.[41]

Digby immunized himself from the virus of Pelagianism by turning to natural philosophy. Digby's first treatise on the nature of bodies had established the inherent coexistence that the body and the soul shared. Further, even though the *separated* soul was divine, it was subject to corruption and damnation as long as it existed within a corporeal body. While the soul was bound to an individual's body, the soul was subject to the reason of the individual, and reason had always been subject to sin and error. Still, the soul could be exalted while existing within the body through the action of knowledge. Digby believed that the soul, like matter, could mutate and increase its knowledge while the body existed. He remarked that "a soule in this life is subject to mutation, and may be perfected in knowledge."[42] However, knowledge did not simply elevate a soul in this world, but *improved* the soul's existence in the next: "That the knowledges which a soule getteth in this life will make her knowledge in the next life more perfect and firme."[43]

Yet it was not simply knowledge but natural philosophy that was particularly efficacious in the elevation of the soul. Digby noted that "the soules of men addicted to science whilst they lived here are more perfect in the next world then the soules of unlearned men."[44] Simply studying human action, Digby wrote, or working to obtain a skill in

---

40 Richard S. Westfall, "The Rise of Science and the Decline of Orthodox Christianity: A Study of Kepler, Descartes and Newton," in *God and Nature: Historical Essays on the Encounter between Christianity and Science*, ed. David C. Lindberg and Ronald L. Numbers (Berkeley: University of California Press, 1986), pp. 218–37, esp. pp. 226–7.

41 Digby was certainly not alone in his concerns. Pierre Gassendi in particular and Thomas Hobbes to a lesser extent tried to understand the role of God in the mechanical philosophy. For more on this issue, see Osler, *Divine Will and the Mechanical Philosophy*, pp. 36–77, and Lisa T. Sarasohn, "Motion and Morality: Pierre Gassendi, Thomas Hobbes and the Mechanical World-View," *Journal of the History of Ideas* 46 (1985): 363–79. Sarasohn notes that theological theories played a relatively small role in Gassendi's thought compared with his detailed studies in physical theories (p. 367). For more on Gassendi's philosophy in particular, see Lynn Sumida Joy, *Gassendi the Atomist: Advocate of History in an Age of Science* (Cambridge: Cambridge University Press, 1987).

42 Digby, *Two Treatises*, p. 433.

43 Ibid., p. 434.

44 Ibid., p. 435.

an art or a trade would not suffice in exalting the soul. Digby noted that, "they who spend their life here in the study and contemplation of the first noble objects, will, in the next, have their universall knowledge (that is the soule) strong and perfect: whiles the others, that played away their thoughts and time upon trifles, and seldome raysed their mindes above the pitch of sense, will be fainte throughout their former laizinesse, like bodies benummed with the palsey."[45] Knowledge obtained through contemplation and experimentalism and not simply through physical activity would exalt the soul. For Digby, natural philosophy combined contemplation and activity perfectly.

The soul was limited precisely because it was bound to the body, and as such the salvation of the soul was largely dependent upon the decisions of an individual's reason and the actions of an individual's body. This, Digby argued, was how individuals were responsible for their salvation and not subject to a predestined fate. Whether a soul was miserable after death was the result of judgments made on this earth by an individual's free will. Digby thought this consequence should not be seen as an injustice because, he wrote,

> the misery which we putt in a soule, proceedeth out of the inequality, not the falsity, of her judgements: for if a man be inclined to a lesser good, more then to a greater, he will in action betake himselfe to the lesser good, and desert the greater, ... and a greater inclination to a greater good: and in pure spirits, these inclinations are nothing else, but the strength of their judgements: which judgements in soules, whiles they are in their bodies, are made by the repetition of more acts from stronger causes, or in more favourable circumstances.[46]

Individuals were subject to tendencies but tendencies which, presumably, could either be overcome or unheeded; assurance did not figure into Digby's idea of salvation. Knowledge could exalt the soul and perhaps even lead it to its eternal salvation. However, knowledge was also intimately related to the corruptible body and therefore was, by definition, limited. Moreover, it was subject to reason, and that too could err. Thus, according to Digby the mechanist, an individual's will was the "motion" required to determine the outcome of one's soul.

The significance of Digby's *Two Treatises* was not in the natural philosophy it expounded but in the theology it sought to preserve. God was more than just a Creator and Arbiter of universal natural laws but had provided a path to redemption as well – a path that a

45 Ibid.
46 Ibid., p. 442.

Catholic or an Arminian could follow. However, while the mechanical philosophy had provided a soul with the potential for salvation, the means of that salvation remained unclear in the natural world. Digby still needed to identify the soteriological role – the Christ centered role – of God in his mechanical philosophy.

Thus, Digby's amalgam of Aristotelianism and Catholicism may have explained the role of free will and Creation in the universe, but a doctrine equally compelling for Digby, the doctrine of the Trinity, remained to be resolved. God the Redeemer and Spirit remained at large. However, he addressed these concerns not through the mechanical philosophy but through a tradition that had an ancient tie to Christianity, alchemy. Still, it was not merely to find these particular aspects of the Trinity that years later Digby developed his particular vision of alchemy. Alchemy also helped him to understand more clearly the theological idea of the resurrection of the dead. Digby's research in the last years of his life was concerned with the ideas of purification and redemption, and he integrated his Catholic beliefs with alchemy to resolve some of his nagging theological concerns.

Alchemy and Redemption in Digby's Natural Philosophy

The 1650s were largely a fallow period for Digby's studies in natural philosophy as he became more involved in his own nation's politics. Cromwell took advantage of Digby's numerous intimacies with the various European diplomatic circles and Digby came to be known by some as "Cromwell's confidant."[47] Upon the Restoration, Digby transmuted his loyalties into a position in Queen Henrietta Maria's court.

However, Digby never completely deserted his studies in natural philosophy. In 1658 he gave an address in Montpellier, France, on his sympathetic powder.[48] In 1661, four years before he died, Digby published the text from an address he gave at Gresham College in London, *A Discourse Concerning the Vegetation of Plants*. Like Digby's other treatises on natural philosophy, this text extended far beyond the study of botany. To Digby, vegetation, and more specifically germination, were cosmic events. He argued that the vegetating seed was a

---

47 *Calendar of State Papers Domestic*, February 1, 11, 1655–6, p. 159.
48 Sir Kenelm Digby, *A Late Discourse Made in a Solemne Assembly of Nobles and Learned Men at Montpelier in France; By Sr. Kenelme Digby, Knight, &c. Touching the Cure of Wounds by the Powder of Sympathy; With Instructions How to Make the Said Powder Whereby Many Other Secrets of Nature Are Unfolded. Rendred Faithfully out of French into English by R. White, Gent.* (London, 1658).

microcosm of a vast, vegetating process that occurred in the universe, a process that had important corollaries in religion and alchemy.

Not only did he use the language of an alchemist to describe a botanical process, but he also clearly had his own mortality in mind as he developed his argument:

> For, to give you a due account of the Vegetation of Plants; I should first examine the natures of *Rarefaction*, of *Condensation*, of *Filtration*, of *Fermentation*, of *Attraction*, of *Imbibition*, of *Concoction*, of *Augmentation*, of *Nourishment*, of *Assimilation*, and of sundry other actions or vertues (as we might term them) of the like strain: Which I should no sooner have made an end of, and have shewed you wherein consist the life and death of a Vegetable; *but presently I should have in my view, the reparation of a decaying life, and the reproduction of a faded one; and so ingulfe my self in the mysterious contemplation of the resurrection of dead and dissipated bodies, and how they may continue the same individuation, and be again the same identicall body, after so many strange changes, and after having put on so many different habits and shapes, as we daily see in the course of Nature.* [49]

Along with vegetation, fermentation was another alchemical process that he used to explain the resurrection of the dead. Digby wrote that fermentation was a process that could lead to a complete "Putrefaction, Dissolution, and Destruction of the compound." However, if the fermentation process was contained within certain limits, "then the body in which it was wrought, is raised to a nobler pitch, and the Ethereall spirits of it are actuated, and put in possession of their native vertue."[50] At this moment, Digby was providing an alchemical description of the resurrection of a vegetative body; presently, however, we will see that this was more a cornerstone for his revealing that these same principles applied to the resurrection of the dead.

Something that had been seen as a miraculous process was, according to Digby, comprehensible without requiring faith or spirituality, much less occult methods or techniques: "And it is want of consideration of judgement, which maketh men fly to occult and imaginary qualities, to shroud their ignorance under inconceiveable termes: Whereas nature in her self is pervious and open to humane discovery, if a due course be taken to dissect and survay her."[51] Digby used alchemy to explain a the-

---

49 Sir Kenelm Digby, *A Discourse Concerning the Vegetation of Plants. Spoken by Sir Kenelme Digby, at Gresham College, on the 23. of January, 1660. At a Meeting of the Society for Promoting Philosophical Knowledge by Experiments* (London, 1661), pp. 2–4 (emphasis added).

50 Ibid., pp. 11–14.

51 Ibid., pp. 46–8.

ological issue such as the resurrection of the dead but avoided the suggestion that he was engaged in occult matters.

The term "occult" as it was understood in the seventeenth century has received careful attention and analysis. Keith Hutchison argues that the word "was part of the technical Peripatetic terminology used to distinguish qualities which were evident to the senses from those which were hidden. In this context it was the antonym of 'manifest.' " Manifest qualities required using the senses, such as taste or sight. Typical occult qualities consisted of the motion of the planets, the magnetic virtue, or the result of a specific medical treatment.[52] Ron Millen has refined Hutchison's argument, noting that Scholastics in the sixteenth and seventeenth centuries contributed to the understanding of occult qualities for contemporaries.[53] John Henry has argued further, that the role of active or vitalistic principles played such a prominent role in seventeenth-century English natural philosophy precisely because mechanical philosophers believed that active but hidden or occult properties were the best way to understand God's creation and therefore served an essential role in defining the true religion. A cadre of English mechanical philosophers – Robert Boyle, Walter Charleton, and Henry More – considered it essential to identify and clarify God as the Prime Mover in the mechanical philosophy.[54] Although Digby shared the concerns and goals of his colleagues, he turned to alchemy to determine the role of God in the mechanical philosophy.[55]

To see universal implications of life and death in vegetation would have been eminently appropriate in the seventeenth century. However, to see religious significance in the process of regeneration would

---

52 Further, Hutchison has recognized the sympathetic attitude that Walter Charleton and Thomas Hobbes had toward occult qualities. Both natural philosophers, Hutchison argues, did not object to the existence or propriety of occult ideas but rather used occultism as an intellectual refuge. Their attacks were not so much pointed at occultism but Aristotelians who used insensible or occult qualities to bring their inquiries to a close rather than an initiation for new questions. Keith Hutchison, "What Happened to Occult Qualities in the Scientific Revolution?" *Isis* 73 (1982): 233–53, esp. 234–5. See also his "Supernaturalism and the Mechanical Philosophy," *History of Science* (1983): 297–333.
53 Ron Millen, "The Manifestation of Occult Qualities in the Scientific Revolution," in *Religion, Science, and Worldview*, ed. Margaret J. Osler and Paul Lawrence Farber (Cambridge: Cambridge University Press, 1985), pp. 185–216.
54 John Henry, "Occult Qualities and the Experimental Philosophy: Active Principles in Pre-Newtonian Matter Theory," *History of Science* 24 (1986): 335–81, esp. 357–8.
55 John Henry calls More's evangelical interpretation of Descartes's philosophy "somewhat idiosyncratic." Surely the same should be said of Digby's efforts as well. Ibid., pp. 338 and 352–8.

have been equally appropriate and Digby was certainly not alone in discussing this issue metaphorically, theologically, and literally. The idea of the resurrection of the dead had exercised generations of theologians. Questions of how, upon Christ's return, the dead would be raised, what the bodies would look like, and how they would be received by God were questions that were continually asked but for which satisfying answers were few.[56] The New Testament was not the only source for Digby in his search for rejuvenation, resurrection, and redemption; alchemy too promised the same for physical bodies. Therefore, it is likely that Digby's interest in this aspect of alchemy had as much to do with his own mortality as it did Venetia's. For it was in the last years of his life that Digby devoted himself almost solely to the study of palingenesis and other alchemical principles and phenomena.

Recall that palingenesis was the notion that plants and animals could be revived or resurrected from their calcined ashes.[57] Digby studied this process and declared that he had failed in his attempt to revive calcined flowers when he followed the process described by the Jesuit natural philosopher Athanasius Kircher. However, he said, when he followed Joseph Duchesne's method, it worked exactly as he had said it would.[58]

Digby guided his audience through Duchesne's palingenisis process, which ultimately led him to consider the resurrection of the dead.[59] Digby said that, "All this leadeth me to speak something of the Resurrection of humane bodyes. There we may find some firm and solid footing. Hitherto we have wandered up and down in the Mazes of Fleeting matter, *qua nunquam in eodem statu permanet* [as a shadow and continueth not]. And with great truth did *Job* apply that expression to the State of men living in this World."[60] Digby proceeded to explain that once humanity's "frail Mortality" had been put to rest, then "a state of permanence and immutability" would exist.[61]

> Not onely whiles the Soul is seperated from the Earthy Companion, but when she shall be cloathed again, that new flesh

---

56 See Caroline Walker Bynum, *The Resurrection of the Body in Western Christianity, 200–1336* (New York: Columbia University Press, 1995), and Norman T. Burns, *Christian Mortalism from Tyndale to Milton* (Cambridge, Mass.: Harvard University Press, 1972).
57 Dobbs, "Digby and Alchemy," p. 147.
58 Digby, *Vegetation*, pp. 72–6.
59 Ibid., pp. 76–85.
60 Ibid., p. 85. The line Digby invoked was from Job 14:2 which, in the Vulgate that he used read in full as follows: "Qui quasi flos egreditur et conteritur, et fugit velut umbra, et nunquam in eodem statu permanet." His contemporaries would have read the same passage from the King James Version as "He commeth forth like a flower, and is cut down: he fleeth also, as a shadow and continueth not."
61 Digby, *Vegetation*, pp. 85–6.

will partake of the constancy of her glorious Mate. But why doe I call it new flesh? I may be pardoned for doing so, when I consider the new qualities and endowments it shall have put on. But otherwise, *in substance and reality, it is the same, the very same, that (for example) accompanyed me in this long and tedious Pilgrimage upon Earth.* How is this? If a Caniball should feed upon my body, and convert it into the substance of his, can both of us rise again with the same bodyes we enjoyed here? Yes, without doubt we may. And I conceive, that the taking away of this difficulty, which hath so highly perplexed even the best Christians, will be so welcome a performance to them who yet have not met with it; that for its sake you will pardon the tediousnesse and coursenesse of all I have hitherto said. And with that, I will cease further troubling you. [emphasis added][62]

Just as Duchesne's process of palingenesis revived the calcined plant back to its original form, so would the resurrection of the dead be a revivification of bodies that would be *both* new and original. The substance of the human form would not be destroyed with death, only its accidental properties. The state of the body at death was not important. Thus, the destruction of a body (in Digby's case study, the particularly gruesome example of a cannibal) still would not destroy the body's potential to be raised in its original form, reassuring news for Christians concerned about the precise details of Judgment Day.

Digby then proceeded to clarify why it was important to understand the natural philosophy and the alchemical properties of vegetation and resurrection.

> But let us first rightly understand one another. I doe not undertake to shew here how this great work is wrought: . . . *But . . . to convince that there is no impossibility nor contradiction in nature, against this great and amazing Mystery.* If there were contradiction in it, it could not be true; it were not the subject of a Miracle. But if I prove that there is no repugnance against the feasibility of it, I am confident I shall not misse of hearty thanks from those sincere believers who have nothing to shake the firmnesse of their Faith, but the suspected impossibility of Mystery. [emphasis added][63]

For Digby, the resurrection of the dead was not a miraculous process but a natural one and therefore one that individuals could comprehend. If the process could be understood, then not only was it not

---

62 Ibid., pp. 86–7.
63 Ibid., pp. 87–9.

miraculous, but it was also reproducible for anyone who understood the process. The implication was that one did not have to be divinely chosen or "elect" to reproduce the alchemical process that imitated the resurrection of the dead. It was a gift from God explicable to those who could devote the time, energy, and resources to achieve the promised result. By eliminating the mystical or divine aspects of alchemy, Digby implicitly separated himself from the Paracelsian tradition that depended fundamentally on the role of God's blessings and the particular spirituality and sometimes even the election of a particular alchemist.[64] That an alchemical procedure be reproducible – no matter from what church a particular philosopher worshiped – was an important requirement for Digby and illustrates his desire to apply more rigorous principles in natural philosophy to the occult tradition of alchemy than previous generations had.[65]

Yet Digby's position on the doctrine of free choice was only one aspect of his theory on the resurrection of the dead. He also related it to his eclectic matter theory and his alchemy. He wrote that matter was the "Capacity to be this or that or any thing whatsoever." Yet what determined that capacity was "Form," the blueprint of matter on earth. "And consequently as long as the Form remaineth the same, the thing is the same, and the matter is the same. Were it not for [Form], how could any body under Heaven remain the same even but for a short Moments space? All sublunary things are in a perpetuall Flux."[66] By the 1660s, Digby's confidence in Aristotelianism had, to a degree at least, eroded. The Forms to which Digby referred were clearly Platonic Forms. Platonic thought dictated that matter was, by definition, mutable, the imperfect representations of immutable and universal Forms. Matter on earth composed all material objects – which were merely the imperfect representations of the perfect and immutable Forms, always "in a perpetuall Flux."[67] At this moment

---

64 There is no better place to begin a study of Paracelsianism than Walter Pagel's ground-breaking study of Paracelsian thought. See his *Paracelsus: An Introduction to the Philosophical Medicine in the Era of the Renaissance*, 2nd rev. ed. (New York: Karger, 1982), esp. pp. 50–1. For the relationship between alchemy and religion, see Herbert Breeger, "Elias Artista – A Precursor of the Messiah in Natural Science," in *Nineteen Eighty-Four: Science between Utopia and Dystopia*, ed. Everett Mendelsohn and Helga Nowotny (Dordrecht: Reidel, 1984), pp. 49–72; Thomas Willard, "Alchemy and the Bible," in *Centre and Labyrinth: Essays in Honour of Northrop Frye*, ed. Eleanor Cook, Chaviva Hošek, Jay Macpherson, Patricia Parker, and Julian Patrick (Toronto: Toronto University Press, 1983), pp. 115–27.
65 For more on the specifics of Digby's alchemical work, see Dobbs, "Digby and Alchemy," 143–63.
66 Digby, *Vegetation*, pp. 89–94.
67 Ibid., pp. 79–82.

Digby clearly departed from Aristotelianism. The Aristotelian conception of Form was founded on the principle that change was the principal task of nature. Digby's Forms are of the immutable, Platonic variety. Digby's willingness to dilute his Aristotelianism was driven by his desire to integrate alchemy into his natural philosophy.

It was also during this time that the mechanical philosophy became less important to Digby's understanding of natural philosophy. The resurrection of the dead did not occur as the result of the mechanical philosophy and indeed was not even relevant in the vegetation process: "For speaking rigorously, I cannot allow Plants to have Life. They are not *Se Moventia*, They have not a principle of motion within them. It is the operation of outward Agents upon them, that seteth on foot all the dance we have above so heedfuly observed, and which so near imitateth the motions of Life."[68] Resurrection occurred because of a "Universall Spirit" that resided in all things: "This Universall Spirit then being Homogeneall to all things, and being in effect the Spirit of Life, not onely to Plants, but to Animals also: were it not worth the labour to render it as usefull to mens bodyes, as to the reparations of Plants?"[69]

The "particles" of matter of which living bodies were composed were guided and organized by the divine force, Digby's "Universall Spirit," a force intimately related to alchemy.[70] For Digby, this spirit was present in all matter but was diminished as a result of the contact it had with base materials – the single exception was the most noble substance of all, gold. Digby noted that gold had the same nature, "as this aethereall Spirit; or rather, it is nothing but it, first corporifyed in a pure place, and then baked to a perfect Fixation. *Raymund Lully* in his excellent Treatise *de intentione operantium*, describeth admirably well the Genealogie of it. If then this perfect body (I mean Gold) could be rendered familiar and digestible to ours, there is no doubt but it would prove a kinde of Tree of Life to us."[71] Then, in the veiled language of the adept, Digby observed, "It [gold] is of it self too firmly composed for any Agent upon Earth to dissolve it. But peradventure the Mother [the Universall Spirit] that bore it, may reincrudate [i.e., make crude again] him [gold] and reduce him back into his first

---

68 Ibid.
69 Ibid., pp. 70. Dobbs called alchemy "medicine for men and metals," a thought that Digby's remark indicates he surely shared as well. See Dobbs, "Newton as Final Cause and First Mover," *Isis* 85 (1994): 633–43 (reprinted as Chapter 2 in this volume).
70 Dobbs was the first to notice this point. See her "Digby and Alchemy," pp. 143–63, esp. p. 157.
71 Digby, *Vegetation*, pp. 70–2.

volatile principles."[72] Following the principle of "like dissolves like," the "Univerall Spirit" would possibly dissolve gold since the two were so closely related. While not the philosophers' stone itself perhaps, Digby identified his "Universall Spirit" and gold at least as components of an alchemical process.[73]

Digby's reference to a Universal Spirit may be a reference to the Stoic *pneuma*, a medium that the matter of the cosmos shaped into its present state and provided the shape of living creatures on earth as well. Stoicism had experienced a revival in the seventeenth century, due largely to the literary excavations accomplished by one of the great late Renaissance humanists, Justus Lipsius.[74] Digby's description of gold appears to be metaphysical, perhaps even reminiscent of the Stoic *pneuma*. Gold was sublimely infused with the Universall Spirit that could heal wounds and extend life. Digby's natural philosophy at this time was as open to Stoic notions as it was to Aristotelian and Neoplatonic ones.

Digby's last alchemical text was his *Chymical Secrets*, published posthumously in 1683. Although this text was largely an alchemical recipe book, his opening preface was a vigorous apologetic both for alchemy and Catholic doctrine, beginning with a statement of his philosophy of the transmutation of metals. It was very brief – just three paragraphs in length – dwarfed by the 270 pages that follow of alchemical and medical recipes. Yet in these opening paragraphs Digby argued that alchemical processes should be replicable, that the Catholic doctrine of free will was a crucial element in the alchemical process, and that he had identified the place of God the Redeemer in the Trinity in alchemy.

The first point, that alchemy ought to be something that can be replicated, was one he had already established in his address on vegetation:

> Having Written so many Processes, and made so many Tryals, and heard so many Discourse of Learned Men upon this Subject, *I will give you an Account of an easie Method that I*

---

72 Ibid. p. 72. I agree with Dobbs's decoding of the term "Mother" as Digby's "Universall Spirit." See her "Digby and Alchemy," p. 157.

73 Dobbs, "Digby and Alchemy," p. 157.

74 Dobbs argued that the Stoic *pneuma* had important alchemical significance. She suggested that in Newton's alchemical treatise, "Of Natures obvious laws & processes in vegetation," Newton described remarkable similarities between his aether (a spirit that existed as a kind of fifth Aristotelian element) and the Stoic *pneuma*. Both were material, both inspired the form of earthly bodies, and both gave bodies the characteristics of life. See *Janus Faces*, pp. 27–9. The role of ancient philosophies in the history of science has been examined in numerous ways in Osler, *Atoms, Pneuma and Tranquillity*. In this collection, see esp. Peter Barker, "Stoic Contributions to Early Modern Science," pp. 135–54.

have resolved upon for accomplishing this Work. Namely, That all imperfect Metals and common [mercury] may be transmuted into [gold] by one and the same Method; to wit, by Maturation and Coction, and not by Generation; for that which is generated, is no more that which it was before it was generated: And that which is Corrupted, is no more that which it was before it was Corrupted.[75]

Digby's introductory statement had important implications both in natural philosophy and theology. This text was an alchemical recipe book; either Digby and his colleagues had attempted the alchemical recipes described in the book or he conferred with other individuals who had.[76] Thus, Digby's statement that all imperfect metals, including mercury, could be transmuted into gold by the same method was not only an observation about the common properties of metals but about the universality and the reliability of his methods. As Dobbs observed, "The experimental nature of the collection [*Chymical Secrets*] cannot be emphasized too strongly.... Starting materials are clearly described, the quantities necessary for each step are given, and requisite 'degrees of fire' are delineated. In many cases there is little or no difficulty in translating the processes into twentieth-century terminology."[77] Clearly, Digby's experiments were intended to be replicated by anyone who performed them.

Digby's second point was that his method was the result of "Maturation and Coction, and not by Generation." Maturation and coction were both alchemical terms. Maturation in the alchemical process described the process of converting a base metal into gold.[78] Francis Bacon, despite his seeming hostility to alchemy, made a specific reference to alchemical maturation in his *Sylva Sylvarum*: "We conceive indeed, that a perfect good Concotion, or Digestion, or Maturation of some Mettals, will produce Gold."[79]

---

75 Sir Kenelm Digby, *Chymical Secrets, And Rare Experiments In Physick & Philosophy, with Figures Collected and Experimented, by the Honourable and Learned Sir Kenelm Digby, Chancellor to the Late Queen-Mother of England. Containing, Many Rare and Unheard of Medicines, Menstruums, and Alkahests; the Philosophical Arcanum of Flamel Artesius, Pontanus and Zachary, with the True Secret of Volatizing the Fixed Salt of Tartar*, published since his Death by George Hartman Chymist, and Steward to the aforesaid Sir Kenelm (London, 1683), pp. 1–2.
76 Dobbs, "Studies in the Natural Philosophy of Sir Kenelm Digby: Part III, Digby's Experimental Alchemy – The Book of *Secrets*," p. 3.
77 Ibid.
78 *Oxford English Dictionary*, s.v. "maturation."
79 Francis Bacon, *The Works of Francis Bacon*, ed. James Spedding et al. (Boston: Brown and Taggard, 1862), vol. 4, § 326, p. 316. I discuss Bacon's complex approach to alchemy in "Alchemical Visions: Piety and Privilege in Early Modern England" (Ph.D.

The term, coction, also had alchemical significance. While coction can mean to cook something and was also a key term in Hippocratic and Galenic physiology, for the alchemist, coction was a preparation through a natural process in which a metal was brought gradually to perfection.[80] The significance of these two terms and why Digby needed to distinguish them from "Generation" was related to his matter theory. Like virtually all other mechanical philosophers, Digby assumed that God made all the matter of the universe ex nihilo, that is, from nothing. The task of the alchemist was to purify the corruptions from the existing matter. The creation or generation of matter was God's provenance, not humanity's.

However, Digby's theological concerns did not end with his reconciliation of mechanism and generation. He then proceeded with a theological discussion of transmutation:

> But the baser Metals after they are transmuted into [silver] or [gold] are still Metals nevertheless as they were before, and the transmutation of their kind is done by changing their accidental form, not their substantial, the perfection whereof is Maturity; for by Maturation the Metal is brought to a higher degree of perfection . . . yea, [gold] it self may be further perfected, and exalted in colour, as when the Stone is made of it, it will comunicate this Maturity to imperfect Metals.[81]

Digby's religious language – metaphors of perfection and imperfection, exaltation and transformation – illustrates the theological significance of alchemy.

It was in *Chymical Secrets* that he was finally able to put in place the remaining pieces of the Trinity that had eluded him in the *Two Treatises*. The mechanical philosophy had not unequivocally explained the presence of the Trinity in the universe. We saw that in the 1640s he could identify God the Creator, but God the Redeemer and God the Spirit eluded the grasp of the mechanical philosopher. In his *Chymical Secrets*, however, he made the following observation:

dissertation, University of California, Davis, 1996), pp. 211–69. For more on Bacon's alchemical studies, see John C. Briggs, *Francis Bacon and the Rhetoric of Nature* (Cambridge, Mass.: Harvard University Press, 1989), pp. 148–50; Joshua Gregory, "Chemistry and Alchemy in the Natural Philosophy of Francis Bacon," *Ambix* 2 (1938): 93–111, esp. p. 106; Stanton J. Linden, "Francis Bacon and Alchemy: The Reformation of Vulcan," *Journal of the History of Ideas* 35 (1974): 547–60, esp. 551; Paolo Rossi, *Francis Bacon: From Magic to Science*, trans. Sacha Rabinovitch (Chicago: University of Chicago Press, 1968), pp. 12–22; Muriel West, "Notes on the Importance of Alchemy to Modern Science in the Writings of Francis Bacon and Robert Boyle," *Ambix* 9 (1961): 102–14, esp. 103–4.

80 *Oxford English Dictionary*, s.v. "coction."
81 Digby, *Chymical Secrets*, p. 2.

[Y]ou shall suddenly see marvellous things, when the Soul of the said [gold] entreth into the Body of the [gold], by means of the Spirit, which is the Solary [mercury], and that by means of the said Soul, the Spirit uniteth with its Body, of three being made one. . . . The Body of the [gold] which was dead before, being by this only and admirable means animated, dignified, and filled with a Vegetative Life, and thereby acquire an inward Power of Multiplication, as well as the Sperms and Seeds of all Animals and Vegetables, and be made fit to grow and produce Fruit, (being sowed in a fit Earth) which it could not do before, because of that default.[82]

The Trinity of Creator ("Soul"), Incarnation ("Body"), and Holy Spirit ("Spirit") of God was present in the alchemical process. A creation by God's hands was dead but revivified by his "Spirit." Digby's alchemy confirmed that the matter theory of the mechanical philosophy needed to be tempered with other philosophies such as Christian Aristotelianism, Neoplatonism, and even Stoicism that were present in the alchemical process. Digby never discarded the mechanical philosophy but neither did he embrace it in its purest form as Descartes or Hobbes had. Like many other alchemists, Digby's natural philosophy was the seemingly contradictory amalgamation of both Aristotelianism and Neoplatonism that allowed him to incorporate alchemy into his argument. Digby believed in the Triune God but he understood and was convinced of the reasoning of the mechanical philosophy. Alchemy allowed him to demonstrate one of the central principles of the Catholic Church and resolve his conflict between faith and reason.

Digby's *Chymical Secrets* ought to be seen as the specific guide to what he outlined philosophically and theologically in his text on vegetation: the resurrection of the dead was a feasible possibility. The remains of animals and vegetables impregnated with the philosophers' stone received a new life that allowed them to grow and reproduce. The alchemical process was not simply concerned with leavening, growing, and developing. The philosophers' stone made that which was once dead, alive.

Conclusion

Digby's Catholicism permeated his natural philosophy.[83] He viewed alchemy in the same way that he viewed natural philosophy: not only

---

82 Ibid., p. 100.
83 The confluence of natural philosophy and transubstantiation has been identified in the work of Galileo as well. Pietro Redondi argues that Galileo's treatise, *The Assayer*,

as a way to resolve theological problems but also as a contribution to Catholic Reformation doctrine.[84] Digby very likely associated the transformation of the spirit through the rituals of the Church with his natural philosophy. If individuals could be transformed into a state of grace through a series of sacred rituals, then surely a process existed in which material bodies could also be purged of their corrupted elements and their pure elements revealed. If the processes of dissolution and putrefaction had alchemical significance, then they also had theological significance. Digby had already explained that his treatise on the nature of bodies could not be understood without understanding his treatise on the nature of the soul. Principles in natural philosophy that were irrefutable and immutable were therefore, by definition, expressions of God.

Digby's natural philosophy, theology, and even his politics need to be understood with his ultimate goal of unity in mind. Digby's Catholic natural philosophy was not developed in order to destroy Protestantism but to unify Christianity. His efforts to use natural philosophy to prove that the "true Church" resided in Catholic doctrine was part of the larger discussion that other natural philosophers as widely different as Robert Fludd, Sir Francis Bacon, and Thomas Hobbes were having: that natural philosophy can have profound religious and political effects that could heal the divisions in their society. For Digby and others, natural philosophy was the irenic tool that could heal the divisions of Christianity that England and Europe had endured since the Reformation.

The unity of Digby's thought was based fundamentally on his Catholicism. Like many natural philosophers, he borrowed freely from Christian, classical, Judaic, and occult influences but his natural philosophy began and ended with his Catholicism. The beauty that Digby saw in the Catholic Church was the unity of its doctrine. Catholic authorities were unanimous in their interpretation of Scripture whereas "scarce any two authors, out of the Romane Church, that have written of matters of faith have agreed in their tenets, but rather have dissented in fundamentall doctrine, and have inveighed against one another in their writings with great vehemence and bitternesse."[85]

---

was a nominalistic and atomistic attack on the doctrine of the Eucharist, the real focus of his trial, not Copernicanism. See his *Galileo: Heretic*, trans. Raymond Rosenthal (Princeton: Princeton University Press, 1987), pp. 203–71, 328–32. However, Redondi's argument has received close criticism. Richard S. Westfall made a careful examination of the argument in his *Essays on the Trial of Galileo* (Notre Dame, Ind.: University of Notre Dame Press, 1989), pp. 84–103.

84 Henry, "Atomism and Eschatology," p. 232.
85 Digby, *Conference with a Lady*, p. 74.

Digby made it clear that certainly there were disagreements within the Catholic Church, but not on matters as fundamental as Church doctrine.[86] Digby had witnessed personally the destructive results that religious divisions had wreaked on England. Individual religious belief had not resulted, in his mind, in a harmonious "true Church" but had led instead to Civil War. The authority of papal rule was perhaps not ideal, but the alternative was clearly worse. Through his natural philosophy Digby spent much of his life trying to reunite the Christian community – one that was particularly divided in England.

The last few years of Digby's life were devoted largely to his belief that natural philosophy and alchemy held universally redemptive powers. He had a laboratory in London and established a salon that became an established center in London for the mathematicians, chemists, philosophers, and writers of its day.[87] Digby's passing in 1665 brought an end to the salon, but it had already inspired Robert Boyle to produce his work on a mechanical explanation for the alchemical process, a work that Sir Isaac Newton read with great interest and which inspired his life's devotion to the secret art.[88]

As a figure in the Scientific Revolution, Sir Kenelm Digby dovetails rather neatly in the joint created by Dobbs's and Westfall's arguments. Digby's religious belief aside, he clearly devoted an enormous amount of his adult life to the study of natural philosophy. His *Two Treatises*, spanning over four hundred folio pages, is proof-positive that Digby's devotion was not directed solely at the altar of the Roman Catholic Church but to the tabernacle of the mechanical philosophy as well. His desire to integrate his belief with his scientific endeavors should not diminish his contribution to seventeenth-century scientific discourse. Indeed, Westfall's observation that Newton based his defense of Genesis in his correspondence with Thomas Burnet on science, not faith, is particularly salient.[89] Digby employed the same rhetorical strategy as Newton and Burnet in both his explicitly theological writings and his natural philosophy. As profoundly influential as Catholicism was to Digby, he too implicitly subordinated theology to science in his writing, illustrating the pervasive role of natural philosophy in even the most devout individuals.

If Westfall is correct, however, did both Digby's and Newton's rhetoric unwittingly undermine their own positions? Given the unde-

---

86 Ibid., pp. 74–5.
87 B. J. T. Dobbs, *The Foundations of Newton's Alchemy or, "The Hunting of the Greene Lyon"* (Cambridge: Cambridge University Press, 1975), pp. 78–9.
88 Ibid., p. 80.
89 Richard S. Westfall, "The Scientific Revolution Reasserted" (Chapter 3 in this volume).

niable and permanent subordination of faith to reason that occurred at the end of the seventeenth century, it seems not only possible but likely. If that is indeed the case, then Digby seems to share the crucial irony that Dobbs identified in Newton: both Digby and Newton contributed to the erosion of the prominence of religious belief in society by dint of their natural philosophy and science.

To assess Digby – and perhaps any other figure in seventeenth-century natural philosophy – solely through the lens of the Scientific Revolution is clearly to perceive him imperfectly, perhaps even unjustly. Digby's concerns – like Gassendi's, Boyle's, and, of course, Newton's – sometimes extended far beyond what we, even in our most generous moments, could consider to be legitimate contributions to natural philosophy, let alone science. Digby's unyielding efforts to integrate his natural philosophy and occult studies with his Catholic belief complicates any effort to identify his contributions to the Scientific Revolution. Dobbs's argument is prescient precisely for that reason. One may disagree with her view that the metaphor of the Scientific Revolution should be interred and yet still acknowledge her plea – one in which Dobbs and Westfall do agree – that we make every effort to locate and place individuals in the richness of *their* society, not ours. This seems to be the least that historians should be obliged to do.

Indeed, it is none other than Digby's alchemy that brings us back to where we began this discussion, with the death of his wife, Venetia. Although he eventually regained his composure several years after her death and even fell in love again, he never remarried. His consolation ultimately seems to have lain in his intellectual and spiritual pursuits. Digby believed that the alchemical process confirmed God's gift of redemption and resurrection. Alchemy reassured him that his faith did not have to reside merely in hope or miracles but that the agency of God could be replicated to remind individuals of his continuing presence on earth.

6

Vital Spirits: Redemption, Artisanship, and the New Philosophy in Early Modern Europe

PAMELA H. SMITH

In his youth the German chemist Johann Rudolph Glauber (1603/4–70) passed through Basel, where he met a philosophically minded man who showed him his own dead child preserved in the water from a nearby spring. Although this child had died some time before, its corpse remained untouched by putrefaction. Glauber never saw the source in Switzerland from which this water came, but he happened upon a similar spring in Vienna-Neustadt. The water in this spring transmuted everything placed in it into stone, preventing all things from rotting. Moreover it cured Glauber of a long and difficult fever.[1] Glauber stayed the whole winter observing this spring and found that it prevented the swamps from freezing, kept the grass green around its edge, and supported numerous turtles and other amphibious animals during the cold winter.[2]

Later, after many years of reading and experimenting, Glauber claimed to have discovered the makeup of the salt that gave this spring it powers. He called it *sal mirabile*, for the wonders it could work, and claimed that he produced it from common salt and sulfuric acid. He first announced his discovery in 1658, publishing a book about the nature of salts in general. In it he praised the manifold virtues of salt: it was the nutriment of all things, a symbol of eternity, the crucial ingredient in alchemical transmutation and in *aurum potabile* (potable gold), the cause of spontaneous generation, and the principle of all life. All salts, including common cooking salt, partook in the wondrous qualities of the elemental salt, but the *sal mirabile* was truly miraculous. It could be taken internally or applied externally on

---

1 Johann Rudolph Glauber, *Tractatus de natura salium. Oder aussführliche Beschreibung/ deren bekanten Salien, ... und absonderlich von Einem/ der Welt noch gantz unbekantem wunderliche Saltze ...* (Amsterdam: Johan Jansson, 1658), p. 71. Glauber says he was in Vienna-Neustadt twenty-one years earlier, which would put the date at 1637; however, some biographers claim he was there in 1625/6.
2 Ibid., p. 74.

wounds, and it caused both poisons and medicines to work with more efficacy.³ It was in fact the universal spirit by which the generating power and heat of the Sun was conveyed to Earth. In earlier writings, Glauber had believed this spirit to be a niter of some sort, but after 1658 he claimed to have discovered the vital spirit in his new salt.

According to Glauber, the medical reformer and chemical philosopher Theophrastus Philippus Aureolus Bombastus von Hohenheim called Paracelsus (ca. 1493–1541) had known of this salt and had called it *sal Enixum*. It had also been described by the author of *De re metallica*, Georgius Agricola (1494–1555), as occurring naturally in the salt springs, but neither of these authorities had known the composition of this salt and had not been able to produce it in the laboratory.⁴ In contrast, Glauber claimed to be able to produce it on a large scale.

In announcing his discovery, Glauber immediately defended his claim to authority in the matter of salts. Despite his lowly origins as a barber's son who had apprenticed as an apothecary and never been to university, he was proud of his capabilities as an artisan: "why should an experienced *Artist* not come closer [to the truth about the salt] when he industriously considers the matter? A scholar is good for preaching but not much else."⁵

Two years after proclaiming his discovery, Glauber wrote at more length about his *sal mirabile*, this time calling it the *sal artis* – the salt of art – that along with its "brother," saltpeter, was the true *Elias artista*, or the return of the prophet Elias, as the artisan *Elias artista*, to earth as a herald of the coming millennium. The prophet's return signaled the beginning of the millennium, instituting a new regime of goodness and plenty,⁶ and in the form of salt, he would bring about a great reformation in medicine and alchemy, making possible alchemical transmutation.⁷ In this work, as in so many others, Glauber coyly

---

3 Ibid., pp. 85, 95. The salt was hydrated sodium sulphate and is still known as a bath salt in Germany under the name of Glaubersalz.
4 Ibid., pp. 67–70.
5 "Warumb solte ein erfahrner *Artist*, wann er der sachen fleissig nachdencket/ nicht weiter damit kommen können? Einem Gelehrten ist guth predigen/ ein mehrers auff dißmal nicht." Ibid., p. 79.
6 On the *Elias artista* (the addition of the *artista* first made by Paracelsus) tradition, see Walter Pagel, "The Paracelsian Elias Artista and the Alchemical Tradition," in *Kreatur und Kosmos*, ed., Rosemarie Dilg-Frank, (Stuttgart: Gustav Fischer Verlag, 1981), pp. 6–19; and Herbert Breger, "*Elias Artista* – A Precursor of the Messiah in Natural Science," in *Nineteen Eighty-Four: Science between Utopia and Dystopia*, ed. Everett Mendelsohn and Helga Nowotny (New York: Reidel, 1984), pp. 49–72.
7 *Miraculi Mundi Ander Theil. Oder Dessen Vorlängst Geprophezeiten Eliae Artistae Triumphirlicher Einritt. Und auch Was der ELIAS ARTISTA für einer sey?* ... (Amsterdam: Johan Jansson, 1660).

stated his intention of discussing only the uses, not the preparation of the salt.

In 1667, Glauber prophesied again that the *Elias artista* would arrive sometime soon after his own death,[8] and a year later, he explained more closely what he had meant by this prediction. *Elias artista* would come to earth as a great light, good would replace evil, great changes in kingdoms would take place, especially in the Holy Roman Empire, and rulers would be overthrown. Glauber was quick to deny the literal meaning of his prophecy, for he had seen the social tumult that had taken place when Paracelsus had made a similar prediction. At that time, some people had taken his prophecy literally and had rioted, apparently trying to bring about themselves the coming overthrow of their governors. No, Glauber assured his readers, his own prophesying should be understood in its anagrammatic meaning, which could be discovered with some rearrangement of the letters, namely, as *Et Artis Salia*, as the "salts of art," of which *sal artis*, or his *sal mirabilis*, was the most important.[9] This salt, he asserted, would change the world because it contained, like all salts but far more powerfully, the vital spirit that caused generation in the cosmos.

Despite the cosmic significance of this salt as a tool of redemption, there was more than a hint of self-advertisement in Glauber's accounts of it, especially after he fell ill in 1662 and became unable to work in his laboratory. This advertisement was important, for Glauber's livelihood depended on the sale of his chemical products. When he moved from Giessen, where he was court apothecary, to Amsterdam in 1640, he entered a world in which it was possible to make one's fortune by selling goods on the market. Until he became incapacitated in the late 1660s, Glauber seems to have been highly successful in marketing his salts, mineral acids, and their recipes.

The story of Glauber's salt gives insight into the emergence of the new philosophy, for Glauber was a practitioner, claiming authority as an "experienced artist," and he advertised his products in the framework of a commercial market. His *sal mirabile* – simultaneously a part of alchemy, an instrument of redemption, and a valuable commercial commodity – represents a particular moment in the development of the new philosophy; a moment brought about by the entry of a new sort of person into the production of knowledge about nature. This

---

8 *Kurtze Erklärung über die Höllische Göttin Proserpinam* ... (Amsterdam: Johan Jansson van Waesberg and Elizaeus Weyerstraet, 1667), pp. 55–6.

9 *De elia artista. Oder Wass Elias artista für einen sey/ und wass Er in der Welt reformiren/ oder verbesseren werde/ wann Er kombt?* ... (Amsterdam: Johan Waesberg and the widow of Elizaeus Weyerstraet, 1668), pp. 3–4.

chapter investigates the particular tensions to which this new sort of philosopher was subjected and how he negotiated a place in his social and intellectual world. Glauber was trained as an artisan or practitioner, but desired to gain acceptance as an independent and disinterested natural philosopher. He sold his products, but was anxious to represent himself as untainted by commerce, understanding his work instead as part of the redemption of the world. It was, however, precisely this commercial context that made his livelihood as independent practitioner possible. And it was the emergence of the new philosophy that made possible Glauber's authority in natural questions as well as the attention paid to him by the *curiosi* of Europe. At the same time, practitioners such as Glauber helped form the practices of natural philosophy. Glauber's work can be used to illustrate the concerns and aims – and ultimately the new epistemology – such new figures brought with them into the making of the new philosophy of nature.

Practitioners and the New Philosophy

One problem with conventional accounts of the (capitalized) Scientific Revolution, such as that provided by Richard Westfall in Chapter 3, is that they focus on a small number of extraordinary individuals, they take for granted a hierarchy of the disciplines (with physics and astronomy at the apex), and they largely ignore the causal forces of social formation and political power. It is not of necessity that these accounts are so written, but they come out of a particular historiographic tradition, as B. J. T. Dobbs makes clear in her essay. Her essay does not deny the "many changes" of the sixteenth and seventeenth centuries, but it calls for a fresh perspective on the *longue durée* event commonly and usefully known as the Scientific Revolution. This *was* a momentous event; it involved an entire transformation in attitudes to the material world, new identities were created, and the significance of natural knowledge and the way in which it was gathered changed forever.

One of the most fruitful new perspectives that has come out of the turn to cultural history and the social construction of knowledge has been the focus on "practices."[10] For the old story of the Scientific

10 The founding text of this movement, at least for the study of early modern European natural philosophy, was Simon Schaffer and Steven Shapin, *Leviathan and the Air Pump* (Princeton: Princeton University Press, 1985), and its influence has been profound. For an overview of the recent works, see the appropriate sections of the bibliographical essay in Steven Shapin, *The Scientific Revolution* (Chicago: University of Chicago Press, 1996).

Revolution was largely the story of theoretical change. Of course, the story of the invention of "experiment" was also important, but it was written very much as the intellectual history of a practice.[11] This story left out the large numbers of individuals who began to show interest in and to practice the "new philosophy," and whose actions brought about the institutionalization of the new philosophy and, more importantly, made the new method of pursuing knowledge part of the habits of mind and action of European scholarly culture. What such a new perspective is yielding, among other things, is a story of how practices shape intellectual movements.[12]

Two important components of this story would be the new emphasis put on the active nature of the pursuit of natural knowledge and the increased attention paid to material, tangible objects – or *realien*. Nature had to be engaged with actively in order to make it yield up its secrets, whether by observation, collection, anatomizing, or work in the laboratory. These methods were practiced upon the objects of nature, but more significantly, one of the aims of this new pursuit of the knowledge of nature consisted in the production of effects – tangible objects or observable phenomena. An important source for this new epistemology was the entry into the knowledge-making process of a new group of people, such as Johann Rudolph Glauber, who were practitioners, often liminal between scholars trained at university in texts and artisans trained in the workshop by imitation. These practitioners often saw the new philosophy as an opportunity to attain an authority not possible for them in any other sphere, for natural philosophy gave them entry into the republic of letters by drawing not upon

11 An example of such an orientation is Antonio Pérez-Ramos's excellent *Francis Bacon's Idea of Science and the Maker's Knowledge Tradition* (Oxford: Clarendon Press, 1988).
12 Although this approach has gained currency in the wake of the "new cultural history" and its emphasis on practices, it does have forebears in the history of science, including the important essay by Edgar Zilsel, "The Sociological Roots of Science," *American Journal of Sociology* 47 (1942): 544–62, and the work of Paolo Rossi, particularly *Philosophy, Technology, and the Arts in the Early Modern Era*, trans. Salvator Attanasio (New York: Harper and Row, 1970). More recent historians pursuing this approach are James A. Bennett, "The Mechanics' Philosophy and the Mechanical Philosophy," *History of Science* 24 (1986): 1–28, and "The Challenge of Practical Mathematics," in *Science, Culture, and Popular Belief in Renaissance Europe*, ed. Stephen Pumfrey, Paolo Rossi, and Maurice Slawinski (Manchester: Manchester University Press, 1991), pp. 176–90; William Eamon, *Science and the Secrets of Nature: Books of Secrets in Medieval and Early Modern Culture* (Princeton: Princeton University Press, 1994); Michael Hunter, *Science and Society in Restoration England* (Cambridge: Cambridge University Press, 1983); Pamela O. Long, "The Contribution of Architectural Writers to a 'Scientific' Outlook in the Fifteenth and Sixteenth Centuries," *Journal of Medieval and Renaissance Studies* 15 (1985): 265–98, and "Power, Patronage, and the Authorship of Ars," *Isis* 88 (1997): 1–41.

their knowledge of the classics, but instead upon their ability to undertake particular practices and produce tangible effects or objects.[13]

Glauber was born the son of a barber. No document attests in any way to a completed school course, university degree, or apprenticeship training. He describes his education as gained by travel, experience, reading, and conversing with scholars. He appears to have learned metalworking and states that he supported himself before the 1630s with mirror making. Documentary evidence for his travels is first found in 1635/6 when he worked in some capacity in the court apothecary in Giessen. In 1641 we find him in Amsterdam, where he married and called himself an apothecary, although he was not in the medical guild in Amsterdam. He may have remained in the service of the Margrave of Hessen-Darmstadt until 1644, perhaps returning to the apothecary in Giessen, but in 1646 it is clear he was living in Amsterdam and had begun the activities that he was to carry on for the rest of his life, namely practicing and publishing works of chemistry and alchemy. His first book on new types of what he called "Philosophical Furnaces" described several new distilling furnaces for the distillation of minerals and organic matter. In the next twenty-four years he published almost thirty books, all containing in different measure practical processes, advertisement of secret preparations, and a theoretical structure for his inventions and recipes. His works are full of self-defense and biblical citation, and he praises and calls constantly upon the authority of Jan Baptista van Helmont and Paracelsus.[14] A list of his library shows him to have been interested in the century's new discoveries, primarily in geography and natural history, but these works are surpassed by the far more numerous chemical and alchemical volumes in his collection.[15]

13 A somewhat younger contemporary of Glauber who used natural philosophy in similar ways was Johann Joachim Becher (1635–82). See Pamela H. Smith, *The Business of Alchemy: Science and Culture in the Holy Roman Empire* (Princeton: Princeton University Press, 1994).
14 For Glauber's biography, see P. Walden, "Glauber," in *Das Buch der grossen Chemiker*, ed., Günther Bugge, vol. 1 (Berlin, 1929), pp. 151–72; Kurt F. Gugel, *Johann Rudolph Glauber 1604–1670, Leben und Werke* (Würzburg: Freunde Mainfränkische Kunst und Geschichte, 1955); Erich Pietsch, *Johann Rudolph Glauber. Der Mensch, sein Werk und seine Zeit* (Munich: R. Oldenbourg, 1956); and Kathleen W. F. Ahonen, "Johann Rudolph Glauber: A Study of Animism in Seventeenth-Century Chemistry" (Ph.D. diss., University of Michigan, 1972). See also Allen Debus, *The Chemical Philosophy*, 2 vols. (New York: Science History Publications, 1977), 2:429–34.
15 This list, entitled *Catalogus librorum*, is bound with some copies of his *Glauberus concentratus oder laboratorium glauberianum* (Amsterdam: Johan Waesberg & the widow of Elizaeus Weyerstraet, 1668). It contains about twenty-seven works of geography, forty-nine of chemistry, and twenty-two religious works, as well as

His published works and the self-advertisement found in them attest to his position in Amsterdam (and later, temporarily, in Wertheim and Kitzingen) as a private practitioner selling his products and his processes on the market instead of producing them for a patron or within a guild. He often quoted approvingly the mottoes taken over by Paracelsus to the effect that a man must be his own master: "Omnia mecum porto" (I carry all things with me) and "Alterius non sit qui suus esse potest" (He who is able to be his own man shall not belong to another). More than anything else, he promised the production of things; he saw himself as an "experienced Artist," one better able to imitate nature in its generative processes and thus to know nature. He maintained that only work in the laboratory could teach about nature: "it is only through fire and not from Aristotle that learning and studying should and must take place. And fire is the correct schoolmaster of all natural arts."[16]

His books are clearly meant to advertise his processes and products. They often contain parts of recipes and only rarely entire processes. Many times we find the statement that he will only reveal the process personally to those who come to his laboratory. Such self-advertisement is a manifestation of his liminality, for his position impelled him to advertise his products and to reveal enough of the recipe to establish his expertise in the subject, but at the same time to keep his processes secret enough to protect his economic interest. This forced Glauber to walk a finer line than either a guild member, who had a clear and overriding economic interest (and often guild regulations on secrecy), or a scholar, the publication of whose ideas established his place and provided his currency in the republic of letters. For a scholar there was usually no reason to hold anything back.

Glauber was clearly an entrepreneur in a commercial society. He even regarded his books as products. When he dissolved his laboratory and library in 1668, he had a large number on hand that he attempted to sell at discount prices.[17] He apparently produced mineral acids and tartaric acid on a large scale, selling them for manufacturing processes.[18] He also offered processes and recipes for sale; for exam-

---

various other works of philology, history, astronomy, medicine, and encyclopedic works. It is remarkable how few classical works his library contains.
16 *Des Teutschlandts Wolfahrt* (Amsterdam: Johan Jansson, 1656), pt. I, p. 79.
17 *Catalogus librorum*, pp. 14–15.
18 In *Testimonium veritatis* (Amsterdam: Johan Jansson, 1657), p. 296, Glauber advertised his process for and the sale of *Spiritus Salis* made according to his inexpensive process.

ple, when Samuel Hartlib sent him a sample of lead ore and tin scoria, Glauber claimed to be able to extract silver and gold (besides lead and tin) from both, but demanded a thousand ducats for the extraction technique for lead ore and two thousand ducats for the tin scoria.[19] Another example of his entrepreneurial activities can be found in the 1649 contract with the natural philosopher Otto Sperling in which Glauber sold the recipes for and privilege to manufacture in Denmark, Norway, Sweden, and "other northerly lands" a brandy half as expensive as French brandy, a long-lasting fruit wine, and the process for making good wine out of bad. In return, Sperling promised to pay Glauber four hundred reichsthaler, and a further one thousand thaler fine if he communicated the processes to anyone outside the area for which he had received the privilege.[20]

Glauber's laboratory in Amsterdam formed a station on many a grand tour. Samuel de Sorbière visited Glauber at the height of his health and prosperity and was impressed with Glauber's modesty and the grandness of the laboratories. Sorbière noted approvingly Glauber's humble but dignified bearing, commenting on Glauber's poor Latin, for which Glauber did not apologize, and Glauber's unembarrassed answers to his questions. Others saw Glauber after 1662 when he was sick and his laboratories had fallen into disrepair. In 1666, when Glauber experienced a fall from a wagon, Robert Boyle reported that Glauber was "spitting much blood, and if the fever prevail upon him he fears his life; which I pray God may be yet continued for giving many good hints, at least to the studiers of nature and arts."[21]

These men interested in natural philosophy regarded Glauber as a producer of knowledge about nature, an experienced artist who could call himself a new philosopher, although he clearly did not possess

19 Hartlib correspondence, Memo from Glauber to Hartlib, undated, fol. 67/15/1A–2B, Hartlib Papers, Sheffield University.
20 Ad Clément and J. W. S. Johnsson, eds., "Briefwechsel zwischen J. R. Glauber und Otto Sperling," *Janus* 29 (1925): 216–17.
21 Samuel de Sorbière, *Relations, lettres, et discours sur diverses matieres curieuses* (Paris: Robert de Ninville, 1660), Letter IV, pp. 103–94, recounted his visit to Glauber, pp. 176–81. Phillip Skippon reported in June 1663 that Glauber was too sick to see him: *An Account of a Journey Made Thro' Part of the Low-Countries, Germany, Italy, and France*, 6 vols. (London, 1732), 6:407. Monsieur de Monconys, *Journal des voyages*, 2 parts (Lyon: Horace Boissat, George Remeus, 1665), pt. 2, p. 179, said of his visit to Glauber in August 1663 that Glauber no longer worked and had no furnaces. In 1668, Edward Brown reported that "the old Glauber the chemist" showed him his laboratory: *Naukeurige en Gedenkwaardige Reysen van Edward Brown*, trans. Jacob Leeuw (Amsterdam: Jan ten Hoor, 1682), p. 23. The passage from Robert Boyle, *Works*, 6:91, quoted in Gugel, *Johann Rudolph Glauber*, p. 29, n. 55.

the same educational background (and thus the same position in the social hierarchy) as they did.[22]

Paracelsus and Artisanal Epistemology

Practitioners such as Glauber were encouraged in their claims to knowledge of nature by the writings of alchemists and chemists, who emphasized the necessity of artisanal expertise in the laboratory, but above all it was Paracelsus who gave these practitioners such confidence. Recall that Paracelsus believed the millennium would be ushered in by *Elias artista*, by artisanal work, and the products of art. Paracelsus's work seems to have struck a chord among artisans as well as radical religious and social reformers.[23]

In Paracelsus's self-proclaimed reform of knowledge, he took the methods of the artisan to be the ideal mode of proceeding in the acquisition of all knowledge, for the artisan worked directly with the objects of nature. Paracelsus considered this unmediated labor in and experience of nature as elevating the artisan both spiritually and intellectually above the scholar. Because the artisan imitated the processes of nature in his creation of works of art from the material of nature, he had a better understanding of nature. And because he was in contact with God's Creation, the artisan also had a better knowledge of God's revelation.

Paracelsus apparently questioned artisans and peasants on their remedies and he spoke frequently of the need for immediate experience of nature in the treatment of patients and the training of physicians. Barber-surgeons were admitted to his courses.[24] He believed knowledge must be gained through actively engaging with the objects

---

22 See also the references to Glauber in the correspondence of Henry Oldenburg and Gottfried Wilhelm Leibniz. The Hartlib papers at Sheffield University contain numerous references to Glauber and some letters from him. Samuel Hartlib and his correspondents were constantly trying to ascertain whether Glauber had communicated his processes in full and if these processes worked. Hartlib's correspondents were sharply divided about whether Glauber was a charlatan. See the interesting analysis of one gentleman's interest in Glauber by Stephen Clucas, "The Correspondence of a XVII-Century 'Chymicall Gentleman': Sir Cheney Culpeper and the Chemical Interests of the Hartlib Circle," *Ambix* 40 (1993): 147–70.
23 See Charles Webster, *The Great Instauration: Science, Medicine, and Reform, 1626–1660* (London: Duckworth, 1975).
24 Walter Pagel, s.v. "Paracelsus," *Dictionary of Scientific Biography*, ed. Charles Gillispie, 18 vols. (New York: Scribner, 1970–86); *Paracelsus: An Introduction to Philosophical Medicine in the Era of the Renaissance*, 2nd ed. (Basel: Karger, 1982); and, *Das medizinische Weltbild des Paracelsus. Seine Zusammenhänge mit Neuplatonismus und Gnosis* (Wiesbaden: Franz Steiner Verlag, 1962).

of nature. His famous statement, "He who would explore [nature], must tread her books with his feet," expresses this belief,[25] and it can be found more clearly in his notion of "experience." "Experience" in the Book of Nature became the central component of the new philosophy's active epistemology in the sixteenth century. Paracelsus used *Erfahrung* to mean the knowledge gained by traveling (i.e., treading the Book of Nature) with open eyes, and he also contrasted *experientia* and *scientia:* "Scientia is in the things given by God [i.e., the objects of nature]; Experientia is a knowledge of those things, in which Scientia is tried."[26] Experience gave knowledge of nature, and it was gained not through a process of reasoning, but by a union of the divine powers of mind and of the entire body with the divine spirit in matter:

> Now note the difference between *experientia* and *scientia*. . . . *Scientia* is inherent in a thing, it is given by God; *experientia* is a knowledge of that in which *scientia* is proven. For instance, the pear tree has *scientia* in itself, and we who see its works have *experientia* of its *scientia*. . . . Thus follows the book of *scientia*, that we experience *scientia*. . . . It is *scientia* to make a patient healthy. But this *scientia* is the medicine, not the physician. . . . These are the books of medicine.[27]

Scientia was the divine virtue in natural things that the physician must "overhear" and with which he must achieve union in order to gain knowledge of medicaments. *Experientia* on the other hand was the final result of this union. This remarkable inversion of theory (*scientia*) and practice (*experientia*) emerged from Paracelsus's contact with hermetic ideas[28] and, I would argue, from his desire to articulate the process by which artisans imitate nature in order to create objects from natural material.[29]

For Paracelsus, knowledge of nature, gained through experience

---

25 Paracelsus, *Four Treatises of Theophrastus von Hohenheim, called Paracelsus* (Baltimore: Johns Hopkins Press, 1941), "The Seven Defensiones," trans. C. Lilian Temkin, 4th defense, p. 29.
26 Pagel, *Paracelsus*, p. 60, quotes Paracelsus: "Scientia ist in dem, in dem sie Gott geben hatt: Experientia ist ein Kuntschafft von dem, in dem Scientia probiert wirt."
27 Paracelsus, *Labyrinthus medicorum errantium* (1538) translated by Nicholas Goodrick-Clarke, in *Paracelsus: Essential Readings* (Wellingborough: Crucible Paperbacks, 1990), p. 104.
28 See Pagel, *Paracelsus*, pp. 218–23.
29 For a fuller treatment of this, see Pamela H. Smith, "Giving Voice to the Hands: The Articulation of an Artisanal Worldview in the Sixteenth Century," in *Popular Literacies*, ed. John Trimbur (Pittsburgh: University of Pittsburgh Press, forthcoming). See also Neil D. Kamil, "War, Natural Philosophy and the Metaphysical Foundations of Artisanal Thought in an American Mid-Atlantic Colony: La Rochelle, New York City, and the Southwestern Huguenot Paradigm, 1517–1730" (Ph.D. diss., Johns Hopkins University, 1988).

and manual labor, was also a form of worship, giving understanding of God's Creation. Moreover, the art of the craftsman "reformed" nature by creating noble objects from the dross of fallen nature. The work of artisans, like the practices of agriculture and medicine, worked to redeem the body and life of humans after the Fall. The labor of refining nature for human needs, common to all work with the hands, brought about the reformation and ultimately the redemption of the world and humankind.[30] The understanding of art as redemptive can be found in artisanal descriptions of their own processes of creation. Jacob Böhme (1575–1624), Benvenuto Cellini (1500–71), and Bernard Palissy (ca. 1510–90) all used language and concepts very similar to that of Paracelsus.[31]

The model for all such redemptive knowledge for Paracelsus was alchemy and chemistry. Alchemical ideas were the basis for his medical-chemical theories[32] and the operations of alchemy were a search for the generative spirit in nature and the means to imitate it by art. The refining processes of alchemy, the model for all the arts, carried out in microcosm the macrocosmic process of human redemption after the Fall.[33]

Glauber took over Paracelsus's chemical theory of three principles, Glauber placing salt as the most important, but more importantly he took over Paracelsus's belief that work in the laboratory, and especially chemistry (for Paracelsus in order to produce medicines, for Glauber in order to produce goods that would contribute to material plenty) was a work of redeeming fallen Nature. Glauber's entire chemical philosophy was shot through with this redemptive understanding of art and alchemy, and the chemical and metallurgical processes he worked and described had significance in the scheme of salvation. He stated several times that knowledge of nature leads to knowledge of God.[34]

Glauber's early practical activity of mirror making appears to have

---

30 On this point, see Kurt Goldammer, *Paracelsus: Natur und Offenbarung* (Hanover: Theodor Oppermann Verlag, 1953).
31 See Kamil, "War, Natural Philosophy."
32 Massimo L. Bianchi, "The Visible and the Invisible: From Alchemy to Paracelsus," in *Alchemy and Chemistry in the Sixteenth and Seventeenth Centuries*, ed. Piyo Rattansi and Antonio Clericuzio (Dordrecht: Kluwer, 1994), pp. 17–50.
33 Owen Hannaway, *The Chemists and the Word: The Didactic Origins of Chemistry* (Baltimore: Johns Hopkins University Press, 1975), esp. pp. 43–5; B. J. T. Dobbs, *The Foundations of Newton's Alchemy* (Cambridge: Cambridge University Press, 1975), and *Alchemical Death and Resurrection: The Significance of Alchemy in the Age of Newton* (Washington, D.C.: Smithsonian Institution Libraries, 1990) develop this point.
34 See for example, *Explicatio oder Ausslegung über die Wohrten Salomonis: In Herbis, Verbis, & Lapidis Magna est Virtus* (Amsterdam: Johan Jansson, 1663), preface; and *Des Teutschlandts Wohlfart*, pt. I, dedication.

provided one underpinning for his theories about metals and about the vital principle. In 1651, he claimed that the sun's warmth penetrates to the center of the earth, which along with the astral powers, heats and vivifies the water there, forcing it to the surface where it appears as healing mineral springs. When the astral powers mix with wetness and earth, they also cause metals to grow.[35] Mirrors could be used to concentrate and capture the rays of the sun and the vital principle they contained.[36]

His later chemical theories also reflect this redemptive understanding of natural knowledge. For example, Glauber believed that the "soul" of metals could be made visible in their color. Glauber claimed the same was true for humans, although this aspect of his thought was less clearly formulated. In the case of metallic souls, when metal ashes were added to glass, the true colors or souls of the metal would manifest themselves. If these "colors" could then be extracted, one would possess the spirit of the metal and hence effect real alchemical transmutation. Working within this framework, Glauber rediscovered the process for red-tinged glass (*Rubin-Glas*), the process for which was believed to be akin to transmutation, and which Glauber claimed contained the soul of gold.[37]

Commerce

Although Glauber took over Paracelsus's understanding of natural knowledge as redemptive and of artisanal labor as a form of worship, his world was not Paracelsus's. Glauber lived in a commercial society, and this subjected him to particular tensions, but also gave him an opportunity to shape himself as a new kind of natural philosopher. As a figure liminal between artisans and scholars with no established qualifications in the republic of letters and no clear estate in society, Glauber was always in danger of slipping down to a purely mechanical or mercantile level. He was anxious to make clear that his activities were not just carried out for the pursuit of wealth. He was particularly sensitive to being called a gold-making alchemist, as he says his neighbors in Amsterdam called him.[38] He argued against this charge, main-

35 *Operis mineralis* (Frankfurt am Main: Heirs of Matthaeus Merianus, 1651), pp. 14–15.
36 In part 4 of the *Furni novi philosophici*, Glauber describes a process for casting mirrors, and in the "sale catalog" of his laboratories and library he published when he could no longer work, there are several burning glasses listed, one of which has a two-foot diameter. *Glauberus concentratus oder laboratorium glauberianum*, p. 33.
37 See Kathleen Ahonen, s.v. "Glauber," *Dictionary of Scientific Biography*, 5:422.
38 *De tribus lapidibus ignium secretorum* (Amsterdam: Johan Waesberg and the widow of Elizaeus Weyerstraet, 1668), esp. p. 10.

taining that "gold-making alchemy" is something completely different from what he does. An alchemist who seeks gold is either a simple laboratory worker who knows nothing significant or someone who has learned about transmutation from reading books. Both activities have nothing to do with true alchemical knowledge. In a passage that reveals his context, he refers to frequent transactions in Amsterdam between the two types of fortune hunter whom he calls "laborer" and "investor," as a kind of cottage industry. Producing gold as a cottage industry results in nothing, however, for

> the *Laborant* knows nothing and the principals or investors know even less. It is not enough that one is a highly schooled, knowledgeable man, for such schooling or learning has nothing to do with alchemy. Alchemy requires people who understand the ways and characteristics of metals.[39]

But gold making was only the worst of fortune hunting; commerce too was distasteful to a natural philosopher. Glauber was also in danger of being seen as a common merchant who sold his goods for private profit, something no true natural philosopher would do.[40] When George Starkey was trying to attract Robert Boyle as his patron, he wrote to Boyle, "I avoid the selling of nature's secrets and for this reason alone is Glauber so antithetical to me."[41] Glauber was anxious to show himself as a disinterested participant in the marketplace. When Monsieur de Sorbière came to visit, he wanted to buy some of Glauber's products, but Glauber said he did not have any at present because he was occupied with plans for performing a "transmutation" in public. Sorbière tells his correspondent that he had news that Glauber had carried out his public performance "without any profit, but only to show the possibility of the thing."[42] What more convincing

---

39 Ibid., p. 24. "Der *Laborant* weiss nichts/ und die *Principalen* oder Verleger noch weniger; Es ist nicht genung/ dass man ein hochgelährter verständiger Mann ist/ solche Gelehrheit oder Welt-verstandt hatt mit der *Alchimia* nichts zu thun/ die *Alchimia* erfordert Leuthe/ welche der *Metall*en arth und eigenschafften verstehen."

40 For other examples of the concerns that those who called themselves natural philosophers had about distancing themselves from mechanics and merchants, see Eamon, *Science and the Secrets of Nature*; Smith, *The Business of Alchemy*; Steven Shapin, *A Social History of Truth: Civility and Science in Seventeenth-Century England* (Chicago: University of Chicago Press, 1994); and Larry Stewart, *The Rise of Public Science* (Cambridge: Cambridge University Press, 1992).

41 William R. Newman, *Gehennical Fire: The Lives of George Starkey, an American Alchemist in the Scientific Revolution* (Cambridge, Mass.: Harvard University Press, 1994), p. 67.

42 Sorbière, *Relations*, pp. 179–80.

proof of his purely philosophical interests (and perhaps better advertisement for his products) than to perform in public?

The report of a conversation to Samuel Hartlib by Henry Appelius in 1647 reveals the interesting mix of commercial entrepreneurship and altruism that characterized Glauber's self-presentation:

> Glauberus <last week when I visited him> was ready as to day to goe to Arnheim with his family, which is a cittie upon the Rhyne not very farr above Utrecht, where the Kufflers dwell & exercise their dying of cloath/ Glauber would faine goe higher, in Germany, & set up their such workes whereby hee might maintaine his family most liberally, by the Inventions especially in metallicis, whereof hy saith hee hath most certaine proofes.... And because hee will bee freed of much acquaintance, leters en the like, hee writes his books, for the benefit of all men.... [H]eertefore he taught the furnaces et the mannour of distilling for monyes, now hy communicats them to the whole world: as soone as there bee men to beare the printing charges, then it <the 2.> may bee printed.... 3 weeks agoe a Doctour from Mets asked leave that his booke might bee translated into Latÿn et French, Glauber answered; there was no necessity to aske leave of him, seing the book was no more his, but all mens.[43]

While Glauber sought to maintain himself by his commerce in mineral products (on the model of the Kufflers who had achieved great commercial success in the cloth trade by the discovery of a tin mordant that fixed red cochineal dye), he at the same time presented himself as making his discoveries freely available to all men (although the contents of his books often belie such intentions).

Glauber wanted to make clear that his activities were of a different order than a merchant's or indeed anyone engaged in commercial transactions, for the possession of his processes and products gave an honest living that prevented a man from being the slave of another. It was wealth that could neither be sinfully hoarded nor stolen by thieves. But most important, when possessed, it made commerce completely unnecessary:

> We do not need to seek the necessaries of life by usury or trickery or by other sinful means. For all commerce, whether large or small, cannot occur without sin, as the highly experienced Jesus Syrach said, the sins between the buyer and the seller stick as tight as a nail in the wall. And not only sin grows in the

---

43 Hartlib Correspondence, August 26, 1647, f. 45/1/33A–34B, f. 33a.

buyer and seller but in all those who are not their own masters, but instead have learned their livelihood under another and must seek [their livelihood] through others. Included in these are physicians, surgeons, apothecaries, jurists, lawyers, procurators and other high or low learned men. When they have only studied and learned an art or craft in order to make money, then sin is rooted in the making of profit and is transformed into an evil habit that will not be extinguished except by death.[44]

Glauber set himself off from the artisans and scholars with whom he might be identified, regarding their work also as tainted. It was in fact only an independent man such as himself who could keep from committing the sin of taking from others by the practice of commerce.

Humans could learn to live without sin if they could recognize and learn how to use the "Universale Sal Mundi," for they would always have plenty, could never slip into poverty, and would remain rich to their graves.[45] Because this salt could give them whatever they desired, be it "health, money, or possessions," they would never harm their fellow humans. Moreover those who were pious would be both visible to all and get their just desserts here on earth because the godless could not possess the art of this salt:

> And this is certainly true, that the godless never will come to knowledge of this salt, and much less to its use. For all philosophers testify to this. Among others, the highly experienced philosopher Bonus Lambertus who affirmed this in his dialogue with Lacinius: *Ars ipsa sancta est, & quam non nisi puros ac sanctos homines habere licet. Nam ut divi Thomae utar [sic] sententia, Ars ista vel reperit hominem sanctum, aut reddet ejus inventio sanctum.*

---

44 *Glauberus concentratus*, pp. 26–7. "Wir haben nicht nöthig durch Wuchen/ und Schinderey/ oder andern zur Sünden-führenden Mitteln unsere Nothdurfft/ oder auffenthalt des Lebens zu suchen. Dann aller Kauffhandell Er sey gleich gross/ oder klein/ ohne Sünde nicht geschehen kan/ wie dann der hocherfahrene Jesus Syrach saget/ dass die Sünde zwischen dem Käuffer/ und Verkäuffer so fäst stäcke/ als ein Nagell in der Wandt; und floririt nicht allein die Sünde bey dem Käuffer/ und Verkäuffer/ sondern auch bey allen dehnen/ welche nicht ihr eygen seyn/ sondern ihren unterhalt durch das ienige so sie gelernt/ von andern zu haben suchen müssen. Darunter auch die *Medici, Chirurgi, Apotheker, Juris*ten/ *Advoca*ten/ Procuratores, und dergleichen Hoch- und Nieder-Gelährten zu rechnen seyn. Wann sie allein darum *studi*ret/ und Kunst/ oder Handwerck gelernet haben/ von andern Geldt dardurch zu gewinnen; dabey die Sünde dann durch das viel Geldt gewinnen also eingewurtzelt/ und in ein tägliche böse gewohnheit verwandelt/ dass sie auch anders nicht wohl auss zu reutten/ als durch den Todt."

45 Ibid., p. 66.

> Art does not lodge by the proud misers, but by those who are satisfied with a little.⁴⁶

Because the knowledge of art was granted by God and only the godly could possess it, Glauber seemed to envision a kind of elect of the artisanal pious being formed. This was, however, not Calvinism, for Glauber was nominally a Catholic, and he had found that in his travels "one must have to do with [different religions], [he] therefore went into the services of Catholics with Catholics, those of Lutherans with Lutherans, and those of Calvinists with Calvinists, and heard of each his opinions, which seemed good to me."⁴⁷ He also read all kinds of religious pamphlets and "over the years heard and saw such quarreling with amazement."⁴⁸ He believed instead in seeking God through nature and salvation through work in the laboratory.

Thus, although Glauber made his living from selling his products and his works are filled with their advertisement, he was at pains to show that it was his labor in the laboratory, rather than his commercial success that gave him his independence and his position. The value of his products was not measured by their worth on the market, but instead by their part in the redemption of the world. This is especially clear in works such as *Des Teutschlandts Wolfahrt oder Prosperitas Germaniae* (1656–61), in which he described many useful processes to make a land fertile (through saltpeter production and composting), to concentrate and preserve surplus agricultural goods, and to protect territories from enemies (by acids that could be fired toward the opposing army). It is telling that part of the stimulus to publish *Prosperitas Germaniae* came directly from the commercial economy in which he found himself. He stated that when there is a surplus harvest, the peasants cannot convert their natural goods into money and thus cannot pay their taxes, for their rulers no longer accept their payments in natural goods.⁴⁹

In his work *Trost der Seefahrenten* (1657), he claimed to want to provide a service to the hard-worked sailors of his adopted Amster-

---

46 Ibid., p. 28. "Und ist dieses gewisslich wahr/ dass die Gottlosen nimmermehr zu der erkändtnüss dieses Saltzes/ und auch noch viel weniger zu dessen nützlichen gebrauch gelangen werden. Wie dann solches alle Philosophi bezeugen: Unter andern auch solches der hoch erfahrene Philosophus, Bonus Lambertus in seinem mit Lacinio gehaltenem Dialogo bekräfftiget/ mit diesen wortten: *Ars ipsa sancta est, & quam non nisi puros ac sanctos homines habere licet. Nam ut divi Thomae utar sententia, Ars ista vel reperit hominem sanctum, aut reddet ejus inventio sanctum.* Die Kunst kehret nicht bey den hofffährtigen Geitzigen ein/ sondern nur bey dehnen/ welche sich mit wenig genügen lassen."
47 *Glauberus Redivivus* (Frankfurt am Main: Thomas Matthias Goetzen, 1656), pp. 78–9.
48 Ibid., p. 82.
49 *Des Teutschlands Wolfahrt*, pp. 50–1.

dam. He set out a method of making seawater potable and keeping drinking water sweet at sea, a process for concentrating grain in a beerlike drink, a recipe for a concentrated and long-lasting bread, and a "spirit of salt" to prevent scurvy. Neither of these books, nor any of his others except one, are dedicated to individuals, but instead are addressed to the "common man" or "the peasants, vine-growers, and gardners," or to his "German Fatherland."[50] Such works proved his devotion to the salvation of humankind.

No doubt sincere piety motivated such works, but it was also the desire to carve out a new space as something more than an artisan and a merchant and different than a scholar. This space was a natural philosopher, both productive and disinterested, a Paracelsian redeemer of the world through the production of material things. The rise of the new philosophy and the rhetoric of Paracelsus made possible Glauber's new identity and formed the means by which he navigated the shoals of making an independent livelihood.

A profound reorientation in attitudes to the material world and material things took place in Europe in the sixteenth and seventeenth centuries. One of the most important markers of this shift was a new understanding of the natural world, which emerged out of a union of scholarly and artisanal ways of thinking about material reality. This new understanding of nature was, however, part of the much larger economic and social transformation of Europe by the growth of an exchange economy. Glauber, working within a commercial market and holding a view of labor and art as redemptive helped to bring about these new attitudes to the natural world and to reshape the concept of salvation as material progress.

50 The only work dedicated to an individual is *Gründliche und warhafftige Beschreibung Wie man auss der Weinhefen einen guten Weinstein in grosser Menge extrahiren sol* (1654), which is addressed to archbishop and elector of Mainz and Wurzburg, Johann Philipp von Schönborn, who had given Glauber a privilege for producing tartaric acid when Glauber was in Wertheim and Kitzingen (1651–4).

7

"The Terriblest Eclipse That Hath Been Seen in Our Days": Black Monday and the Debate on Astrology during the Interregnum

WILLIAM E. BURNS

One result of the dominance of the concept of the "Scientific Revolution" over the study of early modern thought about nature has been that different cultural practices concerning the natural world have been conceived as either on the side of the revolution or against it. Recovery of occult or "unscientific" belief such as astrology or alchemy has either claimed that these activities contributed to the advance of science, as Richard Westfall describes alchemy as part of the Scientific Revolution, or ascribed a positive value to the occult and a negative one to the Scientific Revolution itself. However, to rethink the Scientific Revolution requires the examination of the uses of nature in the early modern period without presupposing either the "scientific" nature of these practices or the "revolutionary" nature of their changes and conflicts. This essay is an attempt to read the English reception of a particular incident – the "Black Monday" solar eclipse of March 29, 1652 – not as an episode in the Scientific Revolution (although such a reading is possible), but as the clash of a variety of positions on natural phenomena and their meaning for humanity.

The English Interregnum, the period between the execution of King Charles I in 1649 and the Restoration of his son Charles II in 1660, saw a passionate debate on both the popular and learned levels on the validity of judicial astrology.[1] The two most influential discussions of the decline of judicial astrology in early modern England have been those of Keith Thomas, in *Religion and the Decline of Magic* (1971), and Patrick Curry, in *Prophecy and Power: Astrology in Early Modern England* (1989). To hark back to the two essays that opened this collection, Thomas's position resembles Dobbs's in its pluralism, and the Scientific Revolution plays only a minor role in his work. He sees the struggle against astrology as mainly carried on in a religious context,

---

1 Nicholas H. Nelson, "Astrology, *Hudibras*, and the Puritans," *Journal of the History of Ideas* 37 (1976): 521–36, includes a useful although incomplete bibliography of Interregnum tracts on the controversy over astrology.

and it and other forms of magic as gradually declining, not suddenly collapsing.[2] Curry's work, by contrast, is organized around a contrast between the antiastrological Restoration and the previous period of the Civil War and Interregnum, characterized as the "halcyon days" of English astrology. Curry's position resembles that of Westfall, in that he sees the decline of the intellectual prestige of astrology among the educated as "a change which is sudden, radical, and complete." Although Curry does not believe that new scientific knowledge, such as Copernicanism, necessarily marginalized astrology, he does believe in the antiastrological impact of the cultural and institutional Scientific Revolution. He sees the Restoration in 1660 and the founding of the Royal Society in 1662 as inaugurating a campaign by the forces of elitist "official" knowledge, particularly natural philosophers, to create an ideological hegemony under the banner of "objective" science. These forces suppressed an alternative astrological tradition that was, at least potentially, one of democratic knowledge. Curry rewrites the history of the Scientific Revolution to feature sympathetic astrologers and villainous natural philosophers, but it is still a revolutionary story he is telling.

A close look at the Interregnum debate, and particularly at the literature surrounding "Black Monday," however, reveals a great variety of conflicting astrological, providential, and satirical uses of natural phenomena in England during the Interregnum, with little effect from the forces of organized natural philosophy. Judicial astrology was, however, intensely contested from other sources, principally the treatises and sermons of ministers and the writings of satirists. Curry's emphasis on the Scientific Revolution as the source of effective opposition to astrology leads him to underrate the strength of antiastrological attitudes during the Interregnum and their continuity across 1660. Change was both less revolutionary and less "scientific" than Curry believes, and Thomas's position seems to fit the evidence better.

On the learned level, the principal polemicists against astrology during the period were ministers, particularly but not exclusively those identified with conservative Puritanism or Presbyterianism.[3] Although these ministers regarded predicting the future by the techniques of judicial astrology as impious, blasphemous, cozening, or

---

2 Michael Hunter has pointed out that science may have been institutionally too weak to challenge astrology directly in the late seventeenth century. Hunter, "Science and Astrology in Seventeenth-Century England: An Unpublished Polemic by John Flamsteed," in *Astrology, Science and Society: Historical Essays*, ed. Patrick Curry (Wolfeboro, N.H., and Woodbridge, Suffolk: Boydell Press, 1987), pp. 261–300.

3 Calvin had written a polemic against judicial astrology, published in English as *An Admonicion against Astrology Judiciall* (London, 1561).

even diabolical, they did not therefore regard all celestial phenomena as lacking meaning for humans.[4] Many who defined themselves and were perceived by contemporary astrologers as being strong opponents of astrology viewed this opposition as perfectly compatible with support for providential interpretations of celestial phenomena. Indeed, complete disbelief in the human meaning of celestial events was simply not an option for orthodox seventeenth-century Christians. There were ample scriptural precedents for providential interpretation of celestial events such as the great darkness at Christ's death or the Star of Bethlehem, which the London minister Nathanael Homes, an Independent and millenarian opponent of astrology, claimed was "more Theologicall than Astrologicall."[5] Strange celestial sights were also prominent in the apocalyptic books of Daniel and Revelation, which claimed that bloody suns and other wonders would be heralds of the apocalypse.[6] The celestial and other prodigies the ancient Jewish historian Flavius Josephus associated with the fall of Jerusalem, including a comet and a star shaped like a broadsword, were constantly cited in early modern prodigy literature at both high and popular levels, and often conceded a special status as having been prophesied by Christ himself.[7] In the controversialists' own time, "prodigies," or strange sights including multiple suns, comets, and apparitions of battles in the sky, were interpreted providentially as signs of God's power and communications to humans by all sides in the conflicts of the English Civil War, and frequently appeared in political and religious propaganda.[8] In addition to being apocalyptic signs, celestial

---

4 Curry's "working definition" of astrology as "any practice or belief that centred on interpreting the human or terrestial meaning of the stars" is too broad, and would lead to the inclusion of the strongest opponents of astrology in the ranks of astrologers. Patrick Curry, *Prophecy and Power* (Princeton: Princeton University Press, 1989), p. 4.
5 Nathaniel Homes, *Daemonologie and Theologie* (London, 1650), p. 171. For Homes's acceptance of the idea that "strange signes in the Heavens" would precede the millennium, see Homes, *The Resurrection Revealed: or the Dawning of the Day-Star* (London, 1653), pp. 548–9.
6 Daniel 7.10, Revelation 8.10, 11; 16.8, 9.
7 *The Jewish War*, 6.300.
8 For discussion of the propaganda role of prodigies during the English Civil War, see Ann Geneva, *Astrology and the Seventeenth-Century Mind: William Lilly and the Language of the Stars* (Manchester: Manchester University Press, 1995); Chris Durston, "Signs and Wonders in the English Civil War," *History Today* 37 (1987): 22–8; Harry Rusche, "Prophecies and Propaganda, 1641–1651," *English Historical Review* 84 (1969): 752–70; and William Burns, "An Age of Wonders: Prodigies, Providence, and Politics in England, 1580–1727" (Ph.D. diss., University of California at Davis, 1994). For a general discussion of prodigies in Early Modern European culture, see Jean Céard, *La nature et les prodiges: L'insolite au XVIe siècle* (Geneva: Droz, 1977).

prodigies were also viewed providentially as warnings from God to repent lest he send greater disasters, or as presages of specific earthly events, a position with many classical as well as scriptural precedents. Unlike judicial astrologers, who treated celestial prodigies as causes as well as signs, claiming that the stars, as well as unusual celestial events such as comets, exerted direct influence on the terrestrial world, providential prodigy interpreters treated God as the only active force, and prodigies merely as signs or warnings exerting no force of their own.[9]

Thomas treats the Interregnum debate as a struggle between the Calvinist predestinarianism of the Presbyterian clergy and the rival celestial determinism of judicial astrology.[10] In doing so, he underrates the importance of the ministers' providential theory of celestial phenomena. Ministers writing antiastrological polemics from a religious position found it necessary to distinguish between (legitimate) providential prodigy-interpretation and (illegitimate) astrology. The veteran moderate presbyterian Thomas Gataker's annotation on the particularly difficult (for astrologers) text Jeremiah 10.2, "Be not afraid at the signs of Heaven, as the heathen are afraid," which touched off his controversy with the famous Parliamentarian astrologer William Lilly, distinguished "signs . . . comming in a constant course, and continued tenor" from "extraordinary dreadfull apparitions, besides the ordinary course of the creature," claiming that the prophet's injunction applied only to the former, and that fear was indeed an appropriate response to God's extraordinary actions.[11] The distinction between providential and astrological interpretation could be drawn in the newspaper press as well as at the learned level where the debate between astrologers and ministers was carried out; one newspaper claimed that a new star in 1653 presaged peace "without the vanity of astrology."[12] Astrologers such as Lilly defended astrology by denying the validity of this distinction, claiming that the very widely held

---

9 For an argument against the view of celestial prodigies as mere signs without independent influence, see Christopher Heydon, *A Defence of Judiciall Astrologie, in Answer to a Treatise Lately Published by M. John Chamber* (Cambridge, 1603), p. 69. This work was particularly influential during the Interregnum. For the centrality of the doctrine of celestial influence in judicial astrology, see John D. North, "Celestial Influence – The Major Premiss of Astrology," in *"Astrologi hallucinati": Stars and the End of the World in Luther's Time*, ed. Paola Zambelli (Berlin: de Gruyter, 1986), 45–100; and "Medieval Concepts of Celestial Influences" in Curry, *Astrology*, 5–18.

10 Keith Thomas, *Religion and the Decline of Magic* (Harmondsworth: Penguin, 1973; first published 1971), pp. 358–71.

11 Thomas Gataker, *Annotations upon all the Books of the Old and New Testament*, 2nd ed., enl. (London, 1651).

12 *The Weekly Intelligencer of the Common-Wealth*, no. 206 (February 21–8, 1653/4): 171.

belief in the significance of unusual celestial events implied the significance of predictable celestial events such as eclipses and conjunctions, and often presenting providential and apocalyptic interpretations alongside more purely astrological ones.[13]

Black Monday was the great event around which much of the learned, popular, and satirical debate on prodigies and astrology was shaped. On Monday, March 29, 1652, there was a total eclipse of the sun visible in England and throughout northern Europe.[14] The eclipse was particularly impressive as it followed a lunar eclipse the same month.[15] Eclipses, like conjunctions, were border phenomena between astrology and providential prodigy-interpretation, being as spectacular as unpredictable events such as comets, but also being predictable events with a known cause. Black Monday was widely publicized before and after it occurred in almanacs, ballads, sermons, and newspapers, attracting a great deal of popular interest and apprehension.[16] Both astrologers and providential interpreters predicted it to be the cause or the sign of great changes in the world, and tried to bend it to establish the intellectual legitimacy of their own positions. One astrologer claimed that the result of the eclipse would confute the "deceitful babling Priests" who opposed astrology.[17] Lilly wrote two works specifically dealing with the eclipse, *Annus tenebrosus* (1652) and *An Easie and Familiar Method whereby to Judge the effects depending on Eclipses Either of the Sun or Moon* (1652). *Annus tenebrosus* was specifically an exposition of the meanings and implications of the eclipse, predicting

13 Some of Lilly's writings in the 1640s such as *The Starry Messenger* (London, 1645), *Supernatural Sights and Apparitions Seen in London June 30, 1644* (London, 1644), and *The World's Catastrophe* (London, 1647) had dealt with prodigies such as triple suns. For discussion of Lilly's role as prodigy monger, see Geneva, *Astrology*.

14 For brief discussions of Black Monday, see Bernard Capp, *English Almanacs: Astrology and the Popular Press, 1500–1800* (Ithaca, N.Y.: Cornell University Press, 1979), pp. 79–80; Derek Parker, *Familiar to All: William Lilly and Astrology in the Seventeenth Century* (London: Jonathan Cape, 1975), p. 184; and Thomas, *Religion*, pp. 299–300. Curry dismisses it in a few sentences, along with other evidence of the strength of opposition to astrology in the 1650s. Curry, *Prophecy and Power*, p. 46.

15 Samuel Thurston, *Angelus anglicanus, or a Generall Judgement of the Three Great Eclipses of the Sun and Moon, Which Will Happen in the Year 1652* (London, 1652).

16 The minister John Swan, in a sermon on the eclipse preached the day before, claimed that his sermon topic had been chosen because of "the great noise which I have heard among the common people concerning that Eclipse of the Sun, which tomorrow in the Forenoon will present itself to us." Swan, *Signa coeli. The Signs of Heaven* (London, 1652), p. 3. Gataker described a dinner at his house before the eclipse at which it had been the major topic of conversation. Thomas Gataker, *His Vindication of the Annotations by Him Published* (London, 1653), p. 148.

17 William Ramesay, *A Short Discourse of the Eclipse of the Sunne, on Monday, Martii 29, 1652* (London, 1651), sig L4i. Ramesey's work was a rare example of a purely astrological reading of the eclipse with no providential or eschatological meaning.

the downfall of Lilly's old enemies, the Presbyterians.[18] *An Easie and Familiar Method* was an older work of astrological theory regarding eclipses in general, which Lilly reissued to take advantage of the current interest in the subject.[19] These works were popular – Gataker claimed that he had not been able to purchase a copy of *Annus tenebrosus* as they had been "sodainly snacht up at their first coming forth."[20] However, not all predictions concerning the consequences of Black Monday were astrological. It was sometimes treated purely as a providential or eschatological prodigy. This was particularly true of writings directed at a less learned audience such as the ballad "England's New Bell-Man," which focused on God's wrath in the Day of Judgment, with the refrain, "Repent therefore O England / The day it draweth near,"[21] or the popular hack ballad writer Laurence Price's *The Shepherd's Prognostication*.[22] One of the earliest writings on the eclipse, called simply *Black Munday*, made much of the alleged parallel between the eclipse and Josephus's prodigies.[23] Some hoped to observe the connection of the eclipse to other abnormal if not prodigious phenomena such as earthquakes and comets.[24]

The eclipse was also relevant to current political conflicts. Some argued that Black Monday was an apocalyptic herald of the Fifth Monarchy. Although one Fifth Monarchy pamphlet, *The Levellers Almanack: For the Year of Wonders 1652*, gave an astrological interpretation of the eclipse, it began by placing it in the context of miracles and prodigies.

> On the 29 of this Moneth, a great Eclipse will happen, in which time the Sun will be almost totally darkened (as at the passion

---

18 Lilly, *Annus tenebrosus* (London, 1652), p. 28. For Lilly's linking Black Monday with the prodigious appearances in the sky discussed in his pamphlets of the 1640s, see *Annus tenebrosus*, 6. For an argument that *Annus tenebrosus* interpreted the eclipse as predicting the death of Charles II, see Geneva, *Astrology*, p. 255. Lilly also discussed the meaning of the eclipse in his almanac, *Merlini anglici ephemeris or Astrologicall Predictions for the Year 1652* (London, 1652), "Astrologicall Predictions."
19 Lilly, *An Easie and Familiar Method*, "To the Reader."
20 Gataker, *His Vindication*, 2.
21 In *The Pepys Ballads in Facsimile*, ed. W. G. Day (Cambridge: Cambridge University Press, 1987), 2:61. This was actually a reprinting of an older ballad with new prefatory material to connect it to the eclipse. An earlier version is in *Pepys Ballads*, 1: 54.
22 Laurence Price, *The Shepherd's Prognostication* (London, 1652). Price worked both sides of the street, publishing after Black Monday *The Astrologer's Bugg-Beare* (London, 1652), which ridiculed astrologers, including Lilly and Nicholas Culpeper, who had predicted disasters to follow the eclipse.
23 *Black Munday* (London, 1651), p. 2. Thomason dated this pamphlet to September 5, 1651.
24 Bodleian Ashmole MS 423, f. 192, Bodleian Library, Oxford.

of our Saviour) and the stars appear in the Firmament in the day time. It happens on a Munday: which may well be called Black-Munday: it is the Precursor of great calamities and mischiefs that shall happen to the people and places concerned in it, especially all Europe, and more especially England: it is the terriblest eclipse that hath been seen in our dayes; and as miraculous as that was at the time of our Saviour's Passion, when there was darknesse on all the Land, from the 6 hour until the 9. and the vail of the Temple was rent in twain; and what horrid calamities hapned to the Jewes after that?[25]

The radical Fifth Monarchist astrologer and herbalist Nicholas Culpeper published two Black Monday tracts, one before the event, *The Year of Wonders*, and one after, *Catastrophe magnatum*. Both explicitly declared that Black Monday would usher in the Fifth Monarchy. In addition, Culpeper devoted much of his 1652 almanac to the eclipse, claiming on the title page that this year would be that of "The Ruine of Monarchy throughout Europe."[26] Culpeper's writings had a particularly broad impact as some of the London newspapers reprinted a long passage from *The Year of Wonders* incorporating predictions of the Fifth Monarchy into their discussions of what Black Monday specifically portended for particular peoples and nations.[27] The third great Parliamentarian astrologer along with Lilly and Culpeper, John Booker, also prominently featured the doom of great ones the eclipse allegedly presaged in his 1652 almanac.[28]

---

25 *The Levellers Almanack: For the Year of Wonders 1652* (London, 1651), p. 4. A similar radical political "Almanack" connecting the eclipse with the fall of European monarchy was "John Napier," *The Bloody Almanack* (London, 1652).

26 Nicholas Culpeper, *An Ephemeris for the Year 1652* (London, 1652).

27 *The French Intelligencer*, no. 16 (March 2–9, 1652): 119; no. 18 (March 16–23): 133–4; and *The Dutch Spy*, no. 1 (March 17–25, 1652): 2–5. The passage reprinted is from *The Year of Wonders* (London, 1652), pp. 9–13 and was also reprinted with additions in *Catastrophe Magnatum* (London, 1652), pp. 68–74.

28 John Booker, *Coelestiall Observations or an Ephemeris of the Planetary Motions . . . 1652* (London, 1652), "March." The idea that the eclipse portended the fall of great ones was not restricted to the almanacs of the high-profile political astrologers Lilly, Booker, and Culpepper. See Vincent Wing, *An Almanack and Prognostication for the Year of Our Lord 1652*, (London, 1652), "Prognostication," and John Rowley, *Speculum perspicuum uranicum 1652* (London, 1652), "Appendix." "N. R., Student in Astrology," *Strange Newes of the Sad Effects of the Fatall Eclipse Happening the 29th of This March, 1652* (London, 1652), combined old prophecies with an analysis of the eclipse to argue both the downfall of great ones and the coming of universal destruction. Perhaps reflecting caution concerning the political situation in England, N. R. focused his analysis of current events on Condé's rebellion in France. The almanac writer Thomas Herbert predicted disaster to Rome, although not in an apocalyptic context. Thomas Herbert, *Speculum anni ab incarnatione verbi 1652* (London, 1652), "Of the Eclipses of the Sun and Moon happening This Year, 1652."

"Doom for great ones" was also interpretable in a Royalist sense, and there were Royalist treatments of the eclipse, including *King Charls His Starre: or Astrologie Defined*, published after Black Monday in 1654. This anonymous Royalist pamphlet (possibly authored by the eccentric Welsh visionary Arise Evans), although mostly focusing on a comet in December 1652, briefly argued astrologically that the eclipse threatened existing governments, including that of England, and was therefore a good sign for the exiled King Charles II.[29] However, few Royalists were as intellectually or personally bold as the anonymous pamphleteer. The best-known Royalist astrologer, George Wharton, was in a vulnerable position in 1652, having been recently released from prison on the promise of avoiding comment on public affairs. He was more cautious than the Parliamentarians Lilly, Booker, or Culpeper, not emphasizing the eclipse in his 1652 almanac. Although Wharton claimed eclipses were "Sad Prodromi of Slaughters and Exile/And Grand Mutations, in Great Britaines Isle," he avowed ignorance as to Black Monday's specific import.[30]

The anonymous tract *The Late Eclipse Unclaspd*, published a few days after the eclipse, put it in a generally apocalyptic, although not specifically Fifth Monarchist, context, displaying the biblical verse "But in those days, after that tribulation, the sun shall be darkened, and the moon shall not give its light" (Mark 13.24) in the position that a biblical text would occupy before a printed sermon. The tract's apocalyptic interpretation of the eclipse did not imply support for judicial astrology. "Foolish people give more credit to judiciary Astrologers then to God's Word, but Gods Ministers may averre, that it is not in the power of any Astrologer to foretell what the Events or Influence of this Eclipse will be in particular."[31]

Some ministers denied both astrological and prodigious meaning to the eclipse in venues that suggest that this view was supported by authorities in the Commonwealth government and the City of London, possibly desiring to minimize the political and economic disruption the eclipse would cause. The minister Fulk Bellers, in a sermon preached the Sunday before the eclipse in front of the lord mayor and aldermen of London and printed with a dedication to them both attacked astrology and claimed that eclipses "are seldome prodigious." Bellers argued

---

29 *King Charls His Starre* (London, 1653), pp. 39–40. For Evans, and a discussion of *King Charls His Starre* in the context of his work, see Christopher Hill, *Change and Continuity in Seventeenth Century England* (London: Weidenfield and Nicolson, 1974), pp. 48–77.

30 George Wharton, *Hemeroscopeion anni intercalaris 1652* (London, 1652), "March."

31 *The Late Eclipse Unclaspd* (n.p., n.d.), p. 7. Thomason added it to his collection on April 12, 1652.

that eclipses were not warned against in Scripture, they were often followed by good rather than bad times, and fear of eclipses was "heathenish."[32] Instead, Bellers claimed that they should be treated as allegorical aids to religious meditation, in that the removal of the sun's light was analogous to the removal of Christ's grace. The kind of meditation Bellers recommended was practiced by the devout Anglican Royalist Alice Thornton, who treated the eclipse not as a herald, but as a type of the Last Judgment. She was moved to "most serious and deep consideration of the day of judgement which would come as sudaine and as certainly uppon all the earth as this eclips fell out."[33] A position similar to Bellers's was taken by "a Reverend Divine of the City of London" whose denunciation of astrologers in connection with Black Monday was printed in the semiofficial newspaper *Several Proceedings in Parliament*.[34] This attack, probably the closest thing to an official or government reaction to the eclipse, emphasized the explainable and predictable nature of eclipses, wavering between the position that such naturally occurring events had no human meaning and the position that this meaning was entirely religious.

Bellers's and the "Reverend Divines'" arguments were directed against the inclusion of eclipses, as regular and predictable events, in the category of the prodigious, not against the existence of the category itself. Few ministers of the time would have been willing to go so far as to attack the providential interpretation of truly anomalous natural phenomena, which would have seemed impious or "atheistic."[35] Culpeper responded to the kind of argument, associated with ministers, that emphasized the importance of the distinction between

---

32 Fulk Bellers, *Jesus Christ the Mysticall or Gospel Sun, Sometimes Seemingly Eclipsed, yet Never Going Down from His People: or, Eclipses Spiritualized* (London, 1652), pp. 28–30. Another sermon preached in London the same day also warned against treating eclipses as prodigies, that of the French Protestant divine Jean Despaigne, with French and English versions published together as *Considerations representée en un sermon Le 28. de Mars, de cette Anee 1652 sur le sujet de l'ECLIPSE qui advint le lendemain* and *Considerations Held Forth in a Sermon the 28 of March, This Year 1653 upon the Eclipse* (London, 1652).

33 Charles Jackson, ed., *The Autobiography of Mrs. Alice Thornton, of East Newton, Co. York* (Durham, 1875), p. 84n.

34 *Several Proceedings in Parliament*, no. 131 (March 25–April 1, 1652), pp. 2033–6. F. A. Inderwick, *The Interregnum* (London, 1891), 132, misattributes this piece to another paper, *Perfect Diurnall*, and claims that it was issued by the Council of State. See Thomas, *Religion*, p. 300n.

35 Nor indeed did all ministers take the opportunity of the eclipse to denounce astrology. Swan's *Signa coeli* was a defense of it, preached on Jeremiah 10.2, the same text Gataker had used to attack astrology in the *Annotations*. For a preacher claiming inspiration and taking a public pulpit to preach incoherently on the significance of the eclipse, see *The Faithful Scout*, no. 64 (April 2–9, 1652): 407–8.

predictable and explainable events and unpredictable and "prodigious" ones in speaking of the next significant solar eclipse after Black Monday, that of August 1654. He claimed: "I do confess [the eclipse] is natural, as the Priest told his people of the last Eclipse: The Eclipse, quoth he, is natural, and therefore can have no such ill effects as the Astrologers tell you of. God help his Calves head; is not meat and drink natural, and yet doth it not nourish? Is not poyson natural, and yet doth it not destroy?"[36] For Culpepper and other judicial astrologers relying on the doctrine of celestial influence, the fact that neither the causes nor the effects of eclipses were supernatural did not mean that they did not have important effects on the human world.

Black Monday itself proved a fiasco, being neither as dark as expected nor immediately leading to the downfall of great ones. Some claimed that the vulgar had believed that the eclipse would cover the earth with a darkness similar to the ninth plague of Egypt, and the earl of Leicester noted that during the hours of the morning the common people had nearly deserted the streets.[37] The reaction to the failure of Black Monday to usher in the millennium, eternal darkness, or indeed any dramatic change whatever was widespread and long-lasting.[38] Observers, supporters, and opponents of astrology all thought Black Monday had been a serious embarrassment to the practice. John Green, a London lawyer and future recorder of the city, claimed that "the Astrologers lost their Reputation exceedingly."[39] The minister John Gaule, in his polemical attack on all forms of magic dedicated to Oliver Cromwell, *The Mag-astro-mancer, or the Magicall-Astrologicall-Diviner Posed, and Puzzled* (1652) punningly claimed that while almanacs had too much influence upon the people, "Only their gross hallucinating in their prodigious portending upon the last eclipse, hath proved not a little to Eclipse their credit with them."[40]

36 Culpeper, *An Ephemeris for 1654* (London, 1654), "Of the Eclipse of the Sun."
37 Price, *The Astrologer's Bugg-Beare*, and HMC, *Dudley and Delisle MSS*, IV, 613.
38 Not everyone hedged their bets as gracefully as N. R., whose title page announced a description of "the direful effects and Prodigies (probably) to be expected in the Aire" during the eclipse.
39 E. M. Symonds, ed., "The Diary of John Green," *English Historical Review* 44 (1929): 112. (Ironically, Green would later defend Lilly in a lawsuit.) For a similar reaction, see *The Diary of John Evelyn*, ed. E. S. de Beer (Oxford: Clarendon, 1955), 3: 63. Charles Hammond, *Englnad's* [sic] *Warning-Piece* (London, 1652), claimed that astrologers had been discredited by the eclipse, and hoped that providential interpretation would not be associated with its embarrassment.
40 John Gaule, *P*us-mantia the mag-astro-mancer, or, The magicall astrologicall-diviner posed, and puzzled* ... (London, 1652), "Dedication." (Booker devoted part of his defense of astrology in his 1653 almanac to a denunciation of Gaule; see *Coelestiall Observations ... 1653* [London, 1653], "To the Reader.") One contemporary writer claimed that if the astrologers had not "proclaimed it [the eclipse] to be such a

Lilly was particularly often claimed to have been discredited by the event, although he had actually minimized the extent or importance of the darkness itself. He clearly regarded the whole affair as a disaster for astrology and himself personally.[41] In the somewhat defensive preface to his 1653 almanac Lilly claimed that after Black Monday he had been victimized by "two round dozen of vinegar pamphlets" as well as thirty Presbyterian sermons, in addition to the works of balladeers and "mercurialists," or newspaper writers. "No Market Town within 20 or 40 miles of London, where Black Monday Ballads were not sung publikely in the streets, and my name only in the Ballad," he complained.[42] Lilly was not the only astrologer forced to defend his 1652 predictions in his 1653 almanac; Henry Harflete had claimed in his 1652 almanac, calculated for and distributed in the city of Sandwich, that the eclipse would be followed by great spring tides, and advised the sewer commissioners to put the waterworks in better repairs.[43] Harflete's 1653 almanac, discussing the absence of the predicted great tides, somewhat testily pointed out that they might have come, if the weather had been worse, and the facilities should be put in better repair anyway.[44] The ex-religious radical turned astrologer John Gadbury defended English astrologers by blaming exaggeration of the darkness of the eclipse on "Italian mountebanks," claiming that "none of our English Astrologers writ any such thing at all."[45]

horrid one, there had not been one man in this City so much as taken notice of any Eclipse at all." William Bromerton, *Confidence Dismounted, or the Astronomers Knavery Anatomized* (London, 1652), p. 6.

41 Lilly, *Annus tenebrosus*, pp. 49–50.
42 Lilly, *Merlini anglici ephemeris or Astrologicall Predictions for the Year 1653* (London, 1653), "To the Reader."
43 Henry Harflete, *Coelorum declaratio 1652* (London, 1652), "April."
44 Harflete, *Coelorum declaratio 1653* (London, 1653), "Prognostication." Wharton, who despite his political differences with Lilly shared with him an interest in defending astrology, devoted a portion of his 1653 almanac to defending Lilly and other astrologers from the charge of making false Black Monday predictions. Wharton, *Hemeroscopeion anni aerae christiani 1653* (London, 1653), pp. 48–50. For Culpeper's defense of *Catastrophe magnatum*, see *An Ephemeris for the Year of Our Lord, 1653* (London, 1653), "April." For more attempts in the years immediately following to defend the significance of the eclipse, see the almanacs of the pseudonymous Johnathan Dove, produced by a group of Cambridge printers, *Speculum anni à partu virginis MDCLIII* (Cambridge, 1652), and *Speculum anni à partu virginis MDCLIV* (Cambridge, 1653).
45 John Gadbury, *Philastrogus Knavery Epitomized* (London, 1652), pp. 13–14. (This tract was a response to an attack on astrologers' focusing on Culpeper, "Philastrogus," *Lillie's Ape Whipt* [Cambridge, 1652]. Although *Lillie's Ape Whipt*, published before Black Monday, referred to it only in general terms, it was still necessary for Gadbury's defense of astrology to deal with Black Monday at some length.) Gadbury claimed several years later that "the famous and memorable Eclipse" had "poured

Veteran antiastrological writers such as Gataker took advantage of the opportunity presented by the astrologers' embarrassment to reiterate their opposition. Gataker's *Vindication* used Black Monday to reaffirm the difference between prodigies "besides, beyond, above, or against the course of nature" and events such as eclipses, which, however striking, were predictable and explainable in terms of the ordinary workings of nature.[46] Quoting Kepler, Gataker ascribed the origin of astrology to the illegitimate extension of techniques legitimate for the interpretation of anomalous prodigies to the interpretation of predictable celestial events. The astrological interpretation of Black Monday was the most recent example of this process.[47]

The only figure principally identifed as a natural philosopher who participated in the debate on astrology following Black Monday was the heliocentric astrologer, Baconian, and conservative divine Joshua Childrey.[48] Childrey, beginning a long career of opposition to providential and eschatological prodigy-interpretation, took a unique position on Black Monday precisely inverting that of Gataker, in his tract setting forth a program for a reformed and heliocentric astrology *Indago astrologica* (1652). Childrey explicitly rejected providential and apocalyptic interpretation in favor of the purely astrological. He claimed the failure of Black Monday to herald the end of the world actually brought religion into contempt, "a help which in these days

> down its influences so violently and cruelly upon Europe, that it set all nations therein, into a confusion." Gadbury, *Ephemeris or a Diary Astronomical and Astrological of the Year of our Lord 1661* (London, 1661), "The State of the Year Predicted." He also subsequently claimed that Gataker's death had been presaged by the eclipse of 1654, suggesting that this made clear the fallaciousness of Gataker's attack on the astrological meaning of eclipses. Gadbury, *Collection Genituarum* (London, 1662), p. 104. For Gadbury and his attempt to reform astrology on a basis of Baconian empiricism, see Mary Ellen Bowden, "The Scientific Revolution in Astrology: The English Reformers, 1558–1686" (Ph.D. diss., Yale University, 1974); Thomas, *Religion*, p. 636; Curry, *Prophecy and Power*, pp. 72–6; and Capp, *English Almanacs*, pp. 184–6.

46 Gataker, *His Vindication*, p. 52.
47 Ibid., pp. 99–100.
48 Childrey would become a Church of England minister, a correspondent of Henry Oldenburg, a contributor to *Philosophical Transactions*, and, along with Gadbury, a zealous opponent of popular and religiously radical providential prodigy-belief after the Restoration. For his unusual and unsuccessful attempt to reconcile astrology, Baconianism, and Copernicanism, see *Dictionary of National Biography*; Bowden "Scientific Revolution in Astrology," pp. 169–76; and Curry, *Prophecy and Power*, pp. 64–7. Neither Bowden nor Curry discuss the relation of Childrey's *Indago astrologica* (London, 1652) to Black Monday. By discussing Childrey in the context of the reaction against astrology during the Restoration, Curry underrates the degree to which Childrey's project of a Baconian astrology was a response to criticism of astrology during the Interregnum.

we need least, since too many already have leaped into Hell without this staff."⁴⁹ He argued that a purely astrological approach to interpreting of eclipses and other strange events was actually more appropriate, and more conducive to religion, than providential or apocalyptic interpretation. If astrological interpretations were erroneous, religion would lose no credit, and public relations disasters such as Black Monday would be limited.⁵⁰ However, Childrey's viewing of Black Monday as primarily a disaster for religious, rather than astrological, interpretation of the skies does not seem to have been widely shared.

The many "vinegar pamphlets" and other satirical writings published after Black Monday reveal the popularity of astrologers as satirical targets and widespread popular opposition to their intellectual pretensions.⁵¹ Appearing in forms available to a broad cross section of the literate public such as broadsheets, these writings treated both well-known astrologers and astrology itself as discredited. One tract, *A Faire in Spittlefields*, for example, depicted Lilly, Culpeper, and Booker as peddlers attempting to sell astrological works no one would buy.⁵² Although astrologers were the main target, some works ridiculed providential prodigy-interpretation as well. The broadsheet *On Bugbear Black-Monday* asked, "Why gaze ye brain-sick People on the Skie / To seek in Heaven some uncouth Prodigie."⁵³

Many accusations of superstition in the English newspaper press, strongly influenced by the Commonwealth government, were aimed at those ever popular targets of abuse, the Scots, in the aftermath of Cromwell's conquest of Scotland. The ordinarily staid government mouthpiece *Perfect Passages of Every Daies Intelligence* as well as other

49 Childrey, *Indago Astrologica*, p. 15.
50 Ibid., pp. 15–16.
51 Curry claims that the use of satire to discredit astrology was particularly characteristic of the Restoration, but it was widespread in the Interregnum as well. Curry, *Prophecy and Power*, pp. 50–1.
52 "J. B. gent.," *A Fair in Spittlefields* (London, 1652). For more attacks on Lilly related to Black Monday, see a work of one of Lilly's old enemies, the lay Presbyterian propagandist John Vicars's posthumously issued broadsheet *On William Lil-lie* (London, 1652); *Black Munday Turn'd White* (London, 1652); and Gataker, *His Vindication*. An anonymous poem, "The Sunn's Eclipse is past, but not the feares," in Huntington Library Ellesmere MSS 8852, San Marino, Calif. also includes mockery of "Prophet Lyly" in a general mockery of popular superstition and astrologers' political predictions of the universal downfall of monarchy.
53 *On Bugbear Black-Monday, March 29, 1652* (London, 1652). For more abuse of the eschatological credulity of astrologers and people in general, see *Weekly Intelligencer of the Common-Wealth*, no. 66 (March 23–30, 1652): 405, and *A Perfect Account of the Daily Intelligence from the Armies*, no. 65 (March 24–31, 1652): 519.

papers claimed that the Scots thought the Black Monday eclipse meant that the sun was going to go out.[54] The government propagandist and newspaperman Marchamont Nedham, in his official newspaper *Mercurius Politicus*, gave a slightly more sophisticated view of the Scots as manipulated by unscrupulous Presbyterian ministers who distorted the true meaning of prodigies.[55] The Welsh, who had supported the Royalist cause in the Civil War and were often associated with credulous belief in ancient prophecies, were also targets of satire.[56]

One event occurring shortly after Black Monday, although no one argued that the eclipse had presaged it, was the reemergence of the comic newspaper press. Tracing their roots to the satirical Royalist newspapers of the late 1640s, such as John Crouch's *The Man in the Moone*, the comic Mercuries printed little news or explicit political commentary, but much humorous and obscene material accompanied by vulgar abuse of whoever had angered the editor.[57] Although there were other reasons for the comic papers edited by the notorious ex-Royalist journalists Crouch and Samuel Sheppard to appear at this time, such as Crouch's recent release from imprisonment, the first issues both of Crouch's infamous *Mercurius Democritus* and Sheppard's short-lived *Mercurius Phreneticus* concentrated heavily on Black Monday.

As these papers ridiculed whatever Puritans and supporters of the Commonwealth held dear without running the risk of openly putting forward Royalist or sectarian alternatives, both providential prodigy-belief and astrology were targets among many others, and Black Monday was a rich source of material. *Mercurius Phreneticus* began its brief career as a lampoon under the pseudo-Arabic pseudonym Galbrion Albumazar entirely devoted to ridiculing Lilly and other astrologers in connection with Black Monday, and only declared itself a periodical in its second issue.[58] The first issue of the most enduring of these

---

54 *Perfect Passages of Every Daies Intelligence*, no. 59 (April 2–9, 1652), *The Faithful Scout*, no. 64 (April 2–9, 1652): 491; and *Mercurius Politicus*, no. 96 (April 1–8, 1652): 1520. For more abuse of the credulity of the Scots, along with preachers and the vulgar, see Bodleian Ashmole MS 423, ff. 192–3.
55 *Mercurius Politicus*, no. 108 (June 24–July 1, 1652): 1703.
56 "The Sunn's Eclipse is Past," lines 45–8.
57 For these papers, see Hyder Rollins, ed., *Cavalier and Puritan* (New York, 1923), pp. 58–62; Joad Raymond, *Making the News: An Anthology of the Newsbooks of Revolutionary England* (New York, 1993), pp. 124, 126; and Joseph Frank, *The Beginnings of the English Newspaper, 1620–1660* (Cambridge: Cambridge University Press, 1961), pp. 195–7, 229–31, 242–3.
58 Sheppard apparently found ridiculing astrologers profitable, as he went on to publish, under the pseudonym Raphael Desmus, a series of parody almanacs titled *Merlinus Anonymus* covering the years 1653–5.

papers, *Mercurius Democritus*, was dated April 8, 1652, less than two weeks after Black Monday, and included a description of "wonderfull fights in the Ayre" during the eclipse, seen by a blind man and heard by a deaf one at Lilly's behest.[59]

Despite the Royalist background of the newspaper satirists and their opposition to Puritan ministers, at least one prominent minister accepted them as allies against judicial astrology. Homes claimed in a sermon preached the first Sunday after Black Monday before the lord mayor of London (as Bellers had preached the last Sunday before it), that "Judiciary Astrology is a LIE. This last Eclipse, March 29. 1652. is the proofe, which not prooving to be so darke, gloomy, and terrible, as they predicted, it hath so eclipsed their credit, that I hope you will forever take them for Liars."[60] In a passage recalling Lilly's enumeration of the forces against him, Homes made explicit the alliance between ministers and satirists, claiming "WITS, POETS, DIURNAL-ISTS, BALAD-MAKERS" as valuable partners in the struggle against judicial astrology.[61] He argued that these writers were more suited than clergymen to reply to astrologers on the astrologers' own level.[62]

The memory of Black Monday persisted, usually as an example of failed astrological prediction, through the rest of the seventeenth century and into the eighteenth century. A newspaper recalled the falsity of astrologers' predictions in 1652 when making its own claims regarding the eclipse of 1654.[63] A 1677 comet tract justified its claims as to the comet's meaning with the argument that "For I think that I have had as good a guess at things and spoke as much Astrological Truth as Lilly or Booker of Black Monday."[64] *The Protestant Almanack for 1685*, speaking of a lunar eclipse, stated "This will be a total and central Eclipse. But you need not fetch home your Cattle as on Black Munday: I hope you will have more wit than so."[65] John Evelyn remembered Black Monday nearly fifty years later, comparing it with a solar eclipse in 1699, and blaming "Lilly the Almanack writer" for

---

59 *Mercurius Democritus*, no. 1 (April 8, 1652): 6. Another attack on Lilly was on pp. 4–5.
60 Nathaniel Homes, *Plain Dealing, or the Cause and Cure of the Present Evils of the Times* (London, 1652), pp. 57–8. Like Bellers, Homes dedicated the printed version of his sermon to the lord mayor and aldermen.
61 Homes, *Plain Dealing*, p. 60.
62 For Booker's counterattack, both against "Phreneticus, Democritus, and other frothy wits, Poets, and Ballad-makers of late dayes" and Homes's attempt to enlist them as allies in the war against judicial astrology, see Booker, *Coelestiall Observations . . . 1653*, "To the Reader."
63 *Weekly Intelligencer of the Common-Wealthe*, no. 216 (July 25–August 1, 1654): 240.
64 *Poor Robin's Opinion of the present Blazing-Star* (London, 1677), p. 2.
65 *The Protestant Almanac 1685* (London, 1685), "Of the Eclipse."

the popular panic.[66] One of the foremost opponents of astrology in the popular press, the almanac writer Richard Saunder, recalled Black Monday to ridicule astrologers as late as 1705.[67]

The controversy over Black Monday reveals an Interregnum marked by extensive debate on natural phenomena and their human meaning with little influence from the "Scientific Revolution." Astrologers, although prominent, did not exert uncontested cultural dominance over interpretation of celestial phenomena. Judicial-astrological positions were fiercely challenged in learned and popular literature, from theological or satirical viewpoints rather than natural-philosophical ones, and many considered strong opposition to astrology perfectly compatible with finding human meaning in the stars.[68] The antiastrological position in England did not make a revolutionary leap into prominence after the Restoration and the creation of institutionalized natural philosophy in the form of the Royal Society, but built on foundations laid by the providential interpreters, ministers, and satirists of the Interregnum.

66 Evelyn, *Diary*, 5:354. For another reference to Black Monday in connection with the 1699 eclipse, see *The True Figure of that Great Eclipse of the Sun* (London, 1699). A possible early eighteenth-century reference to Black Monday was Jonathan Swift's selection of March 29 as the putative date for the death of the astrologer John Partridge in *Predictions for the Year 1708* (London, 1708). The most elaborate discussion of the dating of the pamphlet, George P. Mayhew, "Swift's Bickerstaff Hoax as an April Fools' Joke," *Modern Philology* 61 (1964): 280–7, does not mention the Black Monday connection, but it is an intriguing possibility.

67 Richard Saunder, *1699. Apollo anglicanus, The English Apollo* (London, 1699), "A Type of the Sun's Eclipse, September the 13th" and *1705. Apollo anglicanus, The English Apollo* (London, 1705), "Of the Eclipses of this Year, 1705."

68 The providential tradition, as much or more than horoscope-based "high astrology," was incorporated into late seventeenth- and eighteenth-century natural philosophy. Natural philosophers such as Newton, considering the terrestial effects of celestial phenomena, focused on prodigious events such as comets and meteors, and their catastrophic consequences of famine, plague, and war, rather than the signs, houses, and aspects of judicial astrology. See Sara Schechner Genuth, *Comets, Popular Culture, and the Birth of Modern Cosmology* (Princeton: Princeton University Press, 1997).

8

Arguing about Nothing: Henry More and Robert Boyle on the Theological Implications of the Void

JANE E. JENKINS

Those who are philosophers cannot see [Pascal's vacuum experiments] without wonder; and those who are not, become philosophers when they consider them. There, is observed that brave nothingness against which so many excellent Philosophers have fought for such a long time, that fearful void that frightens all nature, and against which she uses all her forces, that fine nothing which is going to supply arms for its defence, and solid matter to construct discourses in its favour.[1]

As people learned of Robert Boyle's exciting air-pump experiments, published in 1660 as *New Experiments Physico-Mechanical, Touching the Spring of the Air and Its Effects*, the question of whether a vacuum could exist in nature was more keenly debated than ever before. Matthew Hale remarked that the subject of the experiments "is seeingly trivial ... yet it hath exercised the wits and pens of many learned men."[2] Some of the "many excellent Philosophers" debating the possible existence of "that brave nothingness" continued to uphold the view that nature was frightened by empty space, that "fearful void," and would fight to prevent it. This view had long been accepted as a fundamental tenet of Aristotle's, who had predicted that after reading arguments in his *Physics* the concept of "the so-called void will be found to be really vacuous."[3]

In the 1660s, however, the view that nature abhors a vacuum was no longer widely held. Of course, the Cartesian version of the mechanical philosophy of nature reflected the Aristotelian notion that there

---

1 Pierre Guiffard, *Discours du Vuide* (1647), quoted in R. Dugas, *Mechanics in the Seventeenth Century: From the Scholastic Antecedents to Classical Thought*, trans. Freda Joacquot (Neuchatel: Griffon, 1958), p. 211.
2 Matthew Hale, *Difficiles nugae: Or Observations Touching the Torricellian Experiment, and the Weight and Spring of the Air* (London, 1675), "To the Reader."
3 Aristotle, *Physics*, IV.216a.27–8, in *The Complete Works of Aristotle*, ed. Jonathan Barnes, 2 vols. (Princeton: Princeton University Press, 1984).

were no empty spaces in nature. The atomic version of mechanism, however, asserted the existence of void between the atoms. Boyle, ever anxious to avoid confrontation, was reluctant to answer the question of whether vacuum could exist in the world, even though his pneumatic experiments brought the question into the forefront. He understood that people expected that "I should on this occasion interpose my opinion touching that controversy; or at least declare, whether or no, in our engine, the exsuction of the air do prove the place deserted by the air sucked out to be truly empty, that is devoid of all corporeal substance."[4]

However, Boyle was careful to avoid entering the debate surrounding the possible existence of void in nature stating that he had "neither the leisure, nor the ability, to enter into a solemn debate of so nice a question."[5] Rather than indicating lack of interest in the debate, however, Boyle's agnosticism about the existence of the void reflected a keen awareness of the danger, both philosophical and theological, attached to the vacuum debates.[6] While this awareness kept him from endorsing either side, it did not prevent him from writing a lengthy polemic against the well-known Cambridge Platonist, Henry More, concerning the proper interpretation of his early pneumatical experiments.

Boyle knew that his experiments would "be observed and laid hold of, both by the Cartesians and Epicureans, the former of which will endeavour thereby to establish the necessity of the *Materia subtilis*, to maintain the plenitude of the world, . . . and the latter will here triumphantly pretend to have a more illustrious instance than ever of their *vacuum coacervatum* within the world."[7] While Boyle did not get embroiled in the plenist-vacuist debate, he would not sit back silently while More used his experiments to declare that here, finally, was irrefutable proof for the activity of an incorporeal spirit on matter. More had first made passing reference to Boyle's air-pump experiments in the third edition of his *Antidote against Atheism*, published in

---

4 Robert Boyle, *New Experiments Physico-Mechanical, Touching the Spring of the Air, and Its Effects; Made, for the Most Part, in a New Pneumatical Engine*, in *The Works of the Honourable Robert Boyle*, ed. Thomas Birch, 6 vols. (London, 1772; reprinted, Hildesheim: Georg Olms, 1965), 1:37.

5 Ibid.

6 For the argument that Boyle's agnosticism grew out of his fear that open acceptance of atomism would imply association with its atheistic reputation, see J. J. MacIntosh, "Robert Boyle on Epicurean Atheism and Atomism," in *Atoms, Pneuma and Tranquillity: Epicurean and Stoic Themes in European Thought*, ed. Margaret J. Osler (Cambridge: Cambridge University Press, 1991), pp. 197–218.

7 Boyle, *Of the Admirably Differing Extension of the Same Quantity of Air, Rarefied and Compressed*, in *Works*, 3:509.

1662. He presented a more comprehensive discussion of Boyle's pneumatics in *Enchiridion metaphysicum*, published nine years later in 1671. Boyle countered More's appropriation of his hydrostatical and pneumatical experiments in *An Hydrostatical Discourse Occasioned by the Objections of the Learned Dr. Henry More, against Some Explications of New Experiments Made by Mr. Boyle*, published in 1672. Here, Boyle declared himself "determined to write this polemical discourse" even though he had already promised his friends that he would abstain from public debate, believing that the results of his experiments "would be able to make their own way." And he would have kept his promise if "the objections I was to answer had not been by a person of so much fame, proposed with so much confidence, and though with great civility to me, yet with such endeavours to make my opinions appear not only untrue, but irrational and absurd."

Boyle feared that More's "discourse, if unanswered, might pass for unanswerable."[8] Whereas Boyle explained the phenomena exhibited by his air-pump in terms of the mechanical operation of inert matter, More used the same experiments to promote his theory about the operation of an immaterial Spirit of Nature, acting on behalf of God to preserve universal order. Boyle countered this interpretation not simply because he thought More had misinterpreted the details of his newly evolving theory of matter. He was more deeply troubled by the dangerous theological implications of More's account, which, by according agency to nature, threatened to make God unnecessary. Boyle presented his concerns with the kind of interpretation given by More and other "naturists" in *A Free Inquiry into the Vulgarly Received Notion of Nature*, which he wrote in 1666.[9] Here Boyle stated as his explicit goal the refutation of the notion of an intermediary Spirit of Nature, and the assertion of the direct activity of God in nature. This goal would not only promote the use of mechanical explanations but was also considered by Boyle to be "a service to theology."[10] Boyle's project to establish a new, mechanical philosophy of nature was strongly motivated by deep theological assumptions and concerns.[11]

8 Boyle, *An Hydrostatical Discourse Occasioned by the Objections of the Learned Dr. Henry More*, in *Works*, 3:596–7.

9 This treatise of Boyle's was not published, however, until 1686. See Michael Hunter and Edward B. Davis, "The Making of Robert Boyle's *Free Enquiry into the Vulgarly Receiv'd Notion of Nature* (1686)," *Early Science and Medicine* 1 (1996): 204–71.

10 Boyle, *A Free Inquiry into the Vulgarly Received Notion of Nature*, in *Works*, 5:253.

11 For the idea that Boyle's mechanical philosophy of nature was deeply informed by his voluntarist theology, see J. E. McGuire, "Boyle's Conception of Nature," *Journal of the History of Ideas* 33 (1972): 523–42; T. Shanahan, "God and Nature in the Thought of Robert Boyle," *Journal of the History of Philosophy* 26 (1988): 547–69; M. J. Osler, "The Intellectual Sources of Robert Boyle's Philosophy of Nature: Gassendi's Vol-

He believed the new philosophy was beneficial because it provided better explanations of natural phenomena (in terms of matter and motion alone) but also confirmed belief in God's continued and direct activity in the world.[12] Boyle considered that "not the least, service,... our doctrine may do religion, is that it may induce men to pay their admiration, their praises, and their thanks, directly to God himself, who is the true and only Creator of the sun, moon, earth, and those other creatures, that men are wont to call the works of nature."[13] Inert, passive matter interacted to produce the marvels of the world in accordance with God's continued guidance. Nature itself was not an active agent, nor was there a hierarchy of agents between God and the creation. Boyle believed that this kind of philosophy of nature would allow people to praise "God himself for his mundane works, without taking any notice of his pretended vicegerent, nature."[14]

This was not the only time the two prominent Englishmen disagreed nor was it the only issue they disputed. The reasons for their disagreements and the significance of the issues they disputed have been studied by historians and philosophers of science for some time and reflect historiographical trends in interpreting the nature of seventeenth-century science, including reassessments of the Scientific Revolution.[15]

Scholarly interpretations that heralded Boyle as a forerunner of modern science, focused on the theoretical aspects of his natural philosophy, deemphasizing views that have since fallen out of favor. R. A. Greene, writing in 1962 about the controversy between Boyle and More over the proper interpretation of the vacuum experiments, disregarded the merit of More's theological agenda, viewing the controversy in self-evident terms, confirming those aspects of it that adumbrated currently held beliefs. Holding that the air-pump experiments proved the existence of vacuum, Greene lavished rich regard on Boyle, whose "attitude and tone... cannot be too highly or too

untarism and Boyle's Physico-Theological Project," in *Philosophy, Science and Religion in England, 1640–1700*, eds. R. Kroll, R. Ashcraft, and P. Zagorin (Cambridge: Cambridge University Press, 1992), pp. 178–98.

12 For more on the theological underpinnings of Boyle's natural philosophy, see Margaret J. Osler, *Divine Will and the Mechanical Philosophy: Gassendi and Descartes on Contingency and Necessity in the Created World* (Cambridge: Cambridge University Press, 1994).

13 Boyle, *Notion of Nature*, in *Works*, 5:253.

14 Ibid.

15 For an overview of the historiography, see Rose-Mary Sargent, *The Diffident Naturalist: Robert Boyle and the Philosophy of Experiment* (Chicago: University of Chicago Press, 1995), pp. 1–11.

often praised."¹⁶ He saw Boyle as working "with infinite patience" to explain his theories to his opponents, who, with their "continual obtuseness," presented "baseless and gratuitous arguments."¹⁷ He interpreted More's debate with Boyle as "a pathetic spectacle." Although More was "a well-intentioned, learned man," he was nonetheless "unable to adapt his mind to new ideas."¹⁸

This type of interpretation, exemplified in the work of Herbert Butterfield, distinguishes the work of seventeenth-century natural philosophers, such as Boyle and Newton, with the dramatic replacement of Aristotelian assumptions. With the rejection of ancient authorities and methods, the study of nature was said to be transformed through the application of a more modern scientific method and outlook. Emphasizing theoretical transformation, this Scientific Revolution "changed the character of men's habitual mental operations" and marked such a significant turning point in human civilization that it "outshines everything since the rise of Christianity."¹⁹

More recently scholars shifted their focus from the theoretical changes that marked seventeenth-century natural philosophy as the beginning of modern science to an evaluation of the social factors influencing it. This more weakened basic assumptions that identified the Scientific Revolution as an endeavor to uncover universal truths, unencumbered by the politics of power. By highlighting a sociological approach, disagreements between Boyle and More were explained in terms of the dynamic social and political context of Interregnum and Restoration England.²⁰ In their book, *Leviathan and the Air-Pump:*

---

16 Robert A. Greene, "Henry More and Robert Boyle on the Spirit of Nature," *Journal of the History of Ideas* 23 (1962): 462.

17 Ibid., p. 472.

18 Ibid. Other anachronistic studies include C. C. Gillispie, *The Edge of Objectivity* (Princeton: Princeton University Press, 1960); Charles T. Harrison, "Bacon, Hobbes, Boyle, and the Ancient Atomists," *Harvard Studies and Notes in Philology and Literature* 15 (1933): 191–213; J. D. Partington, *A History of Chemistry* (London: Macmillan, 1961); R. P. Multhauf, *The Origins of Chemistry* (London: Oldbourne, 1966); W. E. Knowles Middleton, *The History of the Barometer* (Baltimore: Johns Hopkins Press, 1964). Marie Boas Hall, a major interpreter of Boyle, who wrote at the same time as the authors of these other interpretations, also held him up as a brilliant experimenter, who advanced science. Revealing this perspective, she claimed that Boyle raised chemistry "to a point where physicists recognised it as a real and important science, working in the same spirit as contemporary natural philosophy." Marie Boas Hall, *Robert Boyle and Seventeenth-Century Chemistry* (Cambridge: Cambridge University Press, 1958), p. 232.

19 Herbert Butterfield, *The Origins of Modern Science, 1300–1800* (London: Bell, 1949), p. viii.

20 Studies with a social interpretation of Boyle's work include J. R. Jacob, *Robert Boyle and the English Revolution: A Study in Social and Intellectual Change* (New York: Burt

Hobbes, Boyle, and the Experimental Life, Steven Shapin and Simon Schaffer point to the controversy between More and Boyle without evaluating it in anachronistic or Whiggish terms. They count More as one of Boyle's chief adversaries, critical of his new theory of the air's spring.[21] Rather than explaining this criticism as More's inability to grasp innovative ideas, they frame his response within the social dynamics of the time and consider the dispute between More and Boyle as part of the process leading to the establishment of the new experimental philosophy. Their controversy provided a forum in which Boyle could affirm those elements considered essential to the establishment of an experimental "way of life." These included establishing the proper rules for dispute and identifying which observations and explanations were acceptable as matters of fact.[22] In this view, Boyle's response to More was a deliberate attempt to establish the experimental way of life, thereby settling the strife that had led to civil war. Shapin and Schaffer consider Boyle's experiments on air to be the "centre-piece" of his natural philosophy because they established what could be counted as fact and, thereby, provided consensus and, ultimately, social order.[23] Therefore, they considered any theological differences between the main actors to be only superficial. While "Boyle argued this case on what *might* be called purely theological grounds, ... Boyle's identification of the religious dangers ... was *not* purely theological."[24]

The trend to contextualize Boyle within the broad social and political framework in which he lived has been expanded by other scholars who consider an even wider range of factors, including his theological, ethical, philosophical, and alchemical interests.[25] The importance of Boyle's theological concerns, for instance, appears in John Henry's

Franklin, 1977), and Steven Shapin, *A Social History of Truth: Civility and Science in Seventeenth-Century England* (Chicago: University of Chicago Press, 1994).

21 Steven Shapin and Simon Schaffer, *Leviathan and the Air-Pump: Hobbes, Boyle, and the Experimental Life* (Princeton: Princeton University Press, 1985), pp. 207-24.

22 Ibid., pp. 213, 217.

23 Ibid., pp. 201-2.

24 Ibid., p. 201. The view that Boyle deliberately promoted civility among natural philosophers is also presented in Shapin, *A Social History of Truth*. In this view, Boyle encouraged the acceptance of the mechanical philosophy by constructing an identity founded on his aristocratic standing, his learning, and his piety in order to insure credibility for the new philosophy. A link is drawn between the role of civil, gentlemanly practice and the establishment of truth and facts. See especially chapter 4 (pp. 126-92) for a detailed analysis of Boyle's personal role in the establishment of the mechanical philosophy.

25 Rather than focusing exlusively on political or ideological aspects of Boyle's work, this broader focus is evident in the articles in Michael Hunter, ed., *Robert Boyle Reconsidered* (Cambridge: Cambridge University Press, 1994).

study of a dispute between Boyle and More over the proper interpretation of Boyle's hydrostatical experiments.[26] This eloquent study reveals that their disagreement, which focused superficially on their respective theories of matter, actually revealed a deeper conflict between them, over the nature of God and divine providence.[27] The work of Betty Jo Teeter Dobbs also exemplifies this trend to highlight theological influences in the development of early modern science. Without disregarding or diminishing a wide range of factors, including "intellectual, social, technological, political, economic, and institutional developments," Dobbs put the primary source for developments in early modern science clearly within the theological doctrine of the Judeo-Christian tradition.[28] She thought that issues in theology, particularly debates over the attributes and creative powers of God, promoted and directed the study of nature in the early modern period.

I will situate my study of Boyle's natural philosophy in a theological context. In analyzing the dispute between Boyle and More over the proper interpretation of Boyle's early pneumatical experiments, I demonstrate first that arguments about the existence of the void were not settled on experimental and empirical grounds alone, and therefore should not be held up as symbolic of a dramatic shift to an outlook reminiscent of our modern scientific theories. Instead, I argue that Boyle's interest in issues surrounding the void involved more general theological and philosophical considerations.

I also show that Boyle was able to get around traditional objections to the void (the absurdity of claiming the existence of nothing) by reconceptualizing the problem and thereby allowing it to conform to standards of reasoning in natural philosophy while also keeping it theologically benign. However, the clever epistemological trick used by Boyle to erase conceptual problems with the void should also not

---

26 John Henry, "Henry More versus Robert Boyle: The Spirit of Nature and the Nature of Providence," in *Henry More (1614–1687), Tercentenary Studies*, ed. Sarah Hutton (Dordrecht: Kluwer, 1989), pp. 55–76.

27 The importance of theological differences in disagreements between Boyle and More is also recognized in the work of Barbara Beigun Kaplan. She reveals that the responses of Boyle and More to the healing abilities of Valentine Greatrakes reflected their attitudes to divine providence and the nature of God. More regarded Greatrakes as "an earthly medium transmitting spiritual virtues emanating from God." Boyle worried that viewing his cures to be miraculous would not only diminish the authoritative power of Old Testament miracles but would impute too much power to an autonomous natural world, ignoring God's power. Therefore, he looked instead for mechanical explanations in order to preserve his beliefs about the proper role for divine powers. Barbara Beigun Kaplan, "Greatrakes the Stroker: The Interpretations of His Contemporaries," *Isis* 73 (1982): 178–85.

28 Betty Jo Teeter Dobbs, *The Janus Faces of Genius: The Role of Alchemy in Newton's Thought* (Cambridge: Cambridge University Press, 1994), p. 255.

be held up as evidence that he sought to demolish traditional Aristotelian concepts, a view that would make him a full participant in a Scientific Revolution. Rather than an innovation original with Boyle, his reconceptualization of the vacuum was an idea he found in the work of Johann Alsted, a sixteenth-century scholar firmly commited to Aristotelian thought.

Boyle was able to use a concept of void heuristically in his scientific reasoning, while at the same time not according it any ontological status. He accomplished this shift by suggesting, as Alsted had, that vacuum could be conceptualized in concrete terms, as the privation of a characteristic in a subject with a natural disposition for that characteristic. In this way vacuities in the world were analogous to blindness in a person. Boyle avoided the persistent philosophical problems surrounding the concept of void by referring to it as the absence of matter rather than the presence of a new entity.

Boyle's Air-Pump Experiments

Robert Boyle learned of pneumatic experiments using an air pump from "a book...published by the industrious Jesuit Schottus; wherein,...he related how that ingenious gentleman, Otto Gericke, consul of Magdeburg, had lately practised in Germany a way of emptying glass vessels, by sucking out the air at the mouth of the vessel."[29] This book was Gaspar Schott's *Mechanica hydraulico-pneumatica*, published in 1657, and Boyle remembered how the description of these experiments left him "delighted... since thereby the great force of the external air was rendered more obvious and conspicuous than in any experiment that I had formerly seen."[30] Boyle thought nothing could be "more fit and seasonable" than the "prosecuting and endeavouring to promote that noble experiment of Torricellius."[31]

The results of Boyle's experiments appeared in 1660 as *New Experiments Physico-Mechanical, Touching the Spring of the Air and Its Effects*. This work described forty-three of Boyle's air-pump experiments in order to "insinuate that notion...that most, if not all of them, will prove explicable...that there is a spring, or elastical power in the air we live in."[32] The experiments stirred up interest about both the nature of air and the status of the seemingly empty space left in the machine. Many philosophers such as Huygens and members of the Accademia del Cimento in Florence focused attention and praise on

---

29 Boyle, *New Experiments*, in *Works*, 1:6.
30 Ibid.
31 Ibid.
32 Ibid., 1:11.

Boyle's description of the weight or pressure of air. Others however, such as Franciscus Linus, Thomas Hobbes, and Henry More, challenged Boyle's theory.[33]

Of the forty-three experiments presented by Boyle as evidence for his new theory of the air's spring, More focused on the thirty-third experiment to validate his own theory that an incorporeal Spirit of Nature was active in the world. All the experiments were done with Boyle's air pump (see Figure 8.1). This consisted of two main parts. The upper part was a glass globe (called the receiver) with a thirty-quart volume.[34] At the top of the receiver, a four-inch, round hole, capped with a lid and valve, allowed entry of instruments and experimental objects.

The glass receiver sat atop a pump, used to draw air out. A valve (called a stopcock) between the receiver and pump regulated air flow. The brass pump, which had a three-inch diameter, housed the wooden piston, or "sucker," covered with leather, to fit tightly in the brass cylinder. The piston was cranked up and down by means of a rack-and-pinion device attached to the bottom of the piston. The entire device was framed by a wooden stand.[35]

According to Boyle, the thirty-third experiment "may afford us a nobler instance of the force of the air we live in" because it focused on the action of the piston rather than the receiver itself. Boyle had "not yet been able to empty so great a vessel as our receiver." But the cylinder attached to the bottom of the glass receiver, housing the piston that drew out air, could be emptied more easily and for longer periods of time. Boyle hoped that "by the help of this part of our engine" it would be possible to get "a pretty near guess at the strength of the atmosphere."

It would also serve to confirm Boyle's theory that the air has weight and "a spring, or elastical power." According to this theory "our air consists of, or at least abounds with, parts of such a nature, that in case they be bent or compressed by the weight of the incumbent part

---

33 Boyle responded to Linus's Aristotelian objections in *A Defense of the Doctrine Touching the Spring and Weight of the Air*, in *Works*, 1:118–85, published two years later, in 1662, as an addendum to the second edition of the *New Experiments*. He answered Hobbes in *An Examen of Mr. T. Hobbes His Dialogus Physicus de Natura Aeris; As far as It Concerns Mr. Boyle's Book of New Experiments Touching the Spring of the Air*, in *Works*, 1:186–242.

34 Boyle would have liked a larger receiver but "the glass-men professed themselves unable to blow a larger, of such a thickness and shape as was requisite to our purpose." Boyle, *New Experiments*, in *Works*, 1:7.

35 Boyle's own description of his air pump is in *New Experiments*, ibid. For a more recent description of the pump, see Shapin and Schaffer, *Leviathan and the Air-Pump*, pp. 26–30.

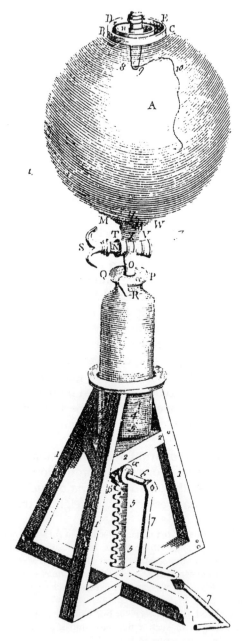

Figure 8.1. Robert Boyle's air pump. From *New Experiments Physico-Mechanical, Touching the Spring of the Air and Its Effects*, in *The Works of the Honourable Robert Boyle*, ed. Thomas Birch, 6 vols. (London, 1672), vol. 1, facing p. 86.

of the atmosphere, or by any other body, they do endeavour, as much as in them lieth, to free themselves from that pressure, by bearing against the contiguous bodies that keep them bent."[36] Boyle considered air to be analogous to "a heap of little bodies, lying one upon another," like so many little pieces of wool. Just like a fleece of wool, air "consists of many slender and flexible hairs; each of which may indeed, like a little spring, be easily bent or rolled up; but will also, like a spring, be still endeavouring to stretch itself out again."[37]

In his thirty-third experiment Boyle set out to measure the strength of the air's tendency to act like a coiled spring and exert a force. The air pump's glass receiver was first removed leaving only the brass cylinder with its internal piston. A valve embedded in the top of the cylinder rendered it airtight when closed. With this upper valve "very carefully and closely stopt," the piston was forcibly pulled to the lowest part of the cylinder and weighted down with hanging weights (see Figure 8.2). Bystanders, with "no small wonder," observed the ascent of the heavily-weighted piston and "could not comprehend, how such a weight could ascend, as it were, of itself."[38] Boyle explained that the action of the piston resulted from the spring exerted by air particles under the piston. The air below the weighted piston was able to push it up because the air particles below, like so many coiled springs, found "little or nothing in the cavity of the evacuated cylinder to resist it." Just like a fleece of wool, compressed smaller by a hand, air particles maintain "an endeavour outwards" causing them to continually push against the pressure exerted upon them.[39]

Boyle's explanation was given in terms of the mechanical actions of tiny particles of air. After initially yielding to the piston drawn forcibly downward, the compressed air below pushed it upward when it found little resistance from the evacuated cylinder above the piston. Speculating that a consideration of all the inferences "that a speculative wit might make" from this experiment would "require almost a volume," Boyle proceeded to discuss only the major consequences of this experiment, including the revision of traditional ideas about vacuum.

36 Boyle, *New Experiments*, in *Works*, 1:11.
37 Ibid. Boyle presented this as one of two possible explanations of the spring's air and did not defend it against the Cartesian hypothesis that the air "is nothing but a congeries or heap of small and ... of flexible particles, of several sizes, and of all kinds of figures ... [with] elastical power ... not made to depend upon their shape or structure." See J. J. MacIntosh, "Robert Boyle's Epistemology: The Interaction between Scientific and Religious Knowledge," *International Studies in the Philosophy of Science* 6 (1992): 104.
38 Boyle, *New Experiments*, in *Works*, 1:72.
39 Ibid.

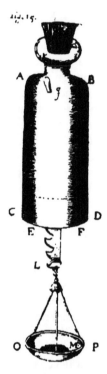

Figure 8.2. Weighted piston used in Experiment thirty-three. From Henry More, *Enchiridion Metaphysicum* (London, 1671), p. 195.

Boyle refuted common opinions regarding nature's abhorrence of vacuum by making reference to a revised theory of what matter is at a fundamental level. The common notion that nature abhors a vacuum was based, in Boyle's view, on appearance alone; it was "but metaphorical and accidental," resulting from "the pressure of the air and of the gravity, and partly also of the fluxility of some other bodies." A vacuum is hindered, yes, but only because of these mechanical responses. With a mechanistic explanation of these experiments, Boyle refuted the Aristotelian dictum that nature abhors a vacuum and showed that any aversion nature had to a void was merely an accident or consequence of the structure and behavior of matter and motion and induced no active response in the world.

Although many natural philosophers applauded Boyle's interpretation of his experiments, Henry More did not agree, fearing that explanations framed in material terms alone would have dangerous theological consequences. More sought to prove the existence of an intermediary, incorporeal Spirit of Nature, operating on the world.

Henry More's Appropriation of Boyle's Thirty-third Experiment

Henry More's interest in the innovative mechanical philosophy of nature was recognized when he was made a member of the Royal Society of London in 1664, although he was never personally involved in carrying out any experiments. More's interest in the mechanical philosophy of nature was ultimately rooted in his desire to solidify Christian theology rather than in promoting empiricism or experimentalism.[40]

Henry More sought to build a system that would ensure the continuation of religious piety while, at the same time, incorporating the innovations of the new mechanical philosophy of nature. He considered it essential to prove the existence of incorporeal entities in the physical world, believing that the demonstration of "the necessity of incorporeal beings" was "a design, that which nothing can be more seasonable in this age; wherein the notion of a spirit is so hooted at by so many for nonsense."[41] Proving the existence of an incorporeal entity would lend credence to the existence of other, more important ones, such as the human soul, angels, and God, thus avoiding materialism.

More sought to ensure the importance of incorporeals in the world because he knew "how prejudicial Des Cartes's mechanical pretensions are to the belief of a God," because "the phaenomena of the world cannot be solved merely mechanically, but . . . there is the necessity of the assistance of a substance distinct from matter, that is, of a spirit, or being incorporeal."[42] More set out to construct a philosophy to replace traditional Aristotelianism and thus prevent the accession of Cartesianism, since he believed that "it is not mere and pure Mechanical motion that causes all these sensible Modifications in Matter."[43] Critical of both the Cartesian and Aristotelian philosophies, he

40 The relationship between More's natural philosophy and theology is presented in Max Jammer, *Concepts of Space: The History of Theories of Space in Physics*, 2nd ed. (Cambridge, Mass.: Harvard University Press, 1969), pp. 40–52, and in Alan Gabbey, "Philosophia Cartesiana Triumphata: Henry More (1646–1671)," in *Problems in Cartesianism*, ed. Thomas M. Lennon, John M. Nicholas, and John W. Davis (Montreal: McGill-Queen's University Press, 1982), pp. 171–250.
41 More to Boyle, December 4, 1671 in *Works*, 6:513.
42 Ibid., 6:513–15.
43 More, *The Immortality of the Soul, so Farre Forth as It Is Demonstrable from the Knowledge of Nature and the Light of Reason* (London, 1659), bk. 3, chap. 3, p. 199.

outlined why "the Atomick philosophy is to be prefered." Firstly, in an age described as "industrious and searching," atomism "is most universally received by free and considering Philosophers." Linking his theology to the new philosophy, More considered it "the most useful for the best ends" because it "serves to support the main parts of natural Religion the best; namely, the Existence of God, of *Genii* or Angels, and the Immortality of the Soul."[44]

More's philosophy, which gave importance to incorporeals by according them dimension, drew on evidence from Boyle's highly acclaimed vacuum experiments for validation. Disagreements over the epistemological and theological underpinnings of More's system led Boyle to respond.

In formulating his own philosophy, More took parts of both Platonic metaphysics (belief in a reality, independent of God) and Cartesian rationalism (reliance on reason as the ultimate criterion for knowledge) and felt that his "interweaving of Platonisme and Cartesianism" would be "making use of these Hypotheses as invincible Bulwarks against the most cunning and most mischievous efforts of Atheism."[45]

More's first step in constructing his new philosophy was to prove the existence of incorporeals in the material world. He saw a definite link between belief in immaterials and belief in God. Those who did not accept immaterial entities in the world, those who considered "the very notion of a spirit or substance immaterial as a perfect incompossibility and pure non-sense," have to consider

> no better consequences then these: That it is impossible that there should be any God, or Soul, or Angel, Good or Bad; or any Immortality or Life to come. That there is no Religion, no Piety nor Impiety, no Vertue nor Vice, Justice nor Injustice, but what it pleases him that has the longest Sword to call so. That there is no Freedome of will, nor consequently any rational remorse of conscience in any being whatsoever, but that all that is, is nothing but matter and corporeal motion; and that therefore every trace of mans life is as necessary as the tracts of lightning and the fallings of thunder; the blind impetus of the

---

44 More, "A Defense of the True Notion of a Spirit," pp. 224–5. This is a translation of chaps. 27 and 28 of More's *Enchiridion metaphysicum* (London, 1671), which appeared in Joseph Glanvill, *Saducismus metaphysicum* (London, 1681). The other reasons More provided for his support of atomism were that it "seems to be the most ancient and nay to have been the old ... Mosaical Philosophy" and that it "is demonstrable to any unprejudiced Reason" (p. 225). More's support of atomism is also clear in *Immortality of the Soul*, preface, p. 3.

45 Henry More, *A Collection of Several Philosophical Writings of Dr. Henry More* (Cambridge, 1662), p. vi.

matter breaking through or being stopt every where, with as certain and determinate necessity as the course of a torrent after mighty storms and showers of rain.[46]

Clearly, the stakes were high.

More looked to Boyle's experiments as an endorsement of his theoretical views about a Spirit of Nature. It was "out of that noble and ingenious Gentleman's Experiments of his Aire-pump" and, in particular, Boyle's thirty-third experiment that More found support for his plan to "plainly demonstrate there must be some Immaterial Being that exercises its directive Activity on the Matter of the World."[47] His interpretation was meant as a corrective to the "rash fancies and false deductions" that had been drawn from these "misunderstood Experiments."[48] More was certain that his interpretations of the experiments were the correct interpretations and that with them he would prove that there was a Spirit of Nature, an incorporeal substance, acting on behalf of God, directing the matter of the universe.

More argued that the phenomenon exhibited in the thirty-third experiment did not arise from "the mere Mechanical powers of Matter" but was "directed and effected by ... some Superintendent Cause."[49] He assumed that "this ascending of the Sucker of the Aire-pump with above an hundred pound weight at it" was sound proof that "there is a Principle transcendng the nature and power of Matter ... an Immaterial Principle, (call it the Spirit of Nature ...) which is the Vicarious Power of God upon this great Automaton, the World."[50] This "absolutely astonishing phenomenon" was not caused by either the piston, the hollow cylinder housing the piston, the air surrounding the piston, or the air within the cylinder. The mechanical explanation put forward by Boyle to explain the ascension of the piston, which relied upon a theory of the ambient air's spring, or pressure, was rejected by More after he first graciously acknowledged that the explanation did have "some verisimilitude."[51] However, "that which charmed us so much at first glance altogether disappoints our hope" for, "there is nothing solid in this mechanical solution," because, "there can be no condensation of the external air in this way produced by solely mechanical causes." More believed that air drawn out of the receiver would be quickly diffused throughout the ambient air, leav-

---

46 More, *Immortality of the Soul*, bk. 1, chap. 9, p. 36.
47 Henry More, *An Antidote against Atheism, or, An appeal to the Natural Faculties of the Minde of Man, Whether There Be Not a God* (London, 1653), bk. 2, chap. 2, p. 43.
48 More, *Immortality of the Soul*, bk. 3, chap. 2, p. 153.
49 More, *Antidote against Atheism*, bk. 2, chap. 2, p. 44.
50 More, *Enchiridion metaphysicum*, chap. 12, p. 2.
51 Ibid., chap. 12, p. 7.

ing it unable to produce any pressure or exert a force upward on the weighted piston. According to Boyle's explanation, the force exerted by the air was exceedingly powerful – able to push up the piston and over one hundred pounds of lead weight. But More concluded that if the air had that much power, "small birds, snails, flies and other similar miniscule animals" would "in a short time be destroyed." Human bodies as well would be "battered and ... crushed ... like wheat between two millstones."[52] The piston's ascension must clearly be caused by "the Effects of the same Immaterial Principle, (call it the Spirit of Nature ...)."[53]

Only "Atheists and Epicureans" would ever imagine that matter could act mechanically or order itself. But since it was evident that the world is ordered, and "attending heedfully to the natural Emanations of unprejudic'd Reason," it is clear that there is a "Knowing Principle, able to move, alter and guide the Matter." It was through the acknowledgment of this "Knowing Principle" that More hoped to achieve his goal of tracing "the visible foot-steps of this Divine Counsel."[54]

Boyle was disturbed by More's apparent misinterpretations of his experiments, worrying that such misrepresentation might contribute to the failure of his mechanical philosophy. Boyle's fears heightened when he read further that More beseeched people to rouse themselves "from this mechanical daydream or torpor" and to free themselves "from the very disease of this age, namely mechanical credulity."[55]

Beyond the threats he felt against mechanism, Boyle was concerned that More was appropriating his experimental work in order to advance a theological agenda that Boyle considered dangerous. Boyle's voluntarist theology held God to be directly responsible for all activity in the world and he worried that inserting an intermediary, which he considered More's Spirit of Nature to be, might eventually interfere with God's free action.[56] It was no secret that Boyle did not approve of More's discussion of his experiments. Anne Conway informed More that "I hear Mr. Boyle sayes you had better never printed it, for you are mistaken in all your experiments."[57] More however, had no

52 Ibid., chap. 12, p. 10.
53 More, *Antidote against Atheism*, bk. 2, chap. 2, p. 46.
54 Ibid., bk. 2, chap. 2, p. 47.
55 More, *Enchiridion metaphysicum*, chap. 13, p. 17.
56 For a discussion of theological differences between Boyle and More, see Margaret J. Osler, "Triangulating Divine Will: Henry More, Robert Boyle, and René Descartes on God's Relationship to the Creation," in *Stoicismo e Origenismo nella Filosofia del Seicento Inglese*, ed. Marialuisa Baldi (Milano: Franco Angeli, 1996), pp. 75–87.
57 Anne Conway to Henry More, February 4, 1675/6. M. H. Nicholson, ed., *Conway Letters: The Correspondence of Anne, Viscountess Conway, Henry More, and Their Friends, 1642–1684* (New Haven: Yale University Press, 1930), p. 420.

regrets, replying to Conway that "If I be mistaken in my Experiments, I suppose Mr. Boyle will show me my mistakes, and all the world besides, which makes me conclude it is better that I have printed them that I and such as I may be undeceived by him or some he may employ against . . . me." More was clearly aware that what "pinches" Boyle is his "exploding that monstrous spring of the ayre."[58] And indeed, Boyle was "pinched" enough to reply at length.

Boyle's Response and Solution: The Absence of Presence

It did not take long for More to learn that "Mr Boyle does not take my dissenting so from him in publick so candidly as I hoped." More confessed to being "very sorry" in light of Boyle's negative response since More stated that he had "a singular honour and esteeme" for Boyle even though More did "meddle with his notions."[59] Nonetheless, More's "meddling" produced a direct response from Boyle in his *Hydrostatical Discourse*.

Boyle worried that More's promotion of an immaterial, incorporeal intermediary, a Spirit of Nature, as the physical cause of certain natural phenomena threatened both his nascent mechanical philosophy of nature and his theology. To have recourse to an immaterial substance "as a physical agent" would, according to Boyle, "prove a work exceeding difficult."[60] Explaining "familiar phaenomena" in terms of an immaterial agent gave only obscure and uncertain explanations by those who could only "pretend knowledge, or disguise ignorance."[61] Boyle disliked the explanations of the "naturists" because he believed reliance on ideas or notions to explain natural phenomena "dangerous to religion in general, and consequently to the Christian."[62] This danger was even greater since "many of the most learned amongst the naturists are Christians"; and, like Henry More, "divines too."[63]

Boyle believed that More's rejection of a totally mechanical explanation of his experiments was motivated by an entirely unacceptable theological position. More's "erroneous conceit" of proposing that an

---

58 Henry More to Anne Conway, February 9, 1675/6; ibid., p. 423.
59 Henry More to Anne Conway, May 11, 1672; ibid., p. 358.
60 Boyle, *Notion of Nature*, in *Works*, 5:191.
61 Ibid.
62 Ibid.
63 Ibid. Although Boyle never mentioned More by name in his *Notion of Nature*, many scholars surmise this is precisely whom Boyle numbered among the "naturists," especially since Boyle focused so much attention in this treatise on a refutation of the notion of a Spirit of Nature. See Richard S. Westfall, *Science and Religion in Seventeenth Century England* (Ann Arbor: University of Michigan Press, 1973), p. 84, and Hunter and Davis, "The Making of Robert Boyle's *Free Enquiry*," p. 246.

incorporeal Spirit of Nature was responsible for the phenomena created in Boyle's experiments was dangerous to religion because of its potential to "defraud the true God of divers acts of veneration and gratitude, that are due to him from men, upon the account of the visible world."[64] Framing explanations in terms of More's Spirit of Nature left God "but very little in the administration of the parts" of the universe. The result of More's maneuver was that "instead of the true God, they have substituted for us a kind of a goddess, with the title of nature; which as they look upon as the immediate agent and director, in all excellent productions, so they ascribe to her the praise and glory of them."[65]

In order to avoid these dangerous theological consequences, Boyle responded to More by reaffirming the mechanical explanations of the experiments in an attempt to diffuse the theological issues that underwrote More's interpretation. The consequences of Boyle's steadfast affirmation of a mechanistic explanation that did not give explanatory power to incorporeals led him to conceive of the void without giving it any independent ontological status. With this reconceptualization he could use a concept of vacuum heuristically in natural philosophical reasoning while, at the same time, denying its use to those promoters of incorporeals, such as More, who might point to the ontological reality of the void as a prime example of an activating incorporeal intermediary, such as the Spirit of Nature.

It is "the mechanical affections of matter" that were sufficient "to produce and account for the phaenomena" that More interpreted, "without recourse to an incorporeal creature."[66] Therefore, Boyle explained the phenomena produced in his thirty-third experiment "upon the laws of the mechanicks." Such a mechanical explanation was sufficient, without "recourse to an incorporeal creature... a *fuga vacui*, or the *anima mundi*, or any such unphysical principle." Boyle wondered "why may not... propositions and accounts that suppose gravity in the air... be looked on as mechanical?"[67]

In the third chapter of his discourse, Boyle responded to More's claim that if the air had weight everything in the world would be crushed, and did so by asserting that experimental evidence would prove More wrong. He argued that "experiments and considerations about the differing pressures of solids, weights, and ambient fluids" would "satisfy you of the invalidity of the proposed objections." This conversion would be especially so, stated Boyle, since "the doctrine it

64 Boyle, *Notion of Nature*, in Works, 5:191.
65 Ibid.
66 Boyle, *Hydrostatical Discourse*, in Works, 3:627.
67 Ibid., 3:601.

impugns, namely the weight and pressure of the atmosphere, is not a bare hypothesis, but a truth made out by divers experiments, by which even professed opposers of it have publickly acknowledged themselves to be convinced."[68] With this evidence he believed he had "answered my learned adversary's objections." Boyle's exasperation with More was evident, repeatedly going over his mechanist explanation, and lamenting that if More had only read some of his other experimental evidence, he would have understood them. For "if the doctor had been pleased to read [Boyle's *Continuation of New Experiments*] he would have received the same satisfaction, that other learned men have done."[69] Declaring More's Spirit of Nature to be "a precarious principle," Boyle contended there was simply no need for it. Mechanical explanations sufficed. There was no reason to explain the phenomena of "the spring and weight of the air, the gravity and fluidity of the water" by an incorporeal agent such as the Spirit of Nature. Such maneuvers Boyle saw as "attempts to accommodate the phaenomena to the hypothesis."[70]

If misgivings still remained after explaining certain natural phenomena mechanically, Boyle suggested that "to remove those remaining scruples ... [one could] range about for other physical helps to solve more completely the problem." The phenomena of compression and air pressure were actions "purely corporeal and mechanical," and so therefore should the explanations of them be. If More thought a mechanical explanation was *not* "a ready and complete solution," he should not "fly to the immediate interposition of an immaterial and intelligent, yet created agent" to explain the phenomena, for an explanation relying on such an agent would involve "a much more difficult task, than the solution of the phaenomenon without it."[71]

If Boyle could not discount More's interpretation by suggesting that he simply failed to understand rather difficult hydrostatical and pneumatic concepts, then he hoped to show that More's interpretation, which involved the postulation of an incorporeal agent, was not only unnecessary but would not work. However, Boyle did not stop there in his goal of quashing the "learned doctor" 's arguments but presented an elegant twist of Aristotelian arguments to deflate More's interpretations.

In claiming that "such mechanical affections of matter," such as the spring of the air could be easily explained "without recourse to an

68 Ibid., 3:603.
69 Ibid.
70 Ibid.
71 Ibid., 3:627.

incorporeal creature," Boyle suggested that even Aristotelian philosophers would concur with Boyle's rejection of More's incorporeal agent. Boyle criticized Aristotelian natural philosophers for relying on immaterial forms and innate tendencies (such as nature's abhorrence of vacuum) to explain natural phenomena, and so, on the surface it appears that he saw More's explanation as simply mirroring Aristotelian ones that also relied on incorporeals. However, Boyle saw a fundamental difference between them. For even though More's recourse to an "incorporeal creature" was much "like the Peripateticks," even Peripatetics themselves would consider More's incorporeal "less qualified" than "their *fuga vacui*" as an explanatory device. This was because Aristotelians considered nature's desire to resist a vacuum to be limited. More's Spirit of Nature, on the other hand, had "unlimited power to execute its functions."[72]

The Aristotelian argument that nature resisted but could not entirely prevent a vacuum was essential to explanations of such classic experiments as the Torricellian mercury experiment. If nature's abhorrence of vacuum was entirely unlimited, then there would be no way to explain the fact that the mercury column was held at a certain level. It was a *limited* horror of vacuum that prevented the complete drop of the mercury in the tube, and the force exerted by this repugnance to vacuum was equal to how far the mercury column fell.[73] Aristotelians therefore would object to More's Spirit of Nature because it had unlimited powers and therefore would not provide adequate explanations of these experiments.

Now Boyle, as usual, was clear that he would "not launch into the controversy, whether nature do any thing *ob fugam vacui*," but showed great delight in suggesting that even Aristotelians with their *fuga vacui* would reject More's incorporeal agent. And Boyle's own purely mechanical interpretations would also alter the received notion about nature's abhorrence of vacuum. Rather than accepting the claim that nature actively seeks to prevent a vacuum, Boyle believed that his thirty-third experiment taught that "the supposed aversion of nature to a vacuum is but accidental, or in consequence, partly of the weight and fluidity, or, at least, fluxility of the bodies here below; and partly, and perhaps principally, of the air, whose restless endeavour to expand itself every way makes it either rush in itself, or compel the interposed bodies into small spaces, where it finds no greater resis-

---

72 Ibid.
73 This is the view held by Pascal before he carried out the Puy-de-Dôme experiment, which led him to conclude that nature has no abhorrence of vacuum and that air pressure alone supports the mercury column.

tance than it can surmount."⁷⁴ This explanation, which held that "nature's hatred of vacuum is but metaphorical and accidental," confirmed Boyle's commitment to mechanical explanations of natural phenomena, a move by which he could avert other explanations that relied on an intermediary Spirit of Nature. The theological problem with More's Spirit of Nature, as Boyle saw it, was that it was an entity intervening between God and his creation, and thereby possessing the potential to interfere with his free exercise of will. Because of his voluntarist theology, Boyle believed that nothing could restrict God's power, that all activity in the world came directly from God rather than some intermediary.⁷⁵

The theological aspect of the disagreement between Boyle and More was recognized by the anonymous reviewer of the *Hydrostatical Discourse*, who described how Boyle, "our Noble Philosopher," believed not only that the world had been created by God but that it is "continually preserv'd by God's divine power and wisdom." This meant it was "his general concourse" that continued to maintain "the Laws by him establish't in it." Therefore, "the phenomena he endeavours to explicate, may be solv'd Mechanically, that is, by the Mechanical affections of Matter, without introducing any precarious Principles, such as he esteems to be Nature's Abhorrence of a Vacuum, Substantial Forms, or Dr. More's Hylarchical Principle, i.e. (in plainer terms,) his Immaterial Director."⁷⁶

Boyle's endless descriptions of entirely mechanical explanations for those experiments appropriated by More displayed his attempt to ensure the success of his mechanical philosophy of nature. Part of his argument against More is the assertion that the Cambridge theologian simply misunderstood his mechanical explanations. However, it is theological misgivings that seem most prominent in Boyle's discourse against More.

At both the beginning and end of his *Discourse* Boyle spoke of God. His introductory comments in which he protected Descartes from More's slanderous charges seem rather curious at first. Acknowledging that there were certain aspects of Cartesian natural philosophy

74 Boyle, *New Experiments*, in *Works*, 1:75.
75 Boyle's voluntarism is presented as the source of disagreements with More in Osler, *Divine Will and the Mechanical Philosophy*, pp. 226–7 and Henry, "Henry More versus Robert Boyle," pp. 63–4. Similar theological debates led to conflict between More and another voluntarist, Richard Baxter. See John Henry, "A Cambridge Platonist's Materialism: Henry More and the Concept of Soul," *Journal of the Warburg and Courtauld Institutes* 49 (1986): 187.
76 Anonymous, "An Account of Two Books," *Philosophical Transactions of the Royal Society of London* 92 (March 25, 1673): 5199.

that he found somewhat theologically unsettling (not ascribing to God a more direct and "immediate" role in the universe), Boyle nonetheless protected Descartes from those, such as More, who suggested that Cartesian principles reeked of irreligion.[77] On the contrary, Boyle pointed to the suggestions he found in Descartes's philosophy for God's continued activity in the world.

Boyle's defense of Descartes seems curious until one reads the end of his Discourse where the topic of God appears again, not having drawn any mention whatsoever through any other part of the long discourse. His final argument against More's "precarious principle" of an incorporeal substance is founded on the claim that even "heathen philosophers" do not wish to incorporate such an immaterial agent into their explanations of natural phenomena. For, the surest way to ensure the unchallenged existence of God in the world and the way that even non-Christian philosophers accomplish the "being of a divine architect of the world" is "by the contemplation of so vast and admirably contrived a fabrick, wherein, yet taking no notice of an immaterial *principium hylarchicum*." They believed things "to be managed in a mere physical way, according to the general laws, settled among things corporeal, acting upon one another."[78]

Therefore, in his philosophy of nature, Boyle gave no explanatory power to noncorporeal entities such as the void. He denied the positive existence of vacuum, as Aristotle had, but not because he considered dimension, and therefore existence, to be linked to material entities only, but because he thought that giving vacuum, an insensible entity, explanatory power would be a return to the same kind of insensible forms and essences used within Aristotelian philosophy. Boyle was concerned that giving dimension and existence to an incorporeal vacuum threatened the fundamentals of mechanistic explanations that relied on matter and motion alone to explain natural phenomena.

Similar concerns challenged his theological sensibilities. Although Boyle did not deny the existence and activity of incorporeal entities such as angels and human souls, he did not accord them universal power to act on matter. To do so would diminish the ability of God to act directly on the matter of the world. With all the debate surrounding the status of the vacuum, Boyle feared that More's desire to give activity to an all-encompassing incorporeal Spirit of Nature might lead him to consider the vacuum to be a prime example of an immaterial

---

77 Other important seventeenth-century philosophers, including Gassendi and Newton, accused Descartes of being insufficiently voluntarist. See Osler, *Divine Will and the Mechanical Philosophy*, p. 149.

78 Boyle, *Hydrostatical Discourse*, in *Works*, 3:628.

force. Not only would this proclaim the death of his mechanical philosophy but it would also lead to atheism.

Boyle refused to declare that vacuum had a positive or independent existence or that it was the efficient cause or agent of any material change. In the same way, he considered the sensation of cold to "be but a privation." Even though cold "affects the organs of feeling, and sometimes causes great pain in them, condenses air and water, and breaks bottles that are too well stopped," these effects are "undeservedly ascribed to cold," and are "rather the occasion, than the true efficient cause of such effects."[79] Such "effects so numerous and great ... are properly to be ascribed to those physical agents, whose actions or operations happen to be otherwise modified, than else they would have been upon the occasion of that imminution or slackness of agitation which they meet with in cold bodies."[80] In the same fashion, darkness is "confessedly a privation of light," deafness a privation of hearing, and ignorance a privation of knowledge. And "death itself is but a privation of life," even when "a man is killed by a bullet [and] his death is effected by a positive, and even impetuous action."[81]

Boyle considered the vacuum to be the privation of air. This "absence, or not touching of the air, be but a privative or negative thing." He recounted what happened when "birds, rats, and cats, and such kind of warm animals" were placed in the receiver of his air pump and "by the operation of that instrument, which withdraws the former air, and keeps out the new, the air, that was wont continually to act upon them, is kept from doing so any longer." The animals die quickly, "without the visible intervention of any positive agent."[82]

Boyle compared this to what happened during war, when enemies divert the stream powering a town's mills. The action of all the mills would cease, "though the besiegers do not produce this change by any positive and direct violence that they offer to the mills, but only by hindering them from receiving the wonted impulses which were requisite to keep them in motion."[83]

Giving significance to negative sense perceptions, such as cold, darkness, and empty space allowed Boyle to justify the importance of incorporeal entities such as vacuum without giving these incorporeals a direct explanatory power. He explained such occurrences as rarefaction in terms of the increased distance between the material corpuscles

79 Boyle, *Of the Positive or Privative Nature of Cold*, in *Works*, 3:737, 744.
80 Ibid.
81 Ibid., 3:750.
82 Ibid., 3:746.
83 Ibid., 3:745. This notion of privation was attractive to Boyle because it upheld the nominalist dictum: "Entia non sunt multiplicanda absque necessitate."

rather than in terms of the increased amount of incorporeal vacuum. In order to avoid declaring himself either for or against the existence of a void, Boyle simply stated that there was an absence of air in his receiver, rather than the presence of void. For,

> to call it void absolutely, would be judged by many as declaring himself a vacuist, who does not yet own the being either of their opinion, or a downright plenist; or else he must be troublesome to the reader and himself, by frequently explaining what sort of vacuum he understands; whereas he declares once for all, that by the *Vacuum Boylianum* he means such a vacuity or absence of common air, as is wont to be effected or produced in the operations of the *Machina Boyliana*.[84]

With this kind of reasoning, Boyle reconceptualized the void as privation. It could be considered in concrete, material terms while not posing the paralogism of saying that nothing existed. A notion of void could be incorporated into explanations without either asserting or denying its actual existence in reality.

The Aristotelian conceptual framework refuted the possibility of void by arguing that a vacuum, which was a denial of existence, could not be asserted to exist.[85] Boyle's reconceptualization shifted emphasis from the logical contradiction implicit in consideration of the existence of nonbeing to the notion that privation was a negation in a subject with a natural disposition for possessing the particular characteristic. This conceptual move allowed Boyle to skirt what he considered to be the dangerous theological implications of giving a nonmaterial entity, such as a Spirit of Nature, an absolute existence.

Although Boyle's concept of the void-as-privation did undermine the traditional Aristotelian denial of vacuum, it should not be counted as evidence for a systematic rejection of the entire Aristotelian system of thought. To do so would number Boyle as an innovative pioneer in the Scientific Revolution, a contributor to the demise of Aristotelianism, laying the groundwork for the objective empiricism of modern science. Those holding such a view consider that as Boyle's stature grew, "the philosophical shackles of earlier times were shaken off," to build "a framework of knowledge to which twentieth-century science is still indebted."[86] Boyle was certainly involved in reworking aspects of the Aristotelian conceptual framework. In his treatise, *The Origin of Forms and Qualities*, Boyle presents the theoretical and experimental basis of his corpuscularian philosophy of nature to refine Aristotelian

---

84 Boyle, *Pneumatical Experiments about Respiration*, in Works, 3:372.
85 Aristotle, *Physics*, IV.213a.11–217b.28.
86 Stuart Fleming, "The Search for Nothing: Boyle's Experiments with a Vacuum," *Archaeology* 40 (1987): 83.

philosophy rather than replace it completely. Just like a miner, whose job is "to discover new mines, and to work those that are already discovered," Boyle sets out "not only to devise hypotheses and experiments, but to examine and improve those that are already found out."[87]

Rather than indicating a dramatic, forward-looking break with traditional thought, marking a transition to the modern secularism of science, Boyle's reconcepualization of the vacuum grew from his desire to address important theological issues concerning the nature and role of God in the world. Such concerns, which were essential and fundamental to the study of natural philosophy in the seventeenth century, have often been overlooked by histories of a Scientific Revolution, which look for precursors of our present, secular world view.[88]

Furthermore, Boyle's reformulation of the concept of vacuum was not an innovation unique to him, one that added another blow to the demise of Aristotelianism. I place the source of this reformulation firmly within the Aristotelian tradition, with the late-Renaissance scholar, Johann Alsted.[89] A confirmed Aristotelian, Alsted asserted in his *Encyclopaedia* that while vacuum was not found in nature, it could be "refered to in concrete terms," as "a place, being without any body."[90] In this way, conceiving of a vacuum would be "much like refering to a blind man, if you define that man as being without sight."[91]

Although the notion of privation was certainly familiar to Aristotelian thinkers, it was ultimately rejected as a valid means to character-

---

87 Boyle, *The Origin of Forms and Qualities according to the Corpuscular Philosophy*, in *Works*, vol. 3, "The Proemial Discourse to the Reader."
88 For a discussion on how definitions of science and natural philosophy determine how histories of science are written, see Andrew Cunningham, "Getting the Game Right: Some Plain Words on the Identity and Invention of Science," *Studies in the History and Philosophy of Science* 19 (1988): 365–89, and Andrew Cunningham and Perry Williams, "De-Centring the 'Big Picture': 'The Origins of Modern Science' and the Modern Origins of Science," *British Journal for the History of Science* 26 (1993): 407–32.
89 Alsted, a professor of theology at Herborn, drew on a wide range of sources (Old Testament, Aristotle, Paracelsus, Scaliger, John Dee) to construct encyclopedic works used commonly throughout Europe. He was "a favorite author" at Cambridge University. See Walter J. Ong, *Ramus: Method and the Decay of Dialogue* (Cambridge, Mass.: Harvard University Press, 1958), pp. 163–4; Leroy E. Loemker, "Leibniz and the Herborn Encyclopedists," *Journal of the History of Ideas* 22 (1961): 323–38; John T. Harwood, ed., *The Early Essays and Ethics of Robert Boyle* (Carbondale: Southern Illinois University Press, 1991).
90 Johann Alsted, *Encyclopaedia septem tomis distincta*, 2 vols. (Herborn, 1630), bk. 13, pt. 1, chap. 12, sec. 8.
91 Ibid.

ize vacuum because of metaphysical assumptions about the nature of place, which equated it with body. Because the concept of dimension was implicit in the concept of place, it was absurd to contend that a dimensionless vacuum could occupy a place.[92] Nonetheless, while the relationship between privatives and positives was not commonly used by Aristotelian thinkers to characterize void space, it was certainly available as a concept.

Although Alsted upheld the Aristotelian tenet that rejected the existence of void, his rejection was not based on the assumption that it could not occupy place since it did not have dimension. Instead, Alsted showed how place and material body did not have to be conflated; how dimension did not have to be considered a primary or essential quality of void. Since "plenum and vacuum are like modes or definitions of different kinds of place," vacuum could be considered to be "place where bodies are not found where they ought to be," just as "a money pouch is called empty when there is no money in it."[93] Alsted offered a conceptualization of void that avoided the ontological difficulty of giving it dimension. Instead of arguing that it had a positive dimension, this approach referred to it as the negation of material dimension, somewhat similar to a negative number. This way of viewing vacuum avoided the problems associated with giving dimension to a nonmaterial entity, and it was precisely this interpretation of void that appears in Boyle's work. He suggested that

> to our confused, and often also to our inadequate conceptions, belong many of those, that may be called negative, which we are wont to employ, when we speak of privations or negations, as blindness, ignorance, death, etc. We have a positive idea of things, that are square and round, and black and white, and in short of other things, whose shapes and colours make them the objects of our sight; but when we say, for instance, that a spirit or an atom is invisible, those words are attended with a negative conception.[94]

The similarities between Boyle's characterization of the void-as-privation and that of Alsted's seem to be more than coincidental. Other studies have traced the roots of Boyle's thought in moral philos-

---

92 As the opposite of positives, privatives make reference to the thing that is natural for the positive term. Therefore, one cannot talk about blindness in something that does not have the possibility of sight, such as a plant. See Aristotle, *Categories*, 11b. 15–13b.36; *De caelo*, II.286a.26; *Metaphysics*, III.996b.14.
93 Alsted, *Encyclopaedia*, bk. 13, pt. 1, chap. 12, sec. 8, and bk. 13, pt. 8, chap. 4, sec. 4.
94 Boyle, *A Discourse of Things above Reason, Inquiring Whether a Philosopher Should Admit There Are Any Such*, in *Works*, 4:421.

ophy to Alsted.⁹⁵ Harwood's study indicates that most of Boyle's quotations were quite likely from Alsted's *Encyclopaedia*.⁹⁶ Although Alsted's book is not on the partial list of Boyle's library, collated by John Warr shortly after Boyle's death, there is strong evidence to suggest that it was indeed part of Boyle's collection because it was listed in a sale catalog of books sold from John Warr's estate. As Boyle's executor, Warr could have bought the *Encyclopaedia* before Boyle's library went to sale, thus explaining its absence from the book listing.⁹⁷ Given that Boyle almost certainly had a copy of Alsted's *Encyclopaedia*, with its discussion of vacuum, it is not surprising that Boyle's account mirrored Alsted's so closely, although he did not follow Alsted in refuting the void.

The consequences of Boyle's move were twofold. First, he avoided persistent philosophical problems surrounding the concept of void so that explanations of the rarefaction and condensation of air were unproblematic. More important though, he did so while not according void any independent ontological status, thereby preventing its use as a surrogate for incorporeal entities such as More's Spirit of Nature. This explains why Boyle broke his silence to argue against More on the proper interpretation of his experiments. Facing the perceived threat of atheism, he sought to solidify and maintain Christian orthodoxy by establishing a natural philosophy that relied on the mechanical operations of matter. Although Boyle's pneumatic experiments can be given freestanding consideration, any perceived scientific advancements attributed to Boyle might better be evaluated within this complexity of theological commitments and with acknowledgment of the contribution of traditional and unrevolutionary Aristotelian concepts.

95 Harwood, *The Early Essays*, p. xxii.
96 Ibid., p. xxiv. For another study acknowledging Alsted's *Encyclopaedia* as the source for Boyle's comments about the distance of the fixed stars from earth, see Michael Hunter, "How Boyle Became a Scientist," *History of Science* 33 (1995): 78.
97 Harwood, *The Early Essays*, pp. 249–52. Four books by Alsted do appear in the catalog of Boyle's library. These are *Compendium grammatica Latina* (Herborn, 1613), *Methodus theologicae* (Frankfurt, 1614), *Definitiones theologicae* (Frankfurt, 1626), and *Loci communes theologici* (Frankfurt, 1630).

PART III

Canonical Figures Reconsidered

9

Pursuing Knowledge: Robert Boyle and Isaac Newton

JAN W. WOJCIK

In the past few years both Boyle and Newton have received considerable scholarly attention, and today we know much more about these two paragons of the "Scientific Revolution" than did previous generations of scholars. The "Newtonian Industry," as Richard S. Westfall has termed it,[1] has undergone a radical transformation in at least two areas. First, the image of Newton as a significant influence in the establishment of Enlightenment deism has been considerably modified by scholars such as James E. Force, Edward B. Davis, and Betty Jo Dobbs, and it is now widely recognized that even though the Newtonian system could be (and indeed was) associated with a deistic world view, this was most definitely not what Newton himself intended.[2] Second, Newton's alchemical pursuits, long kept secret, and even longer misunderstood, have finally been brought to light by J. E. McGuire and P. M. Rattansi, and have been interpreted by Betty Jo Dobbs in such a way that their relationship to Newton's thought as a whole finally makes sense.[3]

Where Boyle studies are concerned, there has been a virtual explosion of activity, due primarily to the efforts of Michael Hunter to make Boyle's unpublished manuscripts and correspondence available to scholars unable to travel to the Royal Society of London, as well as his efforts to establish and maintain a network of communication among

---

1 Richard S. Westfall, "The Changing World of the Newtonian Industry," *Journal of the History of Ideas* 37 (1976): 175–84.
2 James E. Force, "Newton's God of Dominion: The Unity of Newton's Theological, Scientific, and Political Thought," in *Essays on the Context, Nature, and Influence of Isaac Newton's Theology*, ed. James E. Force and Richard H. Popkin (Dordrecht: Kluwer, 1990); Edward B. Davis, "Newton's Rejection of the 'Newtonian World View': The Role of Divine Will in Newton's Natural Philosophy," *Science and Christian Belief* 3 (1991): 103–17; Betty Jo Teeter Dobbs, *The Janus Faces of Genius: The Role of Alchemy in Newton's Thought* (Cambridge: Cambridge University Press, 1991).
3 J. E. McGuire and P. M. Rattansi, "Newton and the Pipes of Pan," *Notes and Records of the Royal Society of London* 21 (1966): 108–43; Dobbs, *Janus Faces*.

the present generation of Boyle scholars.⁴ This activity has resulted in a better understanding of all aspects of Boyle studies, including his own alchemical pursuits, his view of the relationships between theology, natural philosophy, and epistemology, and the details of his experimental program.⁵

The results of this flurry of activity in both Boyle studies and Newton studies have revealed considerable parallels in the thought of these two men. Both, for example, were active experimenters, and each stressed the provisional nature of experimental findings.⁶ Further, each was concerned to explore ways in which alchemy might contribute to his understanding of the world (despite differences in the approaches of the two; on this see the chapter by Lawrence M. Principe in this volume).⁷ And, perhaps most fundamental of all, each was a voluntarist – that is, one who emphasized God's freedom and omnipotence rather than other divine attributes. As voluntarists, each believed that an a priori knowledge of nature was impossible. Because God had freely chosen to create this world rather than some other possible world, the only way to learn exactly what kind of world God had created was to investigate that world empirically.⁸

4 Michael Hunter, ed., *Letters and Papers of Robert Boyle* and its accompanying *Guide to the Manuscripts and Microfilm* (Bethesda, Md.: University Publications of America, 1992). Particularly fruitful in establishing a web of communication among Boyle scholars was the Stalbridge Boyle Conference that Hunter organized in 1991, which took place in Stalbridge, Dorset, and was funded by the Foundation for Intellectual History.

5 See esp. Michael Hunter, ed., *Robert Boyle Reconsidered* (Cambridge: Cambridge University Press, 1995); this volume consists of papers originating at the Stalbridge Boyle Conference. See also Lawrence M. Principe, *Aspiring Adept: Robert Boyle and His Alchemical Quest* (Princeton: Princeton University Press, 1998); Rose-Mary Sargent, *The Diffident Naturalist: Robert Boyle and the Philosophy of Experiment* (Chicago: University of Chicago Press, 1995); Jan W. Wojcik, *Robert Boyle and the Limits of Reason* (Cambridge: Cambridge University Press, 1997).

6 For Boyle, see esp. Wojcik, *Robert Boyle and the Limits of Reason*, esp. chap. 6, 7. The provisional nature of Newton's experimental findings is expressed quite succinctly in the fourth of his "Rules of Reasoning in Philosophy": "In experimental philosophy we are to look upon propositions gathered by general induction from phaenomena as accurately or very nearly true, notwithstanding any contrary hypotheses that may be imagined, till such time as other phaenomena occur, by which they [the propositions] may be either made more accurate, or liable to exceptions" (Isaac Newton, *Mathematical Principles of Natural Philosophy*, trans. Andrew Motte, rev. Florian Cajori [Berkeley: University of California Press, 1934], 2:400).

7 See esp. Principe, *Aspiring Adept*; Dobbs, *Janus Faces*.

8 For Newton's voluntarism, see the sources cited in note 2. For Boyle's, see Edward B. Davis Jr., "Creation, Contingency, and Early Modern Science: The Impact of Voluntaristic Theology on Seventeenth Century Natural Philosophy. An Essay in the History of Scientific Ideas" (Ph.D. diss., Indiana University, 1984), pp. 122–81; Eugene M. Klaaren, *Religious Origins of Modern Science* (Grand Rapids, Mich.: Erdmans, 1977), esp. pp. 85–184; J. E. McGuire, "Boyle's Conception of Nature," *Journal of the His-*

Despite these similarities, there are marked differences in Boyle's and Newton's approach to natural philosophy. Although both were active experimenters, Boyle's experimental program was influenced by Francis Bacon's view that the acquisition of knowledge is a slow and tedious process. As a result, Boyle emphasized the collaborative efforts of many workers (both those in his own laboratory and others throughout the world). Newton, on the other hand, toiled alone, assisted only for a brief period of about five years by his assistant and amanuensis Humphrey Newton (no relation).[9] Newton's masterpiece consisted of a mathematical description of the law of universal gravitation, whereas what later came to be known as "Boyle's Law" was expressed in far less precise terms than those of mathematics, primarily because Boyle's subject matter (the condensation and rarefaction of air) did not lend itself as easily as Newton's to the precision of mathematics.[10]

Indeed, it might seem at first glance that the primary difference

tory of Ideas 33 (1972): 523–42; Margaret J. Osler, "The Intellectual Sources of Robert Boyle's Philosophy of Nature: Gassendi's Voluntarism and Boyle's Physico-Theological Project," in *Philosophy, Science, and Religion in England, 1640–1700*, ed. Richard Kroll, Richard Ashcraft, and Perez Zagorin (Cambridge: Cambridge University Press, 1992), pp. 178–98; Margaret J. Osler, "Triangulating Divine Will: Henry More, Robert Boyle, and René Descartes on God's Relationship to the Creation," in *"Mind Senior to the World": Stoicismo e origenismo nella filosofia platonica del Seicento inglese*, ed. Marialuisa Baldi (Milano: Franco Angeli, 1996), pp. 75–87; Timothy Shanahan, "God and Nature in the Thought of Robert Boyle," *Journal of the History of Philosophy* 26 (1988): 547–69.

9 For Boyle, see Steven Shapin, "The House of Experiment in Seventeenth-Century England," *Isis* 79 (1988): 373–404. On Newton's sole assistant, see Derek Gjertsen, *The Newton Handbook* (London: Routledge and Kegan Paul, 1986), p. 380.

10 Boyle himself refused to predict that his measurements would hold "universally and precisely" (*Defence of the Doctrine*, in *The Works of the Honourable Robert Boyle*, ed. Thomas Birch, 6 vols. (1772; reprinted with an introduction by Douglas McKie, Hildesheim: Georg Olms, 1965), 1:159. For the inherent difference between Newton's Law of Universal Gravitation and "Boyle's Law," see Maureen Christie, "Philosophers versus Chemists Concerning 'Laws of Nature,' " *Studies in History and Philosophy of Science* 25 (1994): 613–29: "There is an essential difference of kind between exact laws, like . . . Newton's law of universal gravitation, and approximate laws, like Boyle's law. . . . Exact laws are realized exactly within the precision of most plausible experimental measurements, and some very fine nit-picking is required to show their slight inadequacies. But for an approximate law any careful measurement might routinely be expected to be a few percent away from the predicted value, by virtue of the approximate nature of the law itself" (p. 624). See also Thomas Kuhn, "Mathematical versus Experimental Traditions in the Development of Physical Science," *Journal of Interdisciplinary History* 7 (1976): 1–31. Steven Shapin takes a penetrating look at Boyle and mathematics in "Robert Boyle and Mathematics: Reality, Representation, and Experimental Practice," *Science in Context* 2 (1988): 23–58 (although I should note that Shapin places far more emphasis on Boyle's concern that "everyday language and the language of the experimental community" overlap [p. 43] than would I).

between these two giants of the seventeenth-century scientific scene was Newton's mathematical approach to physical science, as compared with Boyle's reluctance to mathematize nature. Certainly this was a marked difference and one well worth investigation, but for a number of reasons (which lie outside the scope of the present study), I do not think that it is as significant as it might at first seem.[11] In the present chapter I concentrate on another difference – the different views each of these natural philosophers had concerning the power and scope of human reason. Indeed, I think that the most important difference between these two natural philosophers (and lay theologians) is that they had dramatically different conceptions of God's intentions concerning human understanding in relationship to what *can* be known in both natural philosophy and theology, *how* that knowledge can best be attained, exactly *who* can attain this knowledge, and *when* it might be learned. In this chapter I explore two manifestations of these differences. First, I discuss how each of these thinkers responded to the alleged "mysteriousness" of Christianity and, second, I explore the different way each reacted to the tradition of an "ancient knowledge."

Doctrinal Concerns

Boyle's and Newton's different assumptions about how much God intended human beings to understand can perhaps most easily be discerned in their different positions on two of the Christian doctrines that have posed tremendous (if not insurmountable) obstacles to human understanding – the doctrine of the Trinity and the doctrine of predestination.

Newton's anti-Trinitarianism has been well documented and needs little elaboration.[12] The threat of anti-Trinitarianism of one form or

---

11 In addition to my comments in note 10, it is worth noting that both Boyle and Newton were aware that a mathematical description of nature does not reveal the cause of the phenomena, which is the true goal of natural philosophy. For Boyle, see Sargent, *Diffident Naturalist*, pp. 66–7; for Newton, see Dobbs, *Janus Faces*, pp. 6–13.

12 James E. Force, "Jewish Monotheism, Christian Heresy, and Sir Isaac Newton," in *The Expulsion of the Jews: 1492 and After*, ed. Raymond B. Waddington and Arthur H. Williamson (New York: Garland Publishing, 1994), pp. 259–80; Richard S. Westfall, *Never at Rest: A Biography of Isaac Newton* (Cambridge: Cambridge University Press, 1980); Dobbs, *Janus Faces*. Anti-Trinitarianism comes in a variety of forms. Socinian anti-Trinitarians believed that Jesus Christ was a man miraculously conceived by Mary in fulfillment of Old Testament prophecies but that despite his special status, his nature was fully human (at least prior to his resurrection, at which time he became the adopted son of God while remaining a distinct individual). This should be distinguished from the belief of Arians that Jesus Christ shared in the divinity of God, was the first act of God's creation and was himself the creator of all else, but

another was a major concern of orthodox Anglican churchmen in the second half of the seventeenth century, and Newton was careful to conceal his views from all but the most trusted of confidants. Not until 1690 did he risk sending his views abroad in the form of three letters to John Locke, and then with the stipulation that they not be published.[13]

Those who held anti-Trinitarian beliefs tended to claim that human reason is competent to interpret and understand Christian doctrines. Earlier in the century, Socinian writers had rejected the orthodox interpretation of the Trinity on the grounds that Scripture must be interpreted in such a way as to be consonant with human reason. In the *Racovian Catechism*, for example, Christ's divinity was denied on the grounds that it is repugnant to reason to claim that two persons imbued with opposing qualities (to be both mortal and immortal, for example, or to have a beginning and to be eternal) could constitute one substance. As a result, scriptural passages traditionally invoked in support of orthodox Trinitarianism must be reinterpreted in such a way as to be consonant with human reason.[14]

> was not coessential with God the Father. Both Socinians and Arians should be distinguished from the beliefs of another group of anti-Trinitarians, the Sabellians, that the entire Trinity was incarnated in Jesus Christ, and that the fullness of the Godhead dwells in each believer by virtue of the union of the believer with the Holy Spirit. Arius's views were condemned at the Council of Nicea in A.D. 325, although they continued to be embraced by some of the faithful for another fifty years, until finally they were condemned again at the Council of Constantinople (381). Sabellius taught, during the latter half of the third century, that there is one God who assumes three offices rather than one God in three persons; his views were condemned by an early council, although exactly which council is not known. Socinianism emerged in the late sixteenth century and was an attempt of more radical Protestant reformers to return to the purity of early Christianity.
>
> Newton has traditionally been portrayed as an Arian. It now seems likely, however, that his views were a variant of Arianism; see Thomas C. Pfizenmaier, "Was Isaac Newton an Arian?" *Journal of the History of Ideas* 58 (1997): 57–80.

13 H. W. Turnbull, ed., *The Correspondence of Isaac Newton*, vol. 3, *1688–1694* (Cambridge: Cambridge University Press, 1961), pp. 82–146, for suppression, see p. 123, n. 1; Westfall, *Never at Rest*, pp. 488–91. For the consequences should Newton's anti-Trinitarianism become public knowledge, see James E. Force, *William Whiston: Honest Newtonian* (Cambridge: Cambridge University Press, 1985).

14 *Catechism of the Church of those people who, in the Kingdom of Poland and the Grand Duchy of Lithuania and in other Domains belonging to the Crown, affirm and confess that no other than the Father of our Lord Jesus Christ is the only God of Israel and that the Man Jesus the Nazarene, who was born of the Virgin, and no other besides him, is the only begotten Son of God* (in Polish, Raków, 1605), translated by John Biddle as *The Racovian Catechisme: Wherein you have the substance of the Confession of those Churches, which in the Kingdom of Poland, and Great Dukedome of Lithuania, and other Provinces appertaining to that kingdom, do affirm, that no other save the Father of our Lord Jesus Christ, is that one God of Israel, and that the man Jesus of Nazareth, who was born of the Virgin, and no other*

Newton's anti-Trinitarianism did not rest on the same grounds as those argued in the *Racovian Catechism*. It was based primarily on his conception of the absolute dominion of God the Father rather than on the grounds that belief in the orthodox interpretation of the Trinity is irrational. Having searched Scripture for passages that showed God's dominion over Christ, Newton dismissed passages indicating the equality or consubstantiability of God and Christ not on the grounds that Trinitarians misinterpreted Scripture, but rather on the grounds that they had *corrupted* Scripture. Clearly, Newton's insistence on the absolute dominion of God the Father conditioned his conception of the status of Christ the Son. Nevertheless, it is telling that in his letter to Locke, Newton complained that a (corrupted) verse of Scripture concerning the three witnesses to the coming of the Son of God interfered with the "sense plain & natural & the argument full & strong." "Let them make good sense of it [the witnesses in heaven] who are able," he said; "For my part I can make none." He continued: " 'Tis the temper of the hot and superstitious part of mankind in matters of religion ever to be fond of mysteries, & for that reason to like best what they understand least." This was not Newton's way, however: "In disputable places I love to take up with what I can best understand."[15]

Boyle, on the other hand, was one of what Newton would consider to be "the hot and superstitious part of mankind in matters of religion"; if he was not "fond of mysteries," he was at least willing to accept them. In an early and unpublished "Essay of the Holy Scriptures" he explicitly rejected the claim of the Socinians. Although he acknowledged their being the "incomparable Masters of Reason," he denied that human reason should be the criterion against which the Christian revelation should be judged. "Those that will rightly judge of Truth must be Inquisitive Examiners but not Refractory," Boyle claimed, arguing that this is particularly true where the doctrine of the Trinity is concerned. "I could wish," Boyle added, that "men had bin lesse Positive and Determinate in their Discourses about the Trinity; in which a Reverence is safer than Nicety."[16] In Boyle's published

---

besides, or before him, is the onely begotten Sonne of God (Amsterdam [London], 1652), p. 28.

15 *The Correspondence of Isaac Newton*, 3:108. The verse at issue is I John 5.7 ("For there are three that bear record in heaven, the Father, the Word, and the Holy Ghost: and these three are one" [King James' Version]); Newton interpreted the three witnesses as being the spirit, the baptism, and the passion of Christ. See also Newton's comments on interpreting difficult scriptural passages later in this chapter.

16 Boyle Papers, vol. 7, f. 48, Archives of the Royal Society of London. For a discussion of this essay, see Wojcik, *Robert Boyle and the Limits of Reason*, pp. 55–9; see also

works there is no sustained discussion of the Trinity, although in his discussions concerning reason's limits the Trinity was often invoked as an example. Perhaps his most explicit reference is found in *Seraphic Love*, where he noted that the "strict relations betwixt the persons of the blessed Trinity" had provided God with internal objects prior to the creation.[17] And at least once he referred to the doctrine of the Trinity (in words Newton would have deplored) as an "adorable mystery."[18]

Boyle's most sustained argument concerning reason's limits is in *A Discourse Concerning Things above Reason* and its accompanying *Advices in Judging of Things Said to Transcend Reason* (1681). In these essays he argued that a Christian should have no difficulty acknowledging that some doctrinal matters are "above reason" (and, indeed, that some doctrines even seem to be, from the perspective of finite human understanding, "contrary to reason") on the grounds that many matters pertaining to natural philosophy are equally impervious to human reason. His primary doctrinal concern in *Things above Reason* was the vexing problem of how, if at all, God's foreknowledge of future contingencies might be reconciled with the free will of human agents. *Things above Reason* and *Advices* were composed to address just this question in response to a very heated controversy among his contemporaries concerning predestination. Throughout the essays, his most frequently used example of two true but contradictory propositions – that is, two propositions the consistency of which cannot be grasped by finite human understanding – was the claim that both the proposition that God has foreknowledge of future contingencies *and* the proposition that human beings have free will are true. He recognized

Michael Hunter, "How Boyle Became a Scientist," *History of Science* 33 (1995): 59–103.

17 *Some Motives and Incentives to the Love of God*, usually cited by its running title *Seraphic Love*, in *Works*, 1:269. Although the original version of *Seraphic Love* was composed in 1648, this passage (as well as others dealing with doctrinal matters) was added later, but prior to its publication in 1659 – perhaps reflecting Boyle's growing concern over the threat of anti-Trinitarianism; see Lawrence M. Principe, "Style and Thought of the Early Boyle: Discovery of the 1648 Manuscript of *Seraphic Love*," *Isis* 85 (1994): 247–60, esp. 257–8. Other references to the Trinity may be found in *Advices*, in *Works*, 4:454; *Appendix* to *Christian Virtuoso*, in *Works*, 6:695–6. A lengthy discussion of the Trinity may be found in Boyle's unpublished "Quo in sensu Religio Christiana rationi conformis sit et quo in sensu eidem Contraria dici possit," Boyle Papers, vol. 6, ff. 55–60. This essay will be included (along with an English translation) in the forthcoming "Pickering Masters" edition of Boyle's works, edited by Michael Hunter and Edward B. Davis. The inclusion of this and other previously unpublished writings of Boyle is being funded by the Leverhulme Trust.

18 *The Excellency of Theology Compared with Natural Philosophy*, in *Works*, 4:15.

the implications of admitting contradictory truths into our body of knowledge when he noted that doing so admits "that grand absurdity, which subverts the very foundation of our reasonings, that contradictories may both be true" – that is, may violate the law of noncontradiction.[19] Strictly speaking, Boyle "saved" the law of noncontradiction by arguing that because these two truths cannot be *really* contradictory, God, in his infinite wisdom, must be able to perceive the harmony between them. Despite this stab at a rational explanation, however, he went about as far as it is possible to go in ascribing limits to human reason in order to make room for the mysteries of Christianity.

Newton, on the other hand, nowhere indicated that he found this particularly knotty problem vexing, or even of concern. In his one reference to "Predestinatio" (found in his *Commonplace Book*), he simply quoted at length from Romans (9.14–23) – the same scriptural passage used by Calvin to argue for the doctrine of double predestination.[20] As James E. Force has pointed out, although the presence of this quotation in Newton's *Commonplace Book* does not prove beyond doubt that Newton was a predestinarian, such a belief would be consistent with his emphasis on God's absolute dominion.[21]

A belief in predestination would also be consistent with Newton's opinion that he was one of God's chosen, not only where salvation was concerned but also when it came to understanding passages of Scripture that are difficult to interpret and understand – passages that

---

19 *A Discourse Concerning Things above Reason*, in *Works*, 4:423. For a complete discussion of *Things above Reason* and *Advices in Judging of Things Said to Transcend Reason*, see Wojcik, *Robert Boyle and the Limits of Reason*, chap. 4.

20 Newton, *Commonplace Book*, s.v. "Predestinatio," Keynes MS 2, King's College Library, Cambridge, quoted by James E. Force in "Sir Isaac Newton, 'Gentleman of Wide Swallow'?," in Force and Popkin, *Essays on the Context, Nature, and Influence of Isaac Newton's Theology*, p. 131. Romans 9.14–23 reads: "What shall we say then? Is *there* unrighteousness with God? God forbid. For he saith to Moses, I will have mercy on whom I will have mercy, and I will have compassion on whom I will have compassion. So then *it is* not of him that willeth, nor of him that runneth, but of God that showeth mercy. For the Scripture saith unto Pharaoh, Even for this same purpose have I raised thee up, that I might show my power in thee, and that my name might be declared throughout all the earth. Therefore hath he mercy on whom he will *have mercy*, and whom he will he hardeneth. Thou wilt say then unto me, "Why doth he yet find fault? For who hath resisted his will? Nay but, O man, who art thou that repliest against God? Shall the thing formed say to him that formed *it*, Why hast thou made me thus? Hath not the potter power over the clay, of the same lump to make one vessel unto honor, and another unto dishonor? *What* if God, willing to show *his* wrath, and to make his power known, endured with much long-suffering the vessels of wrath fitted to destruction: And he that might make known the riches of his glory on the vessels of mercy, which he had afore prepared unto glory" (King James Version).

21 Force, "Sir Isaac Newton, 'Gentleman of Wide Swallow'?" p. 131.

Boyle considered to be "mysterious." If the difficult passages "are never to be understood," Newton asked, "to what end did God reveale them?" Clearly, God did so for "the edification of the church," and so, equally clearly, the church must eventually understand their meaning. But, he hastened to add, "I mean not all that call themselves Christians but a remnant, a few scattered persons which God hath chosen" shall understand.[22]

Clearly, where doctrinal matters were concerned, Boyle and Newton differed greatly in their respective views of the power of human reason. Boyle was content to assent to "adorable mysteries" that remain impervious to human reason, whereas Newton insisted that God had revealed Christian doctrine with the intent that it be understood in a "sense plain & natural." Another way in which Boyle and Newton revealed their different views regarding what God had intended human beings to understand and how he had intended them to reach that understanding is reflected in their different attitudes toward the so-called ancient wisdom.

The Wisdom of Antiquity

The relationship of Christianity to pagan philosophy had been explored by many of the church fathers, and during the Reformation the issue was debated anew as Protestants attempted to purge Christianity of Roman corruptions. The principal assumption of the Protestants who appealed to a *prisca sapientia* to provide guidance in the proper interpretation of Christian doctrine was that some pagan philosophers (although certainly not all) had attained partial insight into eternal truth. If properly interpreted and understood, their writings might yield pearls of wisdom about the true (Christian) religion, the creation of the world, and the hidden structure of the world. Attempts to find and interpret these pearls (believed to presage the fullness of revelation that was the Christian religion) were widespread during the sixteenth and seventeenth centuries.

Taken this far but no further, the existence of a *prisca sapientia* testified, at most, to the competence of human reason alone to achieve at least partial knowledge of true religion and true philosophy. In order to preserve the uniqueness and necessity of the Judeo-Christian revelation, the further claim that the sages of antiquity had derived their knowledge from the Hebrews was often added. Sometimes the source of this wisdom was traced back to Moses, or even earlier to

---

22 Yahuda MS, vol. 1, f. 1, Jewish National and University Library, Jerusalem, published as appendix A by Frank E. Manuel, *The Religion of Isaac Newton* (Oxford: Clarendon Press, 1974), pp. 107–8.

Noah and his sons, or earlier yet, to Adam. In any event, it was believed that at some point in the transmission of this wisdom it had become corrupted, thereby necessitating revelations to recall men to the truth.[23]

Newton's belief in and eclectic borrowing from this tradition of ancient wisdom has been well documented.[24] He seems to have begun his search for religious truth in 1672 when he undertook an intensive study of Christian doctrine, apparently in preparation for his approaching ordination as an Anglican clergyman. Ironically, this ordination never occurred, for in his study of the writings of the church fathers he became convinced that the earliest Christians had *not* in fact affirmed Christ's status as a member of the Godhead. The allegedly orthodox doctrine of the Trinity was in fact a heresy, imposed on Christianity by fourth- and fifth-century corrupters of both doctrine and Scripture. "Orthodox" Trinitarianism, therefore, constituted an idolatrous corruption of true Christian doctrine.[25]

Having concluded that allegedly "orthodox" Christianity was instead a form of idolatry, Newton then attempted to identify the origins of idolatry. In *Theologiae gentilis origines philosophicae*, an unpublished work that dates from the early 1680s, he traced the error back to the Egyptians. When Noah had settled in Egypt after the flood, he and his sons had possessed knowledge of true religion from their ancestors, but in time this knowledge had been lost as the Egyptians (and later others) turned their own ancestors into gods. Only a new revelation – that given to Moses – could recall men to the truth, and in time that revelation, too, had been corrupted.[26]

Newton believed that just as true religion had been corrupted by idolatry, truths of natural philosophy, known to Noah and revealed again to Moses, had, after each revelation, been corrupted by false philosophy. Despite this corruption, certain glimmers of the truth

---

23 See, for example, Stephen Gaukroger, "Introduction: The Idea of Antiquity," in *The Uses of Antiquity: The Scientific Revolution and the Classical Tradition*, ed. Gaukroger, (Dordrecht: Kluwer, 1991), pp. ix–xvi; D. P. Walker, *The Ancient Theology: Studies in Christian Platonism from the Fifteenth to the Eighteenth Century* (Ithaca, N.Y.: Cornell University Press, 1972); Charles B. Schmitt, "Perennial Philosophy: From Agostino Steuco to Leibniz," *Journal of the History of Ideas* 27 (1966): 505–32, reprinted (with same pagination) in Schmitt, *Studies in Renaissance Philosophy and Science* (London: Variorum Reprints, 1981).

24 The most concise accounts of Newton's searches for truth in pagan writings are McGuire and Rattansi, "Newton and the Pipes of Pan," and John Gascoigne, " 'The Wisdom of the Egyptians' and the Secularisation of History in the Age of Newton," in Gaukroger, *The Uses of Antiquity*, pp. 171–212. The most comprehensive account is Dobbs, *Janus Faces*.

25 Westfall, *Never at Rest*, pp. 309–15.

26 Ibid., pp. 351–6.

could still be found, he thought, in the writings of antiquity. For example, one of the revelations vouchsafed to Moses had been the command to keep a perpetual fire burning in the tabernacle, an "emblem," as Newton put it, "of the [Copernican] system of the world."[27] And in his search for the cause of gravity, a search that lasted from the time of the first edition of *Principia* until his death, Newton turned again to the wisdom of the ancients. Indeed, he did not consider his theory of universal gravitation to be "new" knowledge at all, but instead yet another manifestation of a truth once lost but now restored. What is more, certain sages had known not only of universal gravitation and of the inverse-square law, but they had also known the *cause* of gravity, something that Newton had not yet been able to rediscover.[28]

Newton's appeal to ancient wisdom in his search for the cause of gravity was discriminating, of course; he did not blindly accept just anything he encountered in his studies, be it theological or pertaining to natural philosophy. In natural philosophy, any claim as to the cause of gravity had to square with observational and experimental phenomena, and he found that a dematerialized Stoic pneuma did so whereas an aetherial mechanism did not.[29] What is important, however, is not that Newton was discriminating in what he selected from the texts of antiquity but that he believed that from the very beginning God had intended human beings to *know* ultimate truth – whether by means of natural reason or supernatural revelation. At least, God had intended that *some* human beings – "a remnant, a few scattered persons which God hath chosen" – achieve ultimate knowledge. This was a belief that Boyle did not share.

Boyle was as aware as was Newton of the tradition of an "ancient wisdom." Scattered throughout his works are references to the same transmission of knowledge appealed to by those who pursued the writings of antiquity in the search for truth, and occasionally he acknowledged finding glimmers of wisdom there. Ultimately, however, Boyle rejected the wisdom of antiquity as being an unreliable source of knowledge about either theology or nature.

Boyle made more than a passing reference to the "ancient wisdom" in three of his works. In *The Sceptical Chymist* he considered Jan Baptista van Helmont's claim that water is the ultimate source of all things. Here Boyle noted that others had held this belief, most notably the Greek Thales. And even Thales had not been the first, Boyle

---

27 Ibid., pp. 353–5; Dobbs, *Janus Faces*, pp. 150–66.
28 Dobbs, *Janus Faces*, pp. 169–212.
29 Newton's search for the cause of gravity was considerably more complicated than I have indicated here; for a full account, see ibid.

claimed, for the Greeks had borrowed their philosophy and theology from the Phoenicians, who in turn, had borrowed from the Hebrews. Hesitant as always to dogmatize, Boyle contented himself with raising objections to the theory, noting that when he and his gardener had repeated van Helmont's experiment with various plants the results had varied, and that even van Helmont's results could be explained in a number of different ways.[30]

In *The Excellency of Theology*, a work dedicated to proving the superiority of revelation over natural human reason, Boyle paid particular attention to the so-called ancient wisdom, and references to various beliefs held in antiquity are scattered throughout. He noted that some heathens had believed in angels, for example, having learned this truth via "imperfect tradition." Without the confirmation of Scripture, however, such knowledge was precarious – and perhaps even dangerous – for Scripture informs us that some angels are evil.[31] Some pagan philosophers had also believed (correctly) that the world had been created in time, but here, too, their knowledge was imperfect. *At best*, Boyle claimed, they had believed the world to have been formed out of preexistent matter. Further, those who had reasoned out the time of the Creation (the Chaldeans), had reasoned wrongly. The Stoics had believed that the world will ultimately be destroyed by fire (having learned this from the Jews), but such knowledge, he warned, was "physically precarious." Scripture, he urged, is the only reliable source of knowledge concerning events that lie outside the scope of human experience.[32]

Even where knowledge of the current frame of the world was concerned, Boyle saw no evidence that the sages of antiquity were in any privileged position to know. Moschus, the Phoenician, may have invented the "atomical hypothesis," but Aristotle had corrupted Greek philosophy by denying atomism.[33] In the past, astronomers had been wrong in their calculations.[34] Indeed, Boyle claimed, many of the chief truths that had emerged over time had been the productions of time and chance, thereby implicitly denying that there *was* even a precariously maintained transmission of truth in such matters.[35]

30 *The Sceptical Chymist*, in *Works*, 1:494–6, 498–509.
31 *The Excellency of Theology*, in *Works*, 4:9–10. For Boyle's ambivalence regarding spirit-contact, see Michael Hunter, "Alchemy, Magic and Moralism in the Thought of Robert Boyle," *British Journal for the History of Science* 23 (1990): 387–410. A more recent and complete account may be found in Principe, *Aspiring Adept*, chap. 6.
32 *Excellency of Theology*, in *Works*, 4:10–11.
33 Ibid., 4:48, 57.
34 Ibid., 4:51.
35 Ibid., 4:48.

Another example involved the naming of animals. Boyle was aware that many of the philosopher-theologians of his day claimed that Adam's naming of the various animals in the Garden of Eden revealed a perfect knowledge of their natures, but he explicitly rejected this view. He had studied the Hebrew language, he explained, and did not see that the Hebrew names for animals revealed any more insight into their natures than, for example, did their names in Greek, or in other languages.[36] Indeed, he claimed, *no* human beings have had (or will have) a perfect knowledge of nature while on earth – not even Adam in the Garden of Eden. Here on earth, he insisted, "we see through a [glass] darkly, but then [in heaven] face to face; now I know in part, but then I shall know even as also I am known," quoting Paul's letter to the Corinthians (I Corinthians 13.12).[37] The sages of antiquity had not been – and indeed could not have been – the possessors of an esoteric and certain knowledge of matters either theological or natural, for Boyle's God had never intended human beings to have such knowledge, at least not while on earth.

Boyle's most sustained attack on the *prisca sapientia* is found in *The Vulgarly Receiv'd Notion of Nature*, a work in which he attacked the Schoolmen's reification of nature, which he considered to be a conception of "Nature" as an "almost divine thing."[38] Here he devoted some eight pages to showing that, historically, *any* conception of nature that includes an appeal to nature as "alive," or posits any kind of entity between God and the world he created has led to polytheism and idolatry. (Interestingly, he included as one of these entities the Stoic world-soul – an entity that, in its dematerialized Platonic version, Newton eventually came to believe offered a means of understanding God's activity in the world in terms of a force capable of explaining gravity.)[39] Citing Maimonides, Boyle suggested that the first idolaters were probably the Sabeans (or Zabeans; i.e., Chaldeans), with whom Abraham had struggled. It may have been from these people, he wrote, that Zeno, the founder of the Stoics, had gotten the idea that the Sun, the Moon, and the rest of the stars had understanding and

---

36 Ibid., 4:45–6. The source of the belief that the names of animals reveal insight into their essences is Genesis 2.19–20, "And out of the ground the Lord God formed every beast of the field, and every fowl of the air; and brought them unto Adam to see what he would call them; and whatsoever Adam called every living creature, that was the name thereof. And Adam gave names to all cattle, and to the fowl of the air, and to every beast of the field" (King James Version).
37 *Excellency of Theology*, in *Works*, 4:46.
38 *The Vulgarly Receiv'd Notion of Nature*, in *Works*, 5:161.
39 Ibid., 5:183, 184. On Newton, see Dobbs, *Janus Faces*, p. 206.

prudence.⁴⁰ The Egyptians of antiquity were equally guilty, he thought, this time citing Eusebius.⁴¹ And Moses had even found that he had to set some of the *Jews* straight on idolatry.⁴² Indeed, many people, Christians and Jews alike, had fallen into idolatry, he noted, going on to indict two of the sources on which Newton had relied – Origen and Maimonides, as well as Menasseh Ben Israel (although he granted that in these cases the malignity of their error was somewhat mitigated by the fact that they held orthodox beliefs as well).⁴³

The different uses to which Boyle and Newton put their respective studies into the (alleged) wisdom of antiquity is striking. Newton believed that there had in fact been an original revelation, and that although this wisdom had become corrupted it still contained pearls of wisdom – well-hidden pearls, to be sure, but nevertheless, pearls. Further, he believed that with perseverence and God's blessing he, Newton, was just the person to find these pearls and string them together in such a way as to restore their original purity. Boyle, on the other hand, believed that although occasionally a philosopher living in the distant past had stumbled upon some insight into the nature of things (as had Moschus the Phoenician, when he had formulated the atomical hypothesis), most of the time most of them had been in error, and these errors were pernicious because they led to polytheism and idolatry. Trying to glean truth from the writings of the ancients was not the way to go about attaining a knowledge of theology or of nature. And he was disturbed by those of his contemporaries who sought wisdom in these writings. He knew, he said, that even in his own time a "sect of men" who professed themselves to be Christians claimed to have discovered "unheard of mysteries in physics and natural theology," but they were mistaken. They had been seduced by the heathenish notion that God is not really distinct from the "animated and intelligent universe."⁴⁴ Boyle acknowledged that from time to time certain truths, having once been known but then lost or corrupted, had been revived in later times (such as the belief of the

---

40 *Vulgarly Receiv'd Notion of Nature*, in *Works*, 5:180–1.
41 Ibid., 5:182.
42 Ibid.
43 Ibid., 5:183. For Newton's reliance on Eusebius, see Westfall, *Never at Rest*, pp. 312, 314; on Origen, pp. 312, 313; on Maimonides, p. 346.
44 *Vulgarly Receiv'd Notion of Nature*, in *Works*, 5:183. James R. Jacob has identified the sect as one led by Henry Stubbe (*Henry Stubbe: Radical Protestantism and the Early Enlightenment* [Cambridge: Cambridge University Press, 1983], pp. 56–7). Michael Hunter and Edward B. Davis reject this interpretation; for their discussion of other possible referents and the difficulty in judging which Boyle had in mind, see Michael Hunter and Edward B. Davis, "The Making of Robert Boyle's *Free Inquiry into the Vulgarly Receiv'd Notion of Nature* (1686)," *Early Science and Medicine* 1 (1996): 246–51.

Pythagoreans that the Earth revolves around the Sun, as well as Moschus's atomical hypothesis). But, he insisted, "though it be often the fate of an oppressed truth, to have at length a resurrection, yet it is not always its peculiar privilege; for obsolete errors are sometimes revived, as well as discredited truths."[45]

Conclusion

At the beginning of this chapter I stated that Boyle and Newton had radically different ideas as to God's intentions concerning human understanding in relationship to what *can* be known in natural philosophy as well as in theology, *how* that knowledge could best be attained, exactly *who* could attain this knowledge, and *when* it might be learned. It is now time to summarize those differences.

Boyle did not think that God intended *any* human beings to attain a complete understanding of either nature or theological truths during this lifetime. He did think that those individuals who struggled patiently to learn more about nature (utilizing a tedious Baconian methodology) would be rewarded with an ever increasing understanding of the natural world. But he did not think that any human beings would attain a complete knowledge of nature's secrets, and he did not think that any human being could comprehend the mysteries of Christianity; complete understanding, Boyle thought, was something that God had reserved for the afterlife. Indeed, a full understanding in heaven of both nature and theology would be the final reward of Christian virtuosi; this was the best incentive God could have offered for a person to be both a Christian and a virtuoso.

Newton, on the other hand, believed that God had in fact intended at least *some* individuals – "a remnant, a few scattered persons which God hath chosen" – to achieve a complete understanding during this lifetime. In Newton's view, God had revealed ultimate truths of both nature and theology in the beginning, and although that knowledge had been lost time and again, God had sent a new prophet each time to recall his people to the original revelation. And even during the times when the truth was lost, glimmers of it remained in the writings of those who had once possessed it. Those individuals who persevered in seeking that truth by a careful and discriminating investigation into the texts of antiquity and the texts of Scripture would be rewarded – if God so willed it – with a complete understanding in this lifetime.

Newton's views concerning what God intended some (but not all) human beings to know (and when, and how) must be gleaned from

45 *Excellency of Theology*, in *Works*, 4:58.

his actual search for complete understanding in this lifetime; so far as I know he wrote nothing, either published or unpublished, in which he argued that this complete understanding is possible. Boyle, on the other hand, was explicit about his views on the limits of human reason. Repeatedly, in a number of passages both published and unpublished, he explained reason's limits in terms of God's will. For example, in an unpublished essay he suggested that

> on divers occasions it may be a usefull supposition to Imagine such a state of things as was that which proceeded the beginning of the Creation for since then there was no being besides God himselfe who is Eternall all Beings of what kind soever that had a beginning must derive those natures & all their faculties from his Arbitrary will.[46]

In *Appendix to the Christian Virtuoso*, first published in 1744, Boyle repeated his view that when God

> was pleased to create other beings, as he did it freely and uncompelled, so he gave to each of those he created such a nature, and such determinate faculties and powers as he pleased. And of the beings, which he created intelligent, among which are the rational minds of men, he endowed each sort or order with such a measure or degree of the intellectual faculty, as he thought fit, for the ends and purposes for which in his infinite wisdom he created them; but to none of these did he impart a knowledge boundless as his own, that infinite knowledge being the essential prerogative of God, and not communicable to any mere creature.[47]

Further, he noted that the relationship of the human mind to the created world affects the ability of human beings to reason about that

---

46 Boyle Papers, vol. 36, f. 46v. Perhaps Boyle meant "preceded" rather than "proceeded." Another passage: "& consequently man himselfe & all intellectual as well as all Corporeall Creatues, were but just such as he thought fit to make them. And as he freely establisht the affections of schemetisms & the Laws of Motion by which the universe was framed, & doth act; so he freely constituted the Reason of Man & other created intellects" (Boyle Papers, vol. 36, f. 46v). Another passage: "thô among his other Creatures he [God] was pleasd to frame some that were of an Intelligent Nature, as Angells & Rational Souls; & to endow these with Intellectual faculties, more or less capacious & inlightned, as he thought fit; yet he reservd to himself, as his prerogative, the Intimate & Compleat Knowledge of his Works; allowing to Humane Minds (to confine our present Discourse to that sort of Intellectual Beings) various degrees of knowledge, according *to* their congenit Aptitudes for knowledge; & *to* the several Circumstances, that from time to time They find themselves in" (Boyle Papers, vol. 8, f. 187).

47 Appendix to *Christian Virtuoso*, in *Works*, 6:697. As Boyle put it in *Advices*, "to discover particular truths is one thing, and to be able to discover the intercourse and

world. Because God created the world before he created man, God did not take human understanding into consideration when he created the world. Instead, God created the world as he thought fit, leaving human beings to speculate about that world as best they can. As Boyle put it,

> I fear we men have too good a conceit of ourselves, when we think that no such thing can have an existence, or at least have a nature or being, as we are not able to comprehend. For if we believe God to be the author of things, it is rational to conceive that he may have made them commensurate, rather to his own designs in them, than to the notions we men may best be able to frame of them. . . . The world itself was first made before the contemplator of it, man: whence we may learn, that the author of nature consulted not, in the production of things, with human capacities; but first made things in such a manner as he was pleased to think fit, and afterwards left human understandings to speculate as well as they could upon those corporeal, as well as other things.[48]

But, it might be argued, perhaps Boyle was talking about the capacities of *most* human beings. Might not God have chosen some few (such as Newton) to whom a full understanding had been revealed? No. Boyle did believe that God, in creating individual human beings, chose to endow some human beings with "a greater measure of intellectual abilities" than others.[49] And Boyle believed that the scriptures had been written both for individuals "with elevated and comprehensive intellects" *and* for individuals "very weak and illiterate."[50] Nevertheless, although God had given "to the intelligent productions of his power and will various degrees of intellectual capacities," he had reserved full understanding for himself.[51]

---

harmony between all truths, is another thing" (*Works*, 4:466). Boyle believed that angels possessed a finer intellect than human beings ("Quo in sensu Religio Christiana rationi conformis sit et quo in sensu eidem Contraria dici possit," Boyle Papers, vol. 6, f. 59).

48 Appendix to *Christian Virtuoso*, in *Works*, 6:694. In the same passage Boyle went on to point out that even if the world was not made by God but is eternal (as the ancient Peripatetics taught) or was made by chance (as the Epicureans taught), "there is yet less reason to believe, that there is any necessity that the nature of primitive things must be commensurate to our understandings; or that in the origin of other things any regard was had, whether they would or would not prove comprehensible to men."

49 *Excellency of Theology*, in *Works*, 4:26.
50 *Style of the Scriptures*, in *Works*, 2:262.
51 *Things above Reason*, in *Works*, 4:466.

Complete understanding, both for those "with elevated and comprehensive intellects" and for individuals "very weak and illiterate" was reserved for the afterlife:

> In heaven our faculties shall not only be gratified with suitable and acceptable objects, but shall be heightened and enlarged, and consequently our capacities of happiness as well increased as filled.... Our then enlarged capacities will enable us, even in objects which were not altogether unknown to us before, to perceive things formerly undiscerned, and derive thence both new and greater satisfactions and delights.[52] ... There, probably, we shall satisfactorily understand those deep and obscure mysteries of religion, ... [and] we shall discern not only a reconcileableness, but a friendship, and perfect harmony, betwixt those texts, that here seem most at variance.[53]

After death, "all that unwelcome darkness, that here surrounded our purblind understandings, will vanish at the dawning of that bright, and ... eternal day; wherein the resolution of all those difficulties, which here exercised (and perhaps distressed) our faith, shall be granted us to reward it."[54]

In this essay I have shown that Boyle and Newton had radically different conceptions of the power and scope of human reason. The most significant conclusion of this study, however, is the absolute priority of theological concerns in the thought of both men. In each case, theological presuppositions informed epistemology, and scholars who think that Boyle and Newton can be understood in terms of their scientific achievements alone are missing important aspects of the influences on the development of modern thought.

---

52 *Seraphic Love*, in *Works*, 1:283.

53 Ibid., 1:289.

54 Ibid., 1:290. And, in *The Second Part of the Christian Virtuoso*, he claimed that "the study of nature, with design to promote piety by our attainments, is useful, not only for other purposes, but to increase our knowledge, even of natural things, if not immediately, and presently, yet in time, and in the issue of affairs. For, at least, in the great renovation of the world, and the future state of things, those corporeal creatures, that will then, be knowable, notwithstanding such a change, as the universe will have been subject to, shall probably be known best by those, that have here made their best use of their former knowledge, which there will, together with their other gifts, be *congruously*, as well as *munificently*, rewarded. And then the attainment of a high degree of knowledge, which here, was so difficult, may, to the enlightened and enlarged mind, become as *easy*, as it will be *satisfactory*: and our improved understandings, will, with joy, perceive, how much all the knowledge, that we can give ourselves of God's works, is inferior to what their divine Author can impart to us, of them" (*Works*, 6:776).

10

The Alchemies of Robert Boyle and Isaac Newton: Alternate Approaches and Divergent Deployments

LAWRENCE M. PRINCIPE

It is clear that at present there is a growing interest in alchemy. This new attention to a subject long dismissed out of hand as a field for serious scholarly inquiry is in large part due to the successful linkage of alchemy with prominent figures of early modern science, especially Sir Isaac Newton. Once the extent of Newton's involvement with traditional alchemy was made manifest by the labors of B. J. T. Dobbs, Karin Figala, Richard S. Westfall, and others, alchemy could no longer be uncomplicatedly rejected as mere fraud or gullibility without impugning Newton himself. Thus, whereas Newton's status as the rationalist par excellence initially made his alchemical involvement seem unbelievable, that status helped to rehabilitate the much-maligned subject of alchemy. Since the revelation of Newton's alchemical interests, other figures of the early modern period have begun to reveal their own alchemical dimensions. Most notable among these is Robert Boyle. Recent studies demonstrate that the "Father of Chemistry" was equally a son of traditional alchemy. Far from repudiating traditional alchemy, as is commonly believed, Boyle pursued it with great avidity. He strove, for example, to discover the secret preparation of the transmutatory Philosophers' Stone, of whose real existence and powers he was certain. Significantly, Boyle's interest in and devotion to alchemy actually increased over the course of his career rather than being repudiated as a youthful whim.[1] These findings in the case of Boyle strongly reaffirm the conclusions regarding the importance of

---

1 Lawrence M. Principe, *The Aspiring Adept: Robert Boyle and His Alchemical Quest* (Princeton: Princeton University Press, 1998), and "Robert Boyle's Alchemical Pursuits," in *Robert Boyle Reconsidered*, ed. Michael Hunter (Cambridge: Cambridge University Press, 1994), pp. 91–105; Michael Hunter, "Alchemy, Magic and Moralism in the Thought of Robert Boyle," *British Journal of the History of Science* 23 (1990): 387–410; William R. Newman, "Boyle's Debt to Corpuscular Alchemy," in Hunter, *Boyle Reconsidered*, pp. 107–18.

alchemy in the early modern period, which were drawn from earlier investigations of Newton.

Undoubtedly, much more remains to be learned about alchemy among important early modern characters. It is virtually certain that the near future will witness additions to the roster of important Scientific Revolution characters who were likewise devotees of the Noble Art. The conditions are thus developing for a wider evaluation of the role of alchemy in early modern thought. We can already see clearly that alchemy continued to attract the attentions of significant figures and to make positive and important contributions until the early eighteenth century. Yet there is a danger of clouding our vision if some historiographic problems deeply entrenched in the secondary literature on alchemy are left unchallenged while this wider evaluation is underway. I refer now in particular to the pervasiveness of an essentialist view of alchemy, a topic that William Newman and I elsewhere discuss.[2]

The essentialist perspective views alchemy as a uniform and unarticulated monolith, largely unchanging over time and place. This essentialist position obscures the diversity and dynamism within alchemy, and portrays it instead as a constant set of beliefs and practices, thus implicitly distinguishing it (in a negative way) from the developmental nature of modern science. Furthermore, an essentialist perspective invites the dangers of oversimplification and over-generalization; this is particularly perilous in studies of a time of such great intellectual complexity as the seventeenth century. There was in fact no one "alchemy" in the fifteen hundred years from Zosimos to Philalethes just as there was not one "physics" from Aristotle to Einstein. At present, then, a chief desideratum for alchemical studies is an exploration of the diversity and latitude of thought *within* it. One route toward understanding the fine structure of alchemy would prescribe the documentation of the theories and practices of the various schools that coexisted within alchemy, paying due attention to their mutual struggles and development. Another route would suggest the investigation of how different thinkers approached and deployed alchemy. It is this latter route that this chapter chooses to pursue – comparing the "alchemies" of Isaac Newton and Robert Boyle.

In regard to alchemy, Boyle and Newton seem to have much in common. Both are major Scientific Revolution figures who devoted huge amounts of time to alchemical pursuits. But although both serve as arguments in favor of the continued appeal of alchemy, that general

---

2 Lawrence M. Principe and William R. Newman, "Some Problems in the Historiography of Alchemy," *Archimedes*, forthcoming.

similarity may obscure some telling differences that can provide a richer, more nuanced view of the late seventeenth-century status of alchemy as well as of the intellectual complexions of these two great thinkers. I shall first examine the respective strategies Boyle and Newton used to uncover the secrets of alchemy, and then compare the expectations and motivations that each held in regard to his alchemy.

Approaches to Alchemy: Textualism, Communication, and Experimentalism

Seekers after the secrets of alchemy can approach the subject through three kinds of sources – the written record left by past *adepti*, direct communication with living sources, and laboratory investigation.[3] Various alchemical seekers employed these three methods in varying proportions, and Newton and Boyle were no exceptions. But the differences in the relative emphases accorded to these sources of knowledge by Boyle and Newton reveal some intriguing differences not only in their own philosophical dispositions, but also in the varied ways in which alchemy could be viewed and deployed in the early modern period.

The reader of Newton's alchemical manuscripts is struck at once by the preponderance of material that is not of Newtonian origin, a fact commented upon by many historians of science. By far the greater part of Newton's alchemical output is in the form of transcriptions, translations, extracts, collations, and compendia of various alchemical authorities. In contrast, among the Boyle papers there are few such items. Most of the alchemical tracts that occur in the Boyle archive are in fact gifts from their authors or copies made by scribes outside of Boyle's employ, rather than copies made specifically by or for Boyle.[4]

---

3 I cannot agree with certain scholars who wish to diminish (or eliminate altogether) the importance of the last, claiming that alchemy did not involve a significant interplay between theory and practice, or that it is to be studied solely as a literary or philosophical phenomenon – the primary text material is overwhelmingly against such a viewpoint. I argue this point in "Apparatus and Reproducibility in Alchemy," in *Instruments and Experiments in the History of Chemistry*, ed. Frederic L. Holmes and Trevor Levere (Cambridge, Mass.: MIT Press, forthcoming).

4 For example, Johann Joachim Becher's *Concordantia purgationis*, vol. 19, ff. 57–77, Boyle Papers, Archives of the Royal Society of London, which is in the same hand as several other Becher documents which were possibly all given to Boyle by Becher himself. Further, Erasmus Rothmaler's *Consilium philosophorum*, endorsed as given by the author and bequeathed by Boyle to "Mr Newton the Mathematian of Cambridge," Boyle Papers, 23: 307–473, and Eirenaeus Philalethes, *De metallorum metamorphosi*, in the hand of a Hartlib circle scribe, Boyle Papers, 44:1–20. An exception is George Ripley's *Medulla philosophiae*, in the hand of Boyle's amanuensis Robin Bacon, Boyle Papers, vol. 30, ff. 1–72. See Principe, *Aspiring Adept*, pp. 139–41.

While this dearth of transcriptions relative to the Newton papers may be quite simply explained by invoking differences in study habits between the two men – Boyle's sickliness and consequent inability to write for himself, or Boyle's greater wealth enabling him to purchase books and manuscripts more readily – the difference regarding collations and compendia cannot be so readily explained away. The making of concordances and compilations characteristic of Newton's alchemical studies is unknown in Boyle's.

A clear example of Newton's work in this regard is the massive *Index chemicus*, a topical index on which he labored for decades (ca. 1680–1700). In its final form, the *Index* contains 879 headings and exceeds one hundred pages and twenty thousand words in length.[5] This long-term exercise was Newton's attempt to gather together in one place the parsimonious, widely scattered revelations of alchemical truths provided by various authors, and to piece these bits together in the hopes of elucidating a greater part of the great alchemical mystery. Another example of such endeavor is seen in the manuscripts of concatenated extracts, such as the "Praxis" manuscript (written in the late 1690s), which in spite of its title gives very little indication of any practical operations, but is instead a series of quotations from numerous alchemical authors laid together.[6] This type of composition – or rather compilation – occurs frequently in such alchemical collections as the *Theatrum chemicum*, and owes its origin to the alchemical belief that "all authors say one thing," a credo that Newton himself must have adopted.[7]

Another example of this genre appears in Newton's *Sententiae notabiles* composed of quotations from seventy-five alchemical texts, both printed and manuscript. When F. Sherwood Taylor published and commented upon this text many years ago, he made the observation that Newton's alchemical "works are, in their turn, a succession of extracts from earlier authors, with a few of his own comments thereon."[8] While the early state of alchemical studies at that time led Taylor to conclude such practice to be "the standard one" of alchemical authors in general (a view now refuted), the subsequent studies of

---

5 Richard S. Westfall, "Isaac Newton's *Index chemicus*," *Ambix* 22 (1975): 174–85; the original is Keynes MS 30, University Library, Cambridge.

6 Babson MS 420, Babson College Archives, Babson Park, Mass.; published in B. J. T. Dobbs, *The Janus Faces of Genius: The Role of Alchemy in Newton's Thought* (Cambridge: Cambridge University Press, 1991), pp. 293–305.

7 As an example, see David Lagneus, *Harmonia chemica*, in *Theatrum chemicum*, ed. Lazarus Zetzner, 6 vols. (Strasburg, 1659–61; reprint, Torino: Bottega d'Erasmo, 1981), 4: 705–804.

8 F. Sherwood Taylor, "An Alchemical Work of Sir Isaac Newton," *Ambix* 5 (1956): 63; the original is Keynes MS 38.

Newton have borne out that this was in fact *Newton's* standard (or at least primary) method. Indeed, Newton deployed this methodology not only for his alchemical studies, but for his theological and chronologic endeavors as well.

It might be pointed out at this juncture that this habit of Newton's has caused problems in the interpretation of his alchemy, simply because it greatly complicates the matter of deciding which documents (often labeled ambiguously or not at all) are actually Newton's own compositions.[9] For example, Keynes MS 18, "The *Clavis*," was long labeled as a Newtonian composition bearing the marks of his methodology and precision, and it formed a cornerstone for some interpretations of his alchemy. Yet it has since been shown to be only another of Newton's transcripts, this time of a circulating manuscript of "Eirenaeus Philalethes," most of the text of which appeared first in a 1651 letter from George Starkey to Robert Boyle.[10] It is possible that other texts currently believed to be Newtonian compositions may in due course likewise be identified as transcripts of or extracts from presently unrecognized originals.[11]

Boyle was certainly familiar with alchemical texts as well, as his references to them in both published and unpublished works make clear. Yet Boyle seems never to have invested texts with the same degree of importance as did Newton. Boyle's emphasis in his study of alchemy was on nonwritten sources – either results from his own laboratories or reports brought to him of the experiments and experiences of others. Moreover, throughout his life he made it abundantly clear that he valued and invited communication with alchemical *adepti*. In the preface to the *Sceptical Chymist* (1661), Boyle remarks on

---

9 On this problem and the archive more generally, see Rob Iliffe, "A 'Connected System'? The Snare of a Beautiful Hand and the Unity of Newton's Archive," in *Archives of the Scientific Revolution*, ed. Michael Hunter (Rochester, N.Y.: Boydell Press, 1998), pp. 137–57, esp. p. 152.

10 B. J. T. Dobbs, *The Foundation of Newton's Alchemy: or, "The Hunting of the Greene Lyon"* (Cambridge: Cambridge University Press, 1975), pp. 132–56, 175–90, and "Newton's 'Clavis': New Evidence on its Dating and Significance," *Ambix* 29 (1982): 198–202. Karin Figala, "Newton as Alchemist," *History of Science* 15 (1977): 102–37; William R. Newman, "Newton's *Clavis* as Starkey's *Key*," *Isis* 78 (1987): 64–74 and see also his *Gehennical Fire: The Lives of George Starkey, an American Alchemist in the Scientific Revolution* (Cambridge, Mass.: Harvard University Press, 1994).

11 One candidate for this status is Keynes MS 12A, a set of alchemical propositions in Latin dating from about 1669, of which ff. 1v–2 were deployed in Dobbs, *Janus Faces*, pp. 22–6; and B. J. T. Dobbs, "Newton's Alchemy and his Theory of Matter," *Isis* 73 (1982): 512–15. The previous page of the document (f. 1) contains very similar aphorisms (some only reworded relative to those on the verso), most of which are immediately followed by the title and page number of the texts from which they were taken.

how he would be "willingly and thankfully" instructed by the *adepti* regarding the nature and generation of metals; this desire is expressed even more forcefully in the preface to the *Producibleness of Chymical Principles* (1680).[12] Additionally, I have elsewhere argued that Boyle's famous but previously problematic paper in the *Philosophical Transactions* on an "incalescent" mercury is in fact a plea for direct advice from alchemical *adepti* regarding the method of employing this mercury to produce the Philosophers' Stone.[13]

Boyle was not disappointed in his hopes of gaining alchemical communications; travelers and correspondents brought him a wealth of information from across Europe and beyond. Boyle himself mentions many such interviews in both published and unpublished papers. Indeed, in the incalescent mercury paper he remarks how in order to learn about philosophical mercuries, he "purposely enquir'd of several prying Alchymists, that have spent much labour, and many Trials ... and have of late years travelled into many parts of Europe to pry into the Secrets of other Seekers of Metalline Transmutations."[14] In some cases, a traveling adept or his disciple visited Boyle and answered various of Boyle's questions, and sometimes presented him with a small sample of a precious alchemical rarity. The most dramatic of such interviews are those in which traveling *adepti* demonstrate projective transmutation before Boyle himself; a detailed first-person account of the most gripping of these transmutations witnessed by Boyle has recently been published.[15] These events were almost certainly more influential than anything else in shaping and promoting Boyle's alchemical beliefs and activities.

A curious fragment among the Boyle papers may represent original notes from just such an interview.[16] These sheets, judging by their small size and format, are the remnant of a disbound memorandum book of the kind that survives intact elsewhere in the archive. They

12 Robert Boyle, *Sceptical Chymist* (London, 1661), preface, [p. xiii]; *Producibleness of Chymical Principles* (London, 1680), pref., [p. xiv].
13 Principe, "Boyle's Alchemical Pursuits," pp. 96–7, and *Aspiring Adept*, pp. 162–5; Boyle, "Of the Incalescence of Quicksilver with Gold," *Philosophical Transactions of the Royal Society of London* 10 (1676): 515–83.
14 Boyle, "Of Incalescence," p. 518.
15 Principe, *Aspiring Adept*, pp. 93–134, 264–70 (the latter is a section of Boyle's *Dialogue on the Transmutation and Melioration of Metals*, published in full for the first time as appendix 1 to *Aspiring Adept*. The emphasis Boyle placed on such experiences is revealed by the amount of space he devoted to them in his autobiographical interview with Bishop Gilbert Burnet as recorded in the "Burnet Memorandum" (where a much briefer version of the transmutation experience mentioned earlier is recounted); see Michael Hunter, *Robert Boyle: By Himself and His Friends* (London: William Pickering, 1994), pp. 26–34.
16 Boyle Papers, vol. 44, ff. 43–9.

are noteworthy also because they are in Boyle's own hand – which is rare in the archive – suggesting that the original volume was something he may have carried around with him. There amid memoranda (e.g., books to be sent for, including Sprat's *History of the Royal Society*, and a clever turn of phrase regarding Aristotle) are several pages of hastily penciled notes as incoherent as they are incomprehensible. These seem to be memoranda jotted down after or during a series of alchemical conversations, one of which is dated as "15 August" of an unspecified year. The interpretation of this document as interview notes is greatly strengthened by two observations. First, there are bracketed sections such as "due proportion of [q. an commune]" where Boyle seems to be reminding himself to "question whether common" gold or not at the next opportunity. Similarly, there are references to anecdotes like those likely to occur in a conversation; for example, "half a coccos kindled and nearly killed him." Second, a fair copy of these scrawled notes exists elsewhere in the archive. In the hand of Boyle's amanuensis Robin Bacon, these notes are copied out in a neater format, and considerably expanded.[17] For example, while the notebook tersely records "The heated Arme. The Chaotic Net as a Recipient. The water reaching to saffron niter, &c.," the recopied notes read more fully "the heated Arm, and the two fair Birds... A Chaos made reticular of Mars & Venus is a far better recipient of the Eben then any of the imperfect Nidorums. As a bason of water may reach to saffron, Niter, &c, and be impregnated differingly by each of them."[18] This fuller, fairer copy would seem to be a more permanent record transcribed from the notes to Boyle Papers, volume 44 (which were hastily jotted down by Boyle immediately after or during the encounter) and apparently augmented by Boyle's recollections at the time of transcription.

Returning to Newton, we find little evidence of his going out of his way to access oral alchemical knowledge or manifesting any interest

17 Ibid., vol. 25, ff. 37–44.
18 Ibid., vol. 44, f. 46; vol. 26, f. 41v. Note the use of Boyle's encoding system here; *Eben*, from the Hebrew, *Stone* [of the Philosophers]; *Nidorum*, in his alphabetic system, for *metals*; see Lawrence M. Principe, "Robert Boyle's Alchemical Secrecy: Codes, Ciphers, and Concealments," *Ambix* 39 (1992): 65. Note that Boyle also uses "Chaos," one of the Philalethean *Decknamen* for antimony (see Newman, *Gehennical Fire*, pp. 235–6). Here the Chaos is "made reticular," that is, with a netlike crystalline pattern on its surface, which occurs when the antimony ore is reduced with iron [Mars] to the metal, and alloyed with copper [Venus]. Thus this line means that the Philosophers' Stone works better when projected upon an antimony-copper alloy rather than upon base metals. The allusion to a "bason of water" being impregnated differently by different substances refers to a comment made by the early seventeenth-century alchemist Michael Sendivogius.

in meeting reputed *adepti*. The few instances of it are scattered and brief – for example, a few bits of correspondence from Fatio de Duillier and Newton's notorious exchange with John Locke regarding the extraction of alchemical processes from the papers of the recently deceased Boyle.[19] One other exception is Newton's apparent contact with the shadowy Cleidophorus Mystagogus, alias William Yarworth.[20] But these are the exceptions rather than the rule, and in any event are briefer and doubtless far less significant to Newton than Boyle's encounters were to Boyle. Indeed, Newton wrote disparagingly of Boyle's circle of alchemical correspondents, informants, and collaborators – as well as of Boyle himself – and declined repeatedly to participate in such communication. A mutilated letter of 1689 to Fatio states that "Mr Boyle has divers times offered to communicate & correspond with me in these matters but I ever declined it because of his [*word cut from paper*] & conversing with all sorts of people & being in my opinion too open and too desirous of fame."[21]

Newton's condemnation of Boyle's putative "openness" recalls his letter to Henry Oldenburg after the publication of the paper on incalescent mercury. In that missive Newton urges that Boyle "preserve high silence" in this matter, for fear of "immense damage to the world."[22] Although Boyle sought alchemical advice regarding the use of this mercury, he certainly agreed with Newton in regard to secrecy, for he refused to answer questions about this mercury, and the next time he mentioned it in print he averred that "for the sake of Mankind, I resolve not to teach the preparation."[23] Indeed, Newton learned directly how unnecessary his advice was, for after Boyle "revealed" a related prize secret to both John Locke and Newton, Newton later complained that Boyle had "reserved a part of it from my knowledge ... what he has told me is imperfect & useless with knowing more than I do."[24] Thus Newton had to quiz Locke to see if Boyle had divulged more of the process to him. Both Boyle and Newton

---

19 *The Correspondence of Isaac Newton*, ed. H. W. Turnbull, 7 vols. (Cambridge: Cambridge University Press, 1960); letters with Locke, 3:92–3, 215–19; letter with Fatio, 3: 265–7.
20 Karin Figala, "Zwei Londoner Alchemisten um 1700: Sir Isaac Newton und Cleidophorus Mystagogus," *Physis* 18 (1976): 245–73; Figala and Ulrich Petzold, "Alchemy in the Newtonian Circle: Personal Acquaintances and the Problem of the Late Phase of Isaac Newton's Alchemy," in *Renaissance and Revolution: Humanists, Scholars, Craftsmen, and Natural Philosophers in Early Modern Europe*, ed. J. V. Field and F. A. J. L. James (Cambridge: Cambridge University Press, 1993), pp. 173–92.
21 Newton to Fatio, October 10, 1689, in *Correspondence*, 3:45.
22 Newton to Oldenburg, April 26, 1676, in ibid., 2:1–2.
23 Boyle, *Producibleness*, p. 214.
24 Newton to Locke, August 2, 1692, in *Correspondence*, 3:217–18.

understood the importance of maintaining alchemical secrecy. At the start of this volume, Westfall claims that maintaining secrecy in alchemy was a midpoint to ignoring it – implying that Newton was somehow ashamed of his own interests. As we can see by their own words, Boyle and Newton were not at all embarrassed, but were instead in fearful awe of the immense power of alchemy.

This exchange of secret recipes brings us to experimentalism; clearly, both Newton and Boyle engaged in this empirical method of uncovering alchemical knowledge. But judging from the written records, Newton's experimental endeavors are lesser than his textual ones. Dobbs estimated that of the "post-*Principia*" alchemical manuscripts (1687–96) only about one-fifth are records of Newton's own laboratory experiments. An even lower percentage marks the "pre-*Principia*" papers.[25] One must admit, however, that when comparing experimental and literary endeavors the relative quantities of surviving pages of notes may not translate accurately into relative quantities of time expended. Yet, upon comparison with the Boyle papers, a clear difference of emphasis between the literary and the experimental is evident. The Boyle archive contains hundreds of pages of extended entries of laboratory trials, many of which are identifiable as relating to traditional chrysopoetic (i.e., gold-making) studies. I have elsewhere mentioned how almost all of these are in various codes that serve to hide their true nature.[26] But Boyle's pursuit of experimental alchemy extended well beyond the work at his own laboratory at his sister Katherine's house on the Pall Mall. Boyle financed alchemical researches at several other work sites. For example, his assistant Ambrose Godfrey Hanckwitz was set up as the overseer for a German alchemist whom Boyle brought to England and supported. Boyle was also closely involved with a "company" of workers endeavoring to bring a process for transmutation to a successful outcome.[27] It is very possible that Boyle financed or advised many other such enterprises that have not yet, or may never, come to light.

In sum, Newton emphasized the written traditions of alchemy, favoring an approach based on close textual analysis and the drawing up of compendia, which were thereafter complemented with laboratory experimentation. Boyle, on the other hand, although he did in

---

25 Dobbs, *Janus Faces*, p. 171, and *Foundations*, pp. 139–67; Newton's chemical experiments are recorded primarily in Cambridge University Library MSS 3973 and 3975. These were examined, but in a programmatically antialchemical fashion, in Marie Boas and A. Rupert Hall, "Newton's Chemical Experiments," *Archives internationales d'histoire des sciences* 11 (1958): 113–52.
26 Principe, "Boyle's Alchemical Secrecy."
27 Principe, *Aspiring Adept*, pp. 134–6, 149–53, 174–5.

fact read many of the same texts as Newton, favored a more empirical approach to learning about alchemy; he gathered information primarily from laboratory experiments and from verbal or epistolary communication with those who had practical experience with such matters, and was highly solicitious to maintain and to develop lines of alchemical communication.

This difference in approach toward uncovering alchemical knowledge undoubtedly originates in several factors, some of them surely dependent upon nothing more than the two men's very different personalities. But one probable contributing factor that is of wider historical interest is traceable to their divergent opinions regarding the *prisca sapientia* and *prisca theologia*, an important topic treated by Jan Wojcik elsewhere in this volume. In a seminal paper, J. E. McGuire and P. M. Rattansi first emphasized the impact of Newton's adherence to the *prisca* tradition, and much of Dobbs's *Janus Faces* pursues this in terms of his alchemy and theology.[28] Newton believed that the ancients were possessors of abstruse truths in both theology and natural philosophy, these having been revealed to ancient biblical characters, perhaps as far back as Adam himself. This knowledge had become successively corrupted over time, but could be retrieved by the close philological study of the most ancient documents. Newton believed that even his inverse-square law of gravitation was known to the ancients, and consequently that he was its restorer, not its discoverer. It has further been argued that much of Newton's post-*Principia* study of ancient texts was based upon his belief that the ancients understood even the cause of gravity, something Newton wished desperately to know.[29]

The notion of the *prisca sapientia* was popular in the Renaissance and early modern period, and it did not fail to take firm root in alchemical circles as well. Thus arose the belief that some of the patriarchs were possessors of the Stone, that Solomon had furnished the Temple with alchemical gold, and that Moses knew grand alchemical arcana. Had Moses not, after all, reduced the golden calf into a form of which he made the rebellious, idolatrous Israelites drink, implying that he knew how to make potable gold? A thinker like Newton, convinced of the possibility of uncovering fragments of this antique wisdom in surviving texts – masked as they might be by corrupting accretions – would naturally gravitate toward the kind of textual analysis that characterized his approach to alchemy.

Boyle, on the other hand, doubted (at least in his maturer years) that

---

28 J. E. McGuire and P. M. Rattansi, "Newton and the 'Pipes of Pan,'" *Notes and Records of the Royal Society* 21 (1966): 108–43.

29 Ibid., p. 109; see the letter of Fatio de Duillier to Christiaan Huygens, February 5, 1691/2, in *Correspondence*, 3:193.

such ancient knowledge ever existed, and even if it had, it could never be recovered. As Wojcik points out, Boyle did not believe that God had ever revealed the entirety of knowledge to man, but had instead provided man with revelation in the Holy Scriptures for his instruction in divinity, and the senses and reason to equip him to examine the physical world, for, as Boyle remarked more than once, "the World is God's Epistle, written to Mankinde."[30] According to Boyle, earlier writers did not have a privileged position in regard to knowledge. Indeed, Boyle had a strong sense of progress – the "future Advancement of Humane Knowledge." While he believed that his was a time of significant advancement in knowledge, he also believed that

> enlightned Posterity will arrive at such attainments, that the Discoveries and Performances, upon which the present Age most values it self, will appear so easy, or so inconsiderable to them, that they will be tempted to wonder, that things to them so obvious, should lye so long conceal'd to us, or be so priz'd by us; *whom* they will, perhaps, look upon with some kind of disdainful Pity, unless they have . . . the generous gratitude to remember the Difficulties this Age surmounted, in breaking the Ice, and smoothing the way for them, thereby contributing to those Advantages, that have enabled them so much to surpass us.[31]

Accordingly, Boyle did not privilege textual sources – he did not seek secrets predominantly in a written text, but rather in the Book of Nature and from living voices, from those "Possessors of Secrets, by whose Friendly Communication [one] may often learn that in a few Moments, which cost the Imparters many a Years toyl and study."[32]

Specific differences between Boyle's and Newton's tastes in alchemical literature help bear out this point about the *prisca sapientia*. While Boyle and Newton shared some favorite alchemical authors – Basil Valentine, George Ripley, and especially Eirenaeus Philalethes (alias George Starkey) – they had very different views of Hermes Trismegistus and the *corpus Hermeticum*. The Hermetic *Tabula smagdarina* or *Emerald Tablet* and commentaries upon it figure prominently in many collections of alchemical tracts throughout the seventeenth century, and Hermes was regularly seen as the ancient founder of alchemy. Hence, the term "Hermetic Art" was often used synonymously with alchemy, even by Boyle himself.[33]

---

30 Boyle, *Usefulnesse of Experimental Naturall Philosophy* (Oxford, 1663), pt. 1, p. 50; *Excellency of the Mechanical Philosophy* (London, 1674), p. 36.
31 Boyle, *A Natural History of Humane Blood* (London, 1684), pp. 287–8.
32 Boyle, *Usefulnesse*, pt. 2, p. 112.
33 On this synonymity, see Jean Beguin, *Tyrocinium chimicum*, p. 1; Boyle lists his alchemical writings as "Tracts relating to the Hermetical Philosophy," Royal Society

Boyle was introduced to the Hermetic corpus very shortly after his attentions first turned seriously toward experimental natural philosophy. In 1650, while he was writing an essay entitled "Of the Study of the Booke of Nature" (intended as the first part of *Occasionall Reflections*), Boyle encountered John Everard's English translation of *Pymander* and *Asclepius*.[34] The young Boyle was sufficiently attracted by the text that he made several insertions into his manuscript treatise entitled the "Booke of Nature" referring to Hermes Trismegistus as a model and authority. Other additions to that essay quote "that excellent Prince De La Mirandola" speaking of the progressive corruption of ancient wisdom over time, and recount the story – important to believers in the *prisca sapientia* – that Seth had engraved ancient secrets of Nature on pillars of stone and brick to preserve them from the Flood.[35] While some of these references resurface nearly verbatim in *Usefulnesse of Experimentall Natural Philosophy* – a text some of whose origins date to this early period – Boyle never thereafter referred to Hermes or the Hermetic corpus either in print or in manuscript.[36] It must be further pointed out that even in these references Boyle uses Hermes as an example of piety, not as a source of ancient wisdom. Paralleling his brief interest in and subsequent abandonment of the Hermetic corpus, in later writings Boyle became increasingly skeptical of the *prisca sapientia* on both practical and theological grounds, as Wojcik shows.

In contrast, Newton maintained a high level of interest in the writings of Hermes throughout his life, and there is evidence that his regard for Hermes (relative to other alchemical authors) actually increased over time.[37] Keynes MSS 27, 28, 29, 60 and lesser parts of several other manuscripts all deal with Hermes, including both his

Manuscript 198, f. 143v, Boyle Papers, and uses the term in his published works. Unfortunately, this usage persisted until this century, and was reified after the enhanced interest in Hermeticism following the publication of Francis Yates's *Giordano Bruno and the Hermetic Tradition* (Chicago: University of Chicago Press, 1964). The term should, however, be restricted to things directly relating to the *Corpus hermeticum*, which really in point of fact is only tangential to alchemy as it was generally practiced in the Latin West.

34 *Hermes Mercurius Trismegistus, His Divine Pymander ... together with ... Asclepius*, trans. John Everard (London, 1650). Michael Hunter, "How Boyle Became a Scientist," *History of Science* 33 (1995): 68. See also Lawrence M. Principe, "Virtuous Romance and Romantic Virtuoso: The Shaping of Robert Boyle's Literary Style," *Journal of the History of Ideas* 56 (1995): 392–3.

35 Boyle, "Of the Study of the Booke of Nature," Boyle Papers, vol. 8, ff. 123–39; on ff. 127–127v, 128.

36 Boyle, *Usefulnesse*, pt. 1, pp. 10–11, 104, 117.

37 J. E. McGuire, "Neoplatonism and Active Principles: Newton and the *Corpus Hermeticum*," in *Hermeticism and the Scientific Revolution*, ed. Robert S. Westman and J. E. McGuire (Los Angeles: William Andrews Clark Memorial Library, 1977), pp. 93–142.

Emerald Tablet and *Seven Chapters*. Dobbs has charted his study of Hermetic tracts from the early 1680s through enhanced activity in the 1690s to the turn of the eighteenth century.[38] In light of the brevity of the *Tablet* (about two hundred words), this is a disproportionate amount of attention, although it parallels the lengthy commentaries upon it written by some other alchemical authors.[39] Newton's enhanced study of this small piece was largely provoked by its supposed antiquity, and therefore its proximity to the sources of ancient wisdom. A similar cause undergirds Newton's reverence for Nicolas Flamel (another author whom Boyle never cites), for Flamel's knowledge of alchemy supposedly came from an ancient book of "Abraham the Jew" whose antiquity and reputed author's nationality placed it closer to the *prisca sapientia*.[40]

The Differing Rewards of Alchemy for Boyle and Newton

Besides examining the differing approaches of Boyle and Newton to alchemical studies, we should also examine what specific benefits these two students of alchemy expected to reap from such activity. In the case of Boyle, the potential rewards can be grouped into three categories: natural philosophical knowledge, medicinal preparations, and the defense of orthodox Christianity.[41] In relation to the first category, alchemical matter theory has been shown to have had an important role in the development of the corpuscularian conception of matter for which Boyle is renowned. Many of Boyle's notions were formulated by drawing upon the conceptions of a long line of alchemical thinkers, from the late medieval Geber through the Aristotelian iatrochemist Daniel Sennert.[42] The same is apparently true for Newton, for both his notions of "mediation" and his "shell theory" of the microstructure of matter – according to which corpuscules are built up in shells from smaller corpuscules – have been shown to owe a considerable debt to alchemical precursors, particularly Eirenaeus Philalethes.[43]

38 Dobbs, *Janus Faces*, pp. 66–73, 178–9, 271–7.
39 See, for example, the exposition of the *Tabula* by Gerard Dorn, *Theatrum chemicum*, 6 vols. (Argentorati, 1659–61), 1:362–87; and the anonymous scholia on the *Tractatus* in *ibid.*, 4:592–705. Many others are cited in John Ferguson, *Bibliotheca chemica*, 2 vols. (Glasgow: Kressinger, 1906), 1:389–94.
40 Dobbs, *Janus Faces*, p. 179.
41 For a fuller exploration of these three alchemical rewards, see Principe, *Aspiring Adept*, pp. 180–213.
42 Newman, "Boyle's Debt to Corpuscular Alchemy," pp. 107–18; William R. Newman, "Art, Nature and Experiment among Some Aristotelian Alchemists," in *Texts and Contexts in Ancient and Medieval Science*, ed. Edith Sylla and Michael McVaugh (Leiden: Brill, 1997), pp. 305–17.
43 Newman, *Gehennical Fire*, pp. 229–39.

Boyle also expected to obtain the alchemical *summum bonum* – the secret of the preparation of the Philosophers' Stone, the adepts' agent of metallic transmutation. There is evidence in the form of laboratory records and allusive references dating from throughout Boyle's life that he was engaged in the long-term pursuit of this goal.[44] This wonderful substance, once prepared, would provide him with stunning chemical phenomena in areas which were of particular interest to him. For example, the action of a minute proportion of the Elixir in transmuting a great quantity of white, fluid quicksilver into solid, yellow gold would play on Boyle's interest in the disproportionate action of "unheeded agents" as explored in his *Tracts about the Cosmical Qualities of Things* (1671), and on his interest in the interconversion of forms and qualities as expressed in his chemical writings from the *Sceptical Chymist* onward.[45] Indeed, Boyle several times used the Stone's transmuting action as an example of material changes supportive of his corpuscularian views.[46] Further, the Stone was widely regarded as a panacea, and so the search for it likewise involved Boyle's interest in medicine (for both personal and philanthropic uses).

But whereas Boyle did not doubt the real existence of the Stone and its powers (especially not after his eyewitness of transmutation in 1678), Newton seems to have entertained doubts about it in spite of his continued study of alchemical texts. In writing to John Locke, Newton closed by forbearing "to say anything against multiplication [i.e., transmutation] in general because you seem perswaded of it, tho there is one argument against it which I could never find an answer to."[47] What this argument was, we shall, unfortunately, never know. Yet it is curious that Newton would express such doubts about so central a feature of the alchemy he was so studiously pursuing. Indeed, such doubts bolster the notion that Newton's interest in alchemy lay elsewhere than in the preparation of what was generally considered the alchemists' chief goal.[48] Specifically, Dobbs's contention that

---

44 Principe, *Aspiring Adept*, and "The Gold Process: Directions in the Study of Robert Boyle's Alchemy," in *Alchemy Revisited*, ed. Z. R. W. M. van Martels (Leiden: Brill, 1990), pp. 200–5.
45 See John Henry, "Boyle and Cosmical Qualities," in Hunter, *Boyle Reconsidered*, pp. 119–38.
46 See, for example, Boyle, *Of the Mechanical Origine or Production of Fixtness* (London, 1675), pp. 15–18, and *Sceptical Chymist*, p. 158.
47 Newton to Locke, August 2, 1692, in *Correspondence*, 3:218–19.
48 On Newton's experimentation, see Peter Spargo, "Newton's Chemical Experiments: An Analysis in the Light of Modern Chemistry," in *Action and Reaction: Proceedings of a Symposium to Commemorate the Tercentenary of Newton's Principia*, ed. Paul Theerman and Adele F. Seeff (Newark: University of Delaware Press, 1993), pp. 123–43.

Newton's study of alchemy was a search for the existence and means of divine activity in the world fits with such a picture. Accordingly, in relation to the *prisca sapientia* tradition discussed earlier, Newton may have felt that the belief in a physical, transmuting Philosophers' Stone was only a "corruption" of the ancient knowledge propagated through alchemy, and that its kernel of true wisdom was imbedded elsewhere. Nonetheless, Newton did carry out alchemical experiments, many of which came directly from the writings of Stone-seeking chrysopoetic alchemists. Thus, it will in fact be difficult to ascertain where Newton's real beliefs on transmutation lie – that is, was he being unusually negative (or cagey) in his comment to Locke, or did his study of originally chrysopoetic processes promise outcomes different from those in the minds of their authors? However this question may eventually be resolved, Newton's seeming ambivalence to the question of the reality of projective transmutation contrasts strongly with Boyle's certainty.

An area of relative commonality between Boyle's and Newton's alchemical expectations lies in the service they believed alchemy could render to religion. Both men saw alchemy as a corrective to an overly mechanized and potentially atheistic world view.[49] For Newton, according to Dobbs, part of this correction came in the form of the positing of forces in matter that reflected the immediate activity and agency of God in the world at every moment. The Newtonian forces, which aroused the suspicion and criticism from stricter mechanists, may well have been a result of Newton's study of alchemical texts.[50]

For Boyle, traditional alchemy provided an interface between the natural, mechanical realm and the supernatural, miraculous realm. The Philosophers' Stone which Boyle sought to prepare was a physical substance that caused physical changes, but which could also, as Boyle came to believe, attract and manifest spirits and angels by some unknown "congruities or magnetisms." Indeed in a remarkable document on the Stone's ability to attract rational spirits, Boyle explicitly

---

Spargo's edition of Newton's laboratory records is forthcoming from Oxford University Press. Spargo's extensive "study of the records of Newton's alchemical experiments certainly fails to reveal any preoccupation with transmutation" (private communication).

49 Dobbs, *Foundations*, esp. pp. 100–5; Principe, "Boyle's Alchemical Pursuits," pp. 100–2, and *Aspiring Adept*, pp. 201–13.

50 Dobbs, *Janus Faces*; "Gravity and Alchemy," in *The Scientific Enterprise: The BarHillel Colloquium*, ed. Edna Ullmann-Margalit (Dordrecht: Kluwer, 1992); and "Newton's Alchemy." Note in partial contrast, however, John Henry, "Occult Qualities and the Experimental Philosophy: Active Principles in Pre-Newtonian Matter Theory," *History of Science* 24 (1986): 335–81, which charts presence of "activity" in English conceptions of matter well before Newton.

relates such manifestations of spirits to the confounding of atheism.[51] Thus this chief alchemical product had a dual potential use for religious apologetics. First, it could provide an ocular demonstration of the existence of spirits, thus effectively silencing the atheists and "wits" who denied their existence and thereby that of God himself. Second, the Stone represented a "boundary" where corporeal and incorporeal realms met. Boyle could continue in his committment to mechanical explanations of physical phenomena without fearing the possible overextension of such a system to the exclusion of God and his workings because the Stone could show exactly where mechanism was valid and where it was not; the incorporeal world was "protected" from mechanical incursions and usurpations that threatened to exile God not only from the running of his creation, but also from existence itself. Thus alchemy could act for Boyle as a mediator between his two potentially conflicting commitments to mechanical natural philosophy and Christian theology. Whereas Newton amended the brute functioning of inert matter with a system of forces (drawn at least partly from alchemical sources) that he held to be indicators of continuous divine activity, Boyle tried to use alchemy to divide the corporeal and incorporeal realms – showing the reality of each and their ability to affect one another by unknown means.

Yet although alchemy could have similar theistic value for both men on the large scale, the specific theological effects were widely different. On the one hand, Boyle thought that although alchemy could be used against atheism, its study could not add anything directly to revealed religion. The totality of God's direct revelation came in the recognized canon of Scripture; Boyle's Christianity was orthodox. Newton on the other hand, while also a biblicist, thought that the *prisca theologia* embedded in ancient texts could be a prop to his Arian-like heresy. Dobbs argued that Newton – denying the consubstantiality of God the Son with God the Father – identified Christ with "God's viceroy" who governed the forces and nonmechanical agents at work in matter, and sought ancient approbations of this belief in alchemical/Hermetic texts.[52] Thus for Boyle, alchemy's supernatural effects on spirits could vindicate theism in general while the Bible suffices to evince orthodox

---

51 The full text of Boyle's "Dialogue on Spirits" is edited and published in Principe, *Aspiring Adept*, pp. 310–17; it was first uncovered and discussed in Hunter, "Alchemy, Magic and Moralism," pp. 396–8.
52 Dobbs, "Newton's Alchemy," pp. 526–8; for a view that opposes the belief that Newton maintained intermediate agencies between God and creation, see McGuire, "Neoplatonism," esp. p. 107. On Newton's heresy, see Steven Snobelen, "Isaac Newton, Heretic," *British Journal of the History of Science*, forthcoming.

Christianity; for Newton in contrast, the texts he chose "corrected" what he saw as blasphemous in orthodox Christianity. The differing orthodoxy of the two (supported by their differing valuations of the *prisca* tradition) manifested itself in their differing theological expectations from alchemy.

Alchemical Diversity, Essentialism, and the Scientific Revolution

The foregoing comparison of the "alchemies" of Robert Boyle and Isaac Newton demonstrates that alchemy, even within the same temporal (late seventeenth century) and cultural (English) domain, is not a uniform object with uniform outlooks, influences, or deployments. Within the heterogenous assemblage of ideas, texts, and philosophies loosely labeled as "alchemical," exists a highly diverse array of divergent and even contradictory notions. Different thinkers chose or developed different things from that array and applied or interpreted them in manifold ways. Thus it is not acceptable for historical analyses to speak imprecisely about a generic "alchemy" and its influences. Likewise, alchemy as a whole cannot be set down uncomplicatedly as one pole of some historiographical dichotomy. In the cases of Boyle and Newton, beyond some general similiarities, we have seen that they gave different emphases to different parts of the alchemical whole, made different uses of it, and expected different rewards from it. Newton was more of a *textual Hermeticist* (follower of the Hermetic corpus) than Boyle; Boyle was more of an *experimental chrysopoeian* (seeker after transmutation) than Newton. Alchemy as a whole remains a difficult topic of study, and if we are to continue successfully to unpuzzle its enigmas, map its contours, and disentangle its lines of influence, we must recognize its own internal articulated structure.

The essentialism/reductionism in regard to "alchemy" with which I take issue in this chapter impels me to question parts of Westfall's opening reassertion. Essentialist, or at least excessively reductionist, perspectives on both alchemy and science in general appear clearly when he attempts to co-opt alchemy into the old blueprint of the Scientific Revolution – an act undoubtedly shocking to the original framers of the Scientific Revolution model. To do so he first claims that "alchemy was not part of the Aristotelian system" and thus should stand alongside Galilean dynamics as a response to Aristotle. In spite of this assertion, there are in fact scores of Aristotelian alchemists from the Middle Ages to the seventeenth century; the diversity of alchemy includes both Aristotelians and anti-Aristotelians.[53] Alchemy as a whole cannot come down cleanly on either side of the debate over Aristotle.

Second, Westfall restricts "alchemy" in such a way as to state that its "great age ... was the late sixteenth and seventeenth centuries." While it is true that alchemical publications burgeoned in that period, are there legitimate criteria for truncating alchemy's late stage from its many previous centuries of development? Even Boyle and Newton themselves concentrated much of their efforts on authors and notions dating from earlier epochs. Boyle's corpuscularianism draws some of its inspiration from as far back as medieval Scholastics such as Geber, and he shows a greater acceptance of the medieval mercury-sulphur theory of metallic composition than of its later developments.[54] Newton's interests as well focused on many early alchemical authors such as Flamel, Ripley, and, of course, "Hermes." These restrictions come, it seems to me, from preconceived and unwarrantably reductionist notions of what "alchemy" is, and a rather unhappy attempt to mask a failure of the revolution paradigm by making it appear a success.

The importance of alchemy – in differing forms – to Boyle and Newton argues sufficiently for its continuing influence in the early modern period. But the long-term continuity and internal dynamism and diversity of alchemy must argue against the notion of "revolution" as it is rightly understood. Alchemy (or, more properly, *chymistry*) did not undergo a revolution comparable with the switch from geocentrism to heliocentrism, nor can it be validly positioned on one or the other side of the Scientific Revolution as that construct is generally understood.[55] Westfall and Dobbs both remind readers that

---

53 The Scholastic system and style is equally clear in the late medieval works of Geber, Albert the Great, and Petrus Bonus, as well as, for example, the late sixteenth-century chrysopoeian Gaston Duclo and the early seventeenth-centruy iatrochemist Daniel Sennert. See Newman, *Summa perfectionis*, and "Art, Nature, and Experiment among some Aristotelian Alchemists," pp. 305–17; also, Lawrence M. Principe, "Diversity in Alchemy: The Case of Gaston 'Claveus' du Clo, a Scholastic Mercurialist Chrysopoeian," in *Reading the Book of Nature: The Other Side of the Scientific Revolution*, ed. Allen G. Debus and Michael Walton (Kirksville, Mo.: Sixteenth Century Press, 1998).
54 Newman, "Boyle's Debt to Corpuscular Alchemy," and Principe, *Aspiring Adept*, pp. 43–6.
55 The use of the archaic chymistry has been suggested as an inclusive term for alchemy/chemistry during the sixteenth and seventeenth centuries – a time in which the domains of "alchemy" and "chemistry" were neither divided nor defined in the modern sense; see Lawrence M. Principe and William R. Newman, "Alchemy vs.

Scientific Revolution is a metaphor; Dobbs would have us abandon it, whereas Westfall wants us to keep a patched-up version, which need contain neither the attributes of suddenness nor thoroughness. But how unhappily chopped up, propped up, and patched up can a metaphor be and remain of value as a descriptor?

The concept of the Scientific Revolution was built largely by scholars who saw physics and astronomy as paradigmatic of scientific thought. Westfall's insistence on the examples of heliocentrism and kinematics continues this trend, and it seems to be the new physics that forms "the common denominator of other topics that come up with the Scientific Revolution" to which "alchemy does not reduce." Part of the problem with the standard notion of Scientific Revolution is that this common denominator is simply insufficiently common. From the start, chymistry did not fit ("reduce"), and for this reason Butterfield was obliged to construct dubious terms of special pleading like the "postponed chemical revolution." The revolution paradigm may continue to have value for describing the development of parts of the sciences – clearly the sixteenth and seventeenth centuries were in fact times of massive changes for physics and astronomy – but our vision of "the sciences" should now be wide and inclusive enough to force us to question the value of an extension of this revolution in *one part* of the sciences to a general, all-encompassing Scientific Revolution.

Further Alchemical Studies

Dobbs's essay identifies some problems in the classical conceptions of the Scientific Revolution and, even though I disagree with certain of her characterizations of alchemy, still points the way toward new and productive ways of approaching the history of early modern science. Clearly, the studies of Newtonian alchemy, and more recently Boylean alchemy, have irreversibly changed the way we look at these two chief figures of early modern English science. But we are by no means at the end of the road. As I mentioned at the outset of this chapter, there are many contemporaries of Boyle and Newton awaiting further study in regard to their alchemical connections. For example, John Locke shared much of his friends' interest in alchemy. He attended chymical lectures at Oxford in the late 1650s, and his involvement was sufficiently maintained that it was to him that Boyle, near the end of

Chemistry: The Etymological Origins of a Historiographic Mistake," *Early Science and Medicine* 3 (1998): 32–65.

his life, revealed the preparation of his philosophical/incalescent mercury – a process he thought crucial to the confection of the Philosophers' Stone – and other chrysopoetic arcana, which he had long kept as closely guarded secrets.[56] Locke was also chosen by Boyle to be one of the executors of his "chymical" papers. How alchemical notions may have shaped Locke's philosophical positions still remains to be explored. Boyle's other chymical executors – Daniel Coxe and Edmund Dickinson – come also to mind. The latter of these was physician to Charles II, with whom he collaborated on alchemical projects in a laboratory reputedly located under the royal bedchamber. Both Coxe and Dickinson were fellows of the Royal Society, as were Boyle, Elias Ashmole, Sir Kenelm Digby, among others. Clearly the views on transmutational alchemy in the early society are worth investigating further.

On the Continent too, there are important characters of the late seventeenth century who merit further investigation in terms of their alchemical connections. One is Gottfried Wilhelm Leibnitz, who was involved in an alchemical society while a young man. Some preliminary investigations seem to bear out that the development of his metaphysical views, particularly in the formative period of his career, were significantly influenced by such alchemical connections.[57] Like the Royal Society, the Académie Royale des Sciences also provides a potentially intriguing venue for the study of the persistence of alchemical notions. Despite the noted antialchemical sentiments of important members like Nicolas and Louis Lémery and Bernard Fontenelle, the Académie also had members, like Wilhelm Homberg, whose work shows strong and sometimes explicit reliance upon traditional alchemical theories and beliefs.[58]

This catalog is far from complete, and gives only a fleeting taste of the studies yet to be undertaken regarding alchemy in the early modern period. Thus, although Boyle and Newton have yielded up much (but far from all) of their own "alchemical secrets," a great deal more new insight on alchemy's role in the early modern period is to be expected in the near future.

56 Principe, *Aspiring Adept*, pp. 175–8.
57 George M. Ross, "Leibniz and the Nuremberg Alchemical Society," *Studia Leibnitiana* 6 (1974): 222–48; "Alchemy and the Development of Leibniz's Metaphysics," p. 58.
58 See, for example, Lawrence M. Principe, "Chacun à son goût: Theory and Practice in Early Modern Chymistry," in *Chemistry and Chemists in Search of an Identity: Perspectives on the Seventeenth-century*, ed. Brigitte van Tiggelen (forthcoming); *Sudhoffs Archiv Beihefte* (forthcoming); and "Wilhelm Homberg: Chymical Corpuscularianism and Chrysopoeia in the Eighteenth Century" (unpublished manuscript).

11

The Janus Faces of Science in the Seventeenth Century: Athanasius Kircher and Isaac Newton

PAULA FINDLEN

> Roberto still did not understand what Father Caspar Wanderdrossel was. A sage? That, certainly, or at least a scholar, a man curious about both natural and divine science. An eccentric? To be sure.
>
> Umberto Eco, *The Island of the Day Before*

During the second half of the seventeenth century two men in particular occupied a prominent place in the world of natural philosophy. Inhabiting respectively the capital cities of Rome and London, they had much in common. Both were deeply religious men, committed to the study of nature as a sure path toward the revelation of divine wisdom, who began their academic careers as professors of mathematics. Both valued the learning of the ancients, searching ever further into the pagan and Christian past in hope of illumination. For significant portions of their respective careers, that knowledge was found in the occult sciences. In all these respects, they typified the Christian encyclopedic approach to nature that humanists had practiced since the fifteenth century. They were quite probably, along with Leibniz, the last great humanist natural philosophers of the early modern period.

Despite these many similarities we do not usually place the German Jesuit Athanasius Kircher (1602–80) and the English philosopher Isaac Newton (1642–1727) side by side. For almost two centuries, they have inhabited the separate worlds created by modern histories of science and, to a lesser degree, articulated by intellectuals at the end of the seventeenth century who had begun to make these distinctions themselves. John Aubrey exemplified the positive reception of Newton's *Principia* (1687), when he rhapsodized that it reflected "the great

---

Thanks to Jo Dobbs, whose first book inspired my interest in the fate of the occult sciences during the Scientific Revolution, and whose conversations and friendship as a colleague in the History and Philosophy of Science Program at the University of California at Davis are reflected in many pages of this essay.

highth of Knowledge that humane nature has yet arrived to."[1] Instead, critics of Kircher heaped scorn upon his many works such as *Loadstone or the Magnetic Art* (1641), *Egyptian Oedipus* (1652–4), and *Subterranean World* (1664). During the final decades of the seventeenth century criticisms of Newton's natural philosophy decreased, especially after the English assault on Leibniz triumphed, but attacks on Kircher only intensified. After hearing about such inventions as Kircher's miraculous sunflower clock, a magnetic heliotrope that amazed virtuosi throughout Europe, Descartes wrote to his Dutch correspondent Christiaan Huyghens in 1643: "The Jesuit has a lot of tricks; he is more charlatan than scholar."[2] Subsequently Descartes refused to read Kircher's books, returning them unopened to his friend Huyghens.[3]

Another half century of intellectual developments transformed Kircher into the prototype of the foolish polymath popularized in Johann Burckhardt Mencke's *The Charlatanry of the Learned* (1715). In Mencke's exposé of learned gullibility, Kircher became the butt of numerous jokes that parodied his search for hieroglyphic wisdom by offering him fake antiquities, which he happily translated, unable to discern the difference between the real and the imaginary. Surely the best was a forged Chinese scroll on which was written, in reverse characters so as to be discernible only in a mirror: *Noli vana sectari et tempus perdere nugis nihil proficientibus* ("Do not seek vain things, or waste time on unprofitable trifles").[4] As Newton became the image of well-tempered and sober reason, deified for his flashes of divine insight, Kircher receded to the margins of the learned world, a foolish, fantastic figure.

Many scholars found Kircher's work disappointing – "nothing but a heap of unreasonable stuff," pronounced Huyghens.[5] But few shunned it entirely. Ironically, more scholars had direct knowledge of

1 In John Fauvel, Raymond Flood, Michael Shortland, and Robin Wilson, eds., *Let Newton Be!* (Oxford: Oxford University Press, 1988), p. 2.
2 Descartes to Huyghens, January 14, 1643, in Martha Baldwin, "Athanasius Kircher and the Magnetic Philosophy" (Ph.D. diss., University of Chicago, 1987), p. 37; see also Thomas L. Hankins and Robert J. Silverman, *Instruments and the Imagination* (Princeton: Princeton University Press, 1995), p. 19.
3 Baldwin, "Athanasius Kircher," p. 154.
4 Johann Burckardt Mencke, *The Charlatanry of the Learned*, trans. Francis E. Litz, ed. H. L. Mencken (New York: Knopf, 1937), p. 86, n. 64. For the general context of this sort of learned joke, see Anthony Grafton, *Forgers and Critics: Creativity and Duplicity in Western Scholarship* (Princeton: Princeton University Press, 1990).
5 This comment was in response to Kircher's *Iter exstaticum* (1656), an imaginary voyage through the Tychonic heavens that Kircher took, guided by angels. Huyghens, *Oeuvres complètes*, XXI, p. 811, in John E. Fletcher, "Astronomy in the Life and Correspondence of Athanasius Kircher," *Isis* 61 (1970): 59.

Kircher's work than Newton's rather inaccessible prose and difficult calculations. His encyclopedias enjoyed enormous popularity, not only in Europe but also in New Spain, and frequently went into multiple editions. Natural philosophers throughout Europe, from the aging Galileo to members of the Accademia del Cimento and the Royal Society to the young Leibniz, consumed Kircher's publications eagerly. The French scholar Marin Mersenne may have thought that his Jesuit friend was almost innumerate after seeing Kircher's calculations to square the circle, but this did not prevent him from reading his books and encouraging his investigations of nature.[6] Mersenne was ever hopeful that Kircher's seemingly boundless erudition would yield pearls among the swine. This attitude toward Kircher was not unique but shared by other prominent intellectuals, such as Nicolas Claude Fabri de Peiresc, whose patronage secured Kircher's position as professor of mathematics at the Roman College in 1634 to prevent the loss of his valuable linguistic and philosophical skills to the imperial court in Vienna.[7] The German Jesuit's seemingly endless array of skills and interests led admiring disciples such as Johann Koestler to describe him as "the prodigious miracle of our age."[8]

Between the 1640s and 1670s, praise as well as criticism of Kircher increased. Experimenters in Florence may have laughed themselves silly over the improbabilities of Kircher's eclectic Aristotelian physics but a wide array of other scholars, among them Robert Boyle, felt that the Jesuit professor was on the verge of discovering some of nature's and humanity's most important secrets. Boyle's view was shared by no less an observer of European academic life than the German polymath Daniel Georg Morhof, whose *Polyhistor* (1688) celebrated

---

6 Saverio Corradino, S. J., "L' 'Ars magna lucis et umbrae' du Athanasius Kircher," *Archivum historicum Societatis Iesu* 62 (1993): 254, n. 30.

7 I have discussed Kircher's career in greater detail in Paula Findlen, *Possessing Nature: Museums, Collecting, and Scientific Culture in Early Modern Italy* (Berkeley: University of California Press, 1994); and "Scientific Spectacle in Baroque Rome: Athanasius Kircher and the Roman College Museum," *Roma moderna e contemporanea* 3, no. 3 (1995): 625–65. English-speaking readers can also consult P. Conor Reilly, S. J., *Athanasius Kircher S. J., Master of a Hundred Arts 1602–1680* (Wiesbaden: Edizioni del Mondo, 1974); and Joscelyn Godwin, *Athanasius Kircher: A Renaissance Man and the Quest for Lost Knowledge* (London: Thames and Hudson, 1979). Other literature on Kircher is cited throughout this essay. For a broader overview of Catholic science, see William J. Ashworth Jr., "Catholicism and Early Modern Science," in *God and Nature: Historical Essays on the Encounter between Christianity and Science*, ed. David Lindberg and Ronald Numbers (Berkeley: University of California Press, 1986), pp. 136–66.

8 Johann Koestler, *Physiologia Kircheriana Experimentalis* (Rome, 1675), pref., in Lynn Thorndike, *A History of Magic and Experimental Science*, 8 vols. (New York: Columbia University Press, 1923–58), 7:569.

Kircher, along with Bacon and Boyle, as one of the great scholars of his generation, a "Hercules ... among writers."[9] Appearing one year after the publication of Newton's *Principia*, Morhof's guide to the learned world celebrated men of letters whose reputations had earned them fame throughout Europe. Kircher, not Newton, was among those immortal scholars.

Others admired Kircher for his generosity with crucial information for their own work. The Jesuit astronomer Giovanni Battista Riccioli named a large lunar crater after him in appreciation for the quantity of astronomical data that Kircher had procured through his missionary contacts.[10] Other authors dedicated their publications to Kircher in acknowledgment of his status as a fantastically erudite scholar. In 1675, when the Tuscan court naturalist Francesco Redi initiated a correspondence with Kircher in order to debate his views on such topics as spontaneous generation, the efficacy of snakestones, and other forms of sympathetic medicine, he paid homage to Kircher as "the most celebrated man of letters in Europe."[11] While Redi disagreed with virtually ever aspect of Kircher's natural philosophy, he nonetheless respected the intellectual authority that his Jesuit opponent commanded.

The wide range of opinion about Kircher's natural philosophy, in conjunction with the shifting views of the importance of Newton's work later in the century, reminds us that the reputation of these two scholars was by no means cemented in their own day but is an artifact of a later period. Conflicting accounts of Newton also circulated. For the first half of his life, he enjoyed a largely local reputation as a talented mathematician; only later did his fame travel outside of England. Contemporaries alternately viewed Newton as a curious, solitary genius or as a proud philosopher who did not hesitate to take credit for others' insights, principally those of Hooke and Leibniz, in order to further his own reputation.

Nagging questions also abounded about the success of Newton's

---

9 Daniel Georg Morhof, *Polyhistor literarius, philosophicus et practicus* (1688; Lubeck, 1747), 2:156. We should certainly see Morhof's work as an example of what Mencken attacked in *De charlatanaria eruditorum*.

10 Fletcher, "Astronomy," p. 67.

11 Clelia Pighetti discusses Boyle's admiration for Kircher (over the members of the Accademia del Cimento) in her *L'influsso scientifico di Robert Boyle nel tardo '600 italiano* (Milan: Franco Angeli, 1988), p. 95. On Redi's views of Kircher, see Pontificia Università Gregoriana, Rome, *Kircher*, MS 566, f. 40 (Redi to Kircher, Florence, June 24, 1675). The debates between these two scholars are discussed in Martha Baldwin, "The Snakestone Experiments: An Early Modern Medicial Debate," *Isis* 85 (1996): 394–418.

natural philosophy. Had Newton introduced occult forces into a mechanistic account of the heavens in such a way as to cast doubt upon his entire system, creating a glaring, logical inconsistency? Or had he indeed solved the problem of understanding the forces governing the universe by allowing that one could describe a phenomenon without knowing its cause? Was his *experimentum crucis* really decisive, as Newton claimed? Many, including Huyghens, said no.[12] William Stukeley's view of Newton in 1752, as "the great solar orb shining with its own light, and diffusing his beamy influence thro' the whole system of the arts and sciences," reflects the mid-eighteenth-century deification of Newton rather than a seventeenth-century perspective. In this earlier period, Newton enjoyed a reputation in the making that did not fully solidify until a decade or so before his death. Surely it is no coincidence that the crystallization of Kircher's reputation as the most ridiculous of the late Renaissance encyclopedists and the emergence of Newton as the first man of science both occurred in the same period.

Until recently, the dominant view of the Scientific Revolution has portrayed it as an epoch that could not, by definition, take seriously the intellectual claims of "losers" in the contest for knowledge.[13] Newton created a new physics, while Kircher tenaciously clung to the old; we have built on the former's work ever since and forgotten the weighty tomes of the latter without any regret. As the centerpiece of the late seventeenth-century Royal Society and as a cultural hero of the Enlightenment, Newton further embodied the forward-looking image of Protestant science embedded in many of the early histories of science. By contrast, Kircher's membership in the Society of Jesus and his position at the center of a network of largely Catholic intellectuals made him the embodiment of a stagnant, post-Galilean culture of science, fit only for discussion in the margins of such works as Frances Yates's *Giordano Bruno and the Hermetic Tradition* and in the pages of Umberto Eco's historical novels.[14]

12 Caspar Hakfoort, "Newton's Optics: The Changing Spectrum of Science," in Fauvel et al., *Let Newton Be*, p. 92.
13 An important exception among surveys of early modern science is Charles Webster, *From Paracelsus to Newton: Magic in the Making of Modern Science* (Cambridge: Cambridge University Press, 1982); see also Allen G. Debus, *Man and Nature in the Renaissance* (Cambridge: Cambridge University Press, 1979).
14 For Yates's discussion of Kircher, see her *Giordano Bruno and the Hermetic Tradition* (Chicago: University of Chicago Press, 1964), pp. 416–23; and *The Rosicrucian Enlightenment* (London: Routledge, 1972), pp. 274–5. See Umberto Eco, *The Name of the Rose*, trans. William Weaver (New York: Harcourt Brace Jovanich, 1983), p. xvi; and *Foucault's Pendulum*, trans. William Weaver (New York: Harcourt Brace Jovanich, 1989),

In this essay, I would like to take seriously the challenge offered by Yates almost thirty years ago. In an often-cited essay on the role of Hermeticism in Renaissance science, Yates made an eloquent plea for a more historical approach to the work of early modern natural philosophers, writing

> I would thus urge that the history of science in this period, instead of being read solely forwards for its premonitions of what was to come, should also be read backwards, seeking its connections with what has gone before. A history of science may emerge from such efforts which will be exaggerated and partly wrong. But then the history of science from the solely forward-looking point of view has also been exaggerated and partly wrong, misinterpreting the old thinkers by picking out from the context of their thought as a whole only what seems to point in the direction of modern developments. Only in the perhaps fairly distant future will a proper balance be established in which the two types of inquiry, both of which are essential, will contribute their quota to a new assessment.[15]

Since the late 1960s, almost three decades of Newton scholarship and a lively debate about the contributions of the occult sciences to the transformation of early modern natural philosophy offer a great deal of material with which to revise our standard portrait of seventeenth-century science. If Newton was the "last of the magicians," as John Maynard Keynes wrote in 1946, "the last great mind which looked out on the visible and intellectual world with the same eyes as those who began to build our intellectual inheritance rather less than 10,000 years ago," then we ought to compare his activities more carefully against those of other Renaissance magi.[16] Kircher provides an exemplary companion to Newton in this task, not only due to his prominence in baroque intellectual life but also due to the commonality of their interests. Understanding in greater detail the similarities as well

---

pp. 284, 439, 441, 525, 580. In *The Island of the Day Before*, trans. William Weaver (New York: Harcourt Brace and Company, 1995), Eco makes Kircher one of the main protagonists on his shipwrecked cabinet of wonders, calling him Father Caspar Wunderdrossel.

15 Frances Yates, "The Hermetic Tradition in Renaissance Science," in *Art, Science and History in the Renaissance*, ed. Charles S. Singleton (Baltimore: Johns Hopkins University Press, 1967), p. 270.

16 John Maynard Keynes, "Newton the Man" (as quoted 1947), in B. J. T. Dobbs, *The Foundations of Newton's Alchemy: or, "The Hunting of the Greene Lyon"* (Cambridge: Cambridge University Press, 1975), p. 13. As is apparent throughout this chapter, my understanding of Newton's work is highly indebted to the work of my colleague Jo Dobbs.

as the divergence between Kircher's and Newton's natural philosophies provides an interesting way to discuss the varieties of late Renaissance encyclopedism and different responses to new methodologies in the seventeenth century.[17]

Sapientia

The original story of the relations between Kircher and Newton begins with a piece of gossip circulating among scholars a few years after Newton's death, as his reputation increased on the continent. In 1738 Voltaire recorded the following anecdote in his *Elements of Newton's Philosophy*: "I have often heard it said that it was from Kircher that Newton had drawn that discovery of light and sound. In effect, Kircher in his *Great Art of Light and Shadow*, and in other books besides, calls sound the ape of light."[18] Voltaire was curious enough to compare passages in Newton's *Opticks* with those in the Jesuit's volume. He ultimately concluded that the story had no validity, because he discerned too many differences between the two philosophers' viewpoints to credit Kircher's qualitative physics with Newton's quantitative discoveries.

Voltaire's quest to understand the origins of scientific knowledge offers us an important glimpse of an intellectual world in which philosophers did not automatically separate the work of Newton from the activities of his less modern predecessors but envisioned greater continuities. At the dawn of the Enlightenment, it was not yet impossible to imagine that a Jesuit polymath might have anticipated a key element of Newton's natural philosophy. Thus Voltaire did not dismiss the story out of hand but felt compelled to place the *Opticks* next to the *Great Art of Light and Shadow* (1646) in order to discern the truth of the matter.

We do not often imagine Voltaire as a reader of Kircher's large, sprawling studies of virtually every form of knowledge. And yet he was not the only eighteenth-century scholar to enter the baroque labyrinth. Leibniz continued to remember Kircher fondly to the end of his life, even as his criticisms of his fellow German's scholarly apparatus increased. Later in the century, Goethe and Priestley both found

17 This approach harmonizes well with the comments on late humanism made by Anthony Grafton in his *Defenders of the Text: The Traditions of Scholarship in an Age of Science, 1450–1800* (Cambridge, Mass.: Harvard University Press, 1991).
18 Voltaire, *Elémens de la philosophie de Neuton* (Amsterdam, 1738), pp. 178–9. This passage was brought to my attention in José Alfredo Bach, "Athanasius Kircher and His Method: A Study in the Relations of the Arts and Sciences in the Seventeenth Century" (Ph.D. diss., University of Oklahoma, 1985), p. 91, n. 47.

much to commend in Kircher's hefty tomes.[19] In the decades following Kircher's death, prominent scholars continued to consider the possibility that he had contributed something to the modern frame of knowledge. It was a legitimate question to ask, even if Voltaire ultimately could not reconcile Kircher's animistic view of nature, which linked sound and light due to their shared qualities, with Newton's mathematical interpretation that established a precise connection between the diatonic scale and the spectrum of light, both of which divided into seven segments.[20] "A naturalist would scearce expect to see ye science of [colors] become mathematicall," Newton informed Henry Oldenburg, "& yet I dare affirm that there is as much certainty in it as in any other part of Opticks."[21] Kircher was precisely the sort of "naturalist" whom Newton had in mind.

Had Voltaire probed the relationship between Kircher and Newton further, beyond the question of a single discovery linking the fields of optics and acoustics, he might have arrived at some very interesting conclusions. Certainly these two natural philosophers had dramatically different perspectives on the nature and meaning of mathematics. It is hard to reconcile Kircher's mystical Neoplatonism, which treated numbers as cultural artifacts concealing divine meaning, with Newton's use of mathematics as a language in which to express precise, observable relationships among physical phenomena. Yet the commonality of their interests, as evidenced in their choice of topics and in the goals they strove to achieve, is striking. Reframing the question brings the work of Kircher and Newton in surprisingly close proximity.

Both natural philosophers emphasized an encyclopedic and syncretic approach to knowledge; they strove to uncover the universal forces of nature that lay hidden in an ancient and somewhat opaque debate about "occult qualities."[22] Kircher and Newton wrote on a

---

19 For Leibniz's relationship with Kircher, see Paul Friedlander, "Athanasius Kircher und Leibniz. Ein Beitrag zur Geschichte der Polyhistorie im XVII. Jahrhundert," *Rendiconti della Pontificia Accademia Romana di Archeologia* 13, nos. 3–4 (1937): 229–47; on Goethe and Priestley, see Bach, "Athanasius Kircher," pp. 61, 65. Goethe's praise of Kircher came in the context of his critique of Newton's optics, while Priestley discussed Kircher's work as part of his history of electricity.

20 See Penelope Gouk, "The Harmonic Roots of Newtonian Science," in Fauvel et al., *Let Newton Be*, pp. 101–25; and Tonino Tornitore, "L'origine delle sinestesie. Mersenne, Kircher e le corrispondenze fra suoni e colori," *Intersezioni* 8 (1987): 21–51.

21 *The Correspondence of Isaac Newton*, ed. H. W. Turnbull, J. P. Scott, A. Rupert Hall, and Laura Tilling, 7 vols. (Cambridge: Cambridge University Press, 1959–77), 1:96.

22 For a broader discussion of this subject, see Keith Hutchison, "What Happened to Occult Qualities in the Scientific Revolution?" *Isis* 73 (1982): 233–53; and "Supernaturalism and the Mechanical Philosophy," *History of Science* 21 (1983): 297–333.

variety of different topics whose totality, in their minds, was more important than any individual part. Just as we cannot fully appreciate the *Principia* without an understanding of Newton's alchemy, optics, and biblical chronologies, we also cannot evaluate works such as Kircher's *Subterranean World, Great Art of Light and Shadow, Noah's Ark* (1675), and *Tower of Babel* (1679) in isolation.[23] Both philosophers engaged in a lifelong project of knowledge that drew on virtually every available discipline that they deemed useful to the enterprise of knowing God through mastery of nature's laws. The end result was a mosaic of ancient and modern knowledge, two fantastic baroque creations that seem to teeter on the edge of oblivion and, at times, were in danger of lapsing into utter chaos. We see this chaos more easily in Kircher than in Newton because the latter was more circumspect in what he chose to publish. Yet the problem nonetheless was there and at times consumed Newton at the height of his scientific work.

During the 1660s, as a young Newton embarked upon his studies at Cambridge that would lead him to such discoveries as his law of universal gravitation and the invention of the calculus, and to probe the mysteries of alchemy for some forty years, Kircher was at the height of his career. Already famed for his development of the Roman College museum, filled with mechanical inventions, natural wonders, and antiquities, Kircher was in the process of completing a series of encyclopedias that would reunify all knowledge. His *Egyptian Oedipus*, tartly labeled "one of the most learned monstrosities of all times" by historian Frank Manuel, had already established Egypt as the font of all wisdom.[24] Drawing on Hermetic and Neoplatonic texts, Kircher posited that the symbolic language of the Egyptians represented the closest remaining approximation of Adamic language. As Piero Valeriano observed in his influential *Hieroglyphica* (1556), "to speak hieroglyphically is nothing else but to disclose the nature of things divine and humane."[25]

According to Kircher, the central proof of the divinity of Hermetic thought lay in the significance of the Trinity and Unity in its philosophy. Undeterred by the pronouncements of Isaac Casaubon, who in 1614 had unmasked the *Corpus Hermeticum* as an early Christian for-

---

23 This point has been made strongly in B. J. T. Dobbs, *The Janus Faces of Genius: The Role of Alchemy in Newton's Thought* (Cambridge: Cambridge University Press, 1991); and in Thomas Leinkauf, *Mundus combinatus. Studien zur Struktur der barocken Universalwissenschaft am Beispiel Athanasius Kirchers S J* (Berlin: Akademie Verlag, 1993).

24 Frank E. Manuel, *The Eighteenth Century Confronts the Gods* (Cambridge, Mass.: Harvard University Press, 1959), p. 190. Also quoted in Baldwin, "Athanasius Kircher," p. 7, n. 1.

25 In Erik Iversen, *The Myth of Egypt and Its Hieroglyphs* (Copenhagen: Gad, 1961), p. 73.

gery purporting to anticipate Christianity, Kircher continued to see these writings as an essential foundation for any pious philosophy. "All knowledge was reduced with good method by that celebrated Mercurius Trismegistus in order to be handed down to posterity."[26] Kircher envisioned Rome as a unique center within which to unlock the mysteries of Hermetic knowledge, inscribed on the obelisks and ruins surrounding the city and recorded by God in nature.

The significance of Kircher's philosophical musings rested on an equally profound claim that had vaulted him to prominence in the 1640s: his ability to decode hieroglyphs.[27] In the eyes of many contemporaries, the project of restoring Egyptian wisdom entailed nothing less than an effort to renovate the lost arts of communication that linked human and divine languages. A philosopher who could read the hieroglyphs spoke directly to the Diety. Holding this precious, fragile knowledge in his hands, Kircher was a towering figure indeed to his amazed colleagues, who had begun to despair that they could ever possess such knowledge. "Already time and blight have almost consumed the Hieroglyphs, and practically destroyed the emblems," wrote Johannes Vondelius to Kircher; "the lack of foreign languages will little by little obfuscate our intellects and deprive us of knowledge."[28] Certainly it is no exaggeration to suggest that, as long as scholars believed that Kircher had succeeded in his quest for Egypt, they credited him with a discovery as significant, in their eyes, as those which Newton subsequently enjoyed. If language was the key to knowledge, then Kircher's command of so many ancient and modern tongues made him a unique resource for the learned community. Such talents earned him the coveted role of obelisk consultant to popes and cardinals, who enlisted Kircher in their efforts to translate the hieroglyphic wisdom of these ancient monuments for the benefit of the entire Catholic world.

---

26 Athanasius Kircher, *Obeliscus Pamphilius* (Rome, 1650), p. 167. For a broader discussion of Kircher's participation in seventeenth-century Egyptomania, see Giovanni Cipriani, *Gli obelischi egizi. Politica e cultura nella Roma barocca* (Florence: Olschki, 1993); Valerio Rivosecchi, *Esotismo in Roma barocca. Studi su Padre Kircher* (Rome: Bulzoni, 1982), esp. pp. 47–76; and J.-F. Marquet, "La quête isiaque d'Athanase Kircher," *Les Etudes philosophiques* 2–3 (1987): 227–41.

27 Kircher could not actually read the hieroglyphs, which he interpreted largely in symbolic and allegorical terms, drawing on Renaissance Neoplatonic traditions that valued this form of communication above all others. And yet we should not simply dismiss him as a dilettante in this endeavor since he had understood the connection between Coptic and ancient Egyptian languages, and the phonetic function of hieroglyphs, as outlined in the *Prodomus coptus sive Aegyptiacus* (Rome, 1636) and discussed more extensively in *Lingua Aegyptiaca restituta* (Rome, 1643).

28 Pontificia Università Gregoriana, *Kircher*, Ms 563 (IX), f. 311r.

Kircher coupled his quest for *prisca theologia* in the writings of Hermes Trismegistus with an equally urgent search for the arcane forces governing the universe. By the 1630s, he had identified magnetism as the key to unlocking all of nature's secrets. He developed his ideas about universal magnetism into a full-blown philosophy in the *Magnetic Kingdom of Nature* (1667), whose frontispiece proclaimed: "The world is bound by secret knots."[29] For Kircher, magnetism encompassed all actions in nature that did not emanate from any visible or manifest cause. Just as the unseen hand of God inscribed the hieroglyphs, divine wisdom governed nature in ways that the human intellect found difficult to comprehend. The loadstone became both the source of many forces in the natural world, and an analogy for other unseen powers that did not emanate directly from the loadstone but imitated its occult properties. It was, as Kircher wrote in his *Loadstone*, "the epitome of all nature."[30]

Like many early modern natural philosophers, Kircher was deeply committed to identifying the primal forces in nature through which the Deity molded the universe. The great chain of being was no convenient metaphor but a literal description of the connections between macrocosm and microcosm. In choosing magnetism as the central force in the universe – "the only guide and key to all motion whatsoever"[31] – Kircher identified himself as a follower of such sixteenth-century philosophers as William Gilbert, Giovan Battista della Porta, and Johannes Kepler. While repudiating Gilbert's vision of the Earth as a magnet and Kepler's heliocentrism, and attacking della Porta's impiety as a magus who put human invention before divine creation, Kircher nonetheless identified strongly with the tradition of natural magic that imagined the universe to be controlled by animated forces that shared a common source. *Natura clavis una est*, he proclaimed in the *Subterranean World*: nature has but one key.[32] Attraction and repul-

29 My discussion of Kircher's views on magnetism is indebted to the excellent work of Martha Baldwin on this subject. See her "Athanasius Kircher"; and also "Magnetism and the Anti-Copernican Polemic," *Journal of the History of Astronomy* 16 (1985): 155–74. The frontispiece to Kircher's *Magneticum regnum naturae* (Rome, 1667) is reproduced and discussed in John Henry, "Newton, Matter, and Magic," in Fauvel et al., *Let Newton Be*, p. 131. Henry does not elaborate on the comparison between Kircher's and Newton's views of active principles, but it is certainly intriguing that he chose this illustration as a complement to a discussion of Newton. Piyo Rattansi also does the same in his chapter on "Newton and the Wisdom of the Ancients," in Fauvel et al., *Let Newton Be*, where he reproduces Kircher's image of Pan in the *Oedipus Aegyptiacus* (p. 184).
30 Kircher, *Magnes sive de arte magnetica*, proem, in Baldwin, "Athanasius Kircher," p. 463.
31 Kircher, *Magnes* (Cologne, 1643), proem.
32 Kircher, *Mundus subterraneus* (Amsterdam, 1665), II, MS, II, p. 111.

sion, sympathy and antipathy, structured the relations among all objects in the universe. God himself, Kircher proclaimed, was a great magnet.[33] Accordingly, the discovery of nature's occult virtues entailed nothing less than the discernment of divine activity in the cosmos.

The close relationship between nature and divinity further underscored the importance of religious belief in the pursuit of natural philosophy. As a Jesuit, Kircher presented faith as an important prerequisite for reading the Book of Nature. Languages, objects, and experiments provided the tools of science but knowledge of God defined the reason to study nature. Kircher's image of natural philosophy as a religious calling influenced his talented Danish friend, the anatomist Nicolaus Steno, who recorded the following passage from Kircher's *Loadstone* in his 1659 commonplace book, as a medical student in Copenhagen: "Only he whom God and nature have ordained for it should be regarded as fitted and destined for this study."[34] Religion delimited an important boundary in Kircher's conception of science. It provided the ingredients for a correct interpretation of nature while also demarcating a realm of knowledge that was especially difficult for humans to know. Writing in the decades following Galileo's condemnation for openly advocating Copernicanism, Kircher staunchly affirmed the primacy of Scripture over nature. "Whosoever glorifies the dignity of the name of Christ, let him be especially careful not to yearn to philosophize beyond the limits of sacred canonical proscriptions."[35]

Kircher's pious conception of natural philosophy influenced not only his selection of Egypt and universal magnetism as two of his primary subjects but also shaped his investigations of light. "Moreover, light itself is in God, and is God," he wrote in the *Great Art of Light and Shadow*.[36] Light was the symbol of the intellect, yielding a spectrum of colors that ranged from the bright purity of God (white) to the darkness of all insensible creations (black), with humanity painted the color of blood (red) to recall its sensible yet impure status. It also functioned as a "celestial loadstone," moving the heavenly bodies as the magnet propelled earthly bodies.[37]

Kircher's explorations of light cannot be viewed within the narrow

---

33 Baldwin, "Athanasius Kircher," p. 454.
34 Niels Stensen, *A Danish Student in His Chaos-Manuscript 1659*, ed. H. D. Scheplern, Acta Historica Scientiarum Naturalium et Medicinalium 38 (Copenhagen: University of Copenhagen Library, 1987), p. 15.
35 Kircher, *Oedipus Aegyptiacus*, in Baldwin, "Athanasius Kircher," p. 470.
36 Kircher, *Ars magna lucis et umbrae*, p. 800.
37 Baldwin, "Athanasius Kircher," pp. 24, 145.

confines of the science of optics, which occupied only a minor portion of his treatise.[38] He juxtaposed the science of light with a fascination with optical illusions and shadowy demonstrations. Using his famous magic lanterns, he created a vast morality play in the rooms of the Roman College, where students and visitors watched images of Christ and Satan, death and resurrection, dance across the walls of the Jesuit compound. While interested in the mathematical properties of light, Kircher did not make them a central topic in his investigations. More important were light's emblematic qualities, which underscored its importance as a universal symbol of the varied forms of perception that allowed the Christian philosopher to scale the ladder of wisdom.[39]

Kircher saw the ancient and natural worlds as unique repositories of God's messages. Each science provided a different means of understanding the wisdom of the Deity through discernment of the patterns that linked all forms of knowledge to their original source. "Unity is the essence of God," proclaimed Kircher in the *Egyptian Oedipus*.[40] Thus the search for the first language and attempts to create a universal language fueled the quest for nature's essential qualities. Light and shadow, universal magnetism and universal harmony, and theories of subterranean fires at the center of the Earth that shaped the activity on its surface all combined to form a natural philosophy that was Aristotelian and Neoplatonic in origin but capable of accommodating many new forms of learning.[41] This grand synthesis reflected many themes common to seventeenth-century science. As we shall see, they were very much on Newton's mind as he began to develop his own natural philosophy.

Experientia

In many respects, Isaac Newton shared the intellectual world of Kircher in the early 1660s. Yet within a decade he had launched a series of investigations that would ultimately make him a rather different sort of natural philosopher. Like Kircher and the English Renaissance magus John Dee, Newton enjoyed a reputation as a talented

38 Corradino, "Ars magna lucis et umbrae," pp. 252–3. See also Bach, "Athanasius Kircher," which focuses on Kircher's relationship to early modern optical theories.
39 This subject is discussed well in William B. Ashworth Jr., "Light of Reason, Light of Nature: Catholic and Protestant Metaphors of Scientific Knowledge," *Science in Context* 3 (1989): 89–107, esp. 96. For an interesting discussion of the religious significance of the metaphysics of light, see Dino Pastine, *La nascita dell'idiolatria. L'Oriente religioso di Athanasius Kircher* (Florence: La Nuova Italia, 1978), pp. 27–30.
40 Kircher, *Oedipus Aegyptiacus* (Rome, 1652–4), 3:6.
41 Leinkauf's *Mundus combinatus* provides an excellent discussion of the ancient, medieval, and early Renaissance origins of Kircher's natural philosophy.

inventor, sketching perpetual-motion machines in his notebooks and creating clocks, sundials, "speaking trumphets," and "ground and polished glasses for...all kind of optical purposes." Newton's "strange inventions," as one youthful companion dubbed them, mirrored well the experimental machines collected and displayed by Kircher in the Roman College museum, although they eventually yielded a much more famous invention: the reflecting telescope.[42] Such artifacts belonged primarily to the domain of mathematical magic in which the Jesuits figured prominently; during this period we find Newton reading John Wilkins's *Mathematical Magick*.

Natural magic was an essential starting point for Newton's introduction to the study of nature. While flirting with judicial astrology and exploring the project of creating a universal language that was as popular in Cambridge as in Rome, Newton briefly shared Kircher's fascination with the works of the Renaissance magus della Porta. This interest led him to explore some of the classic topics that occupied natural philosophers in the mid-seventeenth century. It was probably his encounter with della Porta's *Natural Magic* that led Newton to include the following queries in his student notebook of 1664–5: "Whether magnetic rays will blow a candle" and "Whether a lodestone will not turn around a red hot iron fashioned like a windmill sails, as the wind does to them."[43] About five years later, in the midst of his alchemical investigations, Newton identified an agent called *magnesia* as the universal life principle. "This & only this is the vital agent diffused through all things that exist in the world."[44]

Ultimately, Newton did not find the secrets of the universe in the operations of the loadstone. Magnetism, like gravity and electricity, counted as one of the "various kinds of natural forces" that organized the cosmos. By 1687, he had concluded that magnetism was a *less universal* force than gravity and less powerful in its ability to act at a distance. "Some bodies are attracted more by the magnet; others less; most bodies not at all."[45] Magnetism attracted roughly at the cube of

---

42 William Stukeley, *Memoirs of Sir Isaac Newton's Life* (London: Taylor and Francis, 1936), p. 57; J. E. McGuire and Martin Tamny, *Certain Philosophical Questions: Newton's Trinity Notebook* (Cambridge: Cambridge University Press, 1983), pp. 313, 378; and Richard S. Westfall, *Never at Rest: A Biography of Isaac Newton* (Cambridge: Cambridge University Press, 1980), pp. 60, 91–2.

43 McGuire and Tamny, *Certain Philosophical Questions*, pp. 377, 379.

44 Newton, "Propositions" (ca. 1669), in Dobbs, *Janus Faces*, p. 24.

45 Newton, "Conclusio" (written around spring 1687), in *Unpublished Scientific Papers of Isaac Newton*, ed. A. Rupert Hall and Marie Boas Hall (Cambridge: Cambridge University Press, 1962), p. 330; *Sir Isaac Newton's Mathematical Principles of Natural Philosophy and His System of the World*, trans. Andrew Motte, ed. Florian Cajori

the distance in contrast to gravity attracting at the square (according to "some rude observations," Newton wrote apologetically in his *Principia*).[46] This conclusion diminished magnetism in the eyes of Newton, who subsequently concentrated his efforts on explaining gravity and his laws of optics. Yet, while rejecting magnetism, Newton continued to explore the power of alchemical *magnesia*, a cornerstone of his theory of universal vegetation, and upheld the idea of spontaneous generation. "All things are corruptible, all things are generable," he wrote in the 1670s.[47] Thus in his nonmathematical investigations, Newton developed a philosophy that echoed important aspects of Kircher's ideas.

The commonality between Kircher's and Newton's philosophies of nature is most clearly evident in their alchemical investigations. Toward the mid-1670s, Newton recorded the following description of the *cauda pavonis*, an elegant alchemical creation known as the peacock's tail: "They grow in these glasses in the form of a tree, and by a continued circulation the trees are dissolved again with the work into a new mercury." Describing his fascination with this process, he wrote: "I reckon this a great secret in Alchemy."[48] While Newton privately re-created the secrets of alchemy, Kircher publicly exposed them in the laboratory of the Roman College museum. He entertained visitors such as Queen Christina of Sweden with his famous "Hermetic experiments." Among them was an experiment to create the peacock's tail.[49] Kircher used this demonstration to support his claims about the possibility of spontaneous generation. Both he and Newton believed it to be a great secret of nature.

The location of both philosophers' alchemical activities, however, delimited an important difference in their approach to this ancient discipline. Deeply steeped in ancient alchemical and Hermetic writings, Kircher and Newton arrived at fairly different conclusions regarding the power of this discipline to yield insight into nature. For Newton, alchemy remained a privileged secret which he shared charily with a limited circle of friends, among them, John Locke and Robert Boyle.[50] He deeply respected the power of this ancient wisdom.

    (Berkeley: University of California Press, 1947), p. 414 (bk. III, prop. VI, theorem VI, cor. V).
46 Newton, *Principia*, p. 414 (trans. Motte).
47 In Dobbs, *Janus Faces*, p. 56.
48 Newton, *Clavis* (mid-1670s), in Gale E. Christianson, *In the Presence of the Creator: Isaac Newton and His Times* (New York: Free Press, 1984), p. 231.
49 Giorgio de Sepi, *Romani Collegii Societatis Iesu Musaeum Celeberrimum* (Amsterdam, 1678), p. 45.
50 Richard S. Westfall, "Newton and Alchemy," in *Occult and Scientific Mentalities in the Renaissance*, ed. Brian Vickers (Cambridge: Cambridge University Press, 1984), p. 315;

Kircher instead enjoyed publicizing alchemists' fraudulent claims. He included lengthy criticisms of alchemy in his *Egyptian Oedipus* and *Subterranean World*. Newton created an alchemical dictionary for his personal use; Kircher published one as part of his campaign to reform alchemy. Denying that the creation of the Philosophers' Stone and the transmutation of base metals into gold could ever really occur, Kircher announced to his readers that he had witnessed the activities of many alchemists and never seen them come to fruition.[51] Alchemy, he argued, bordered on being a personal form of revelation, which was incompatible with the teachings of the Catholic Church. This, of course, was precisely why Newton found it so appealing.

Method also provided an important point of departure for both natural philosophers. During the 1660s Newton embarked on a careful reading of the works of Descartes and Boyle, both of whom were known and cited by Kircher.[52] Yet Kircher's comparison of his *Great Art of Knowledge* (1669) to Descartes's *Discourse on Method*, and his insistence that his experiments adhered to the rigorous procedures to authenticate knowledge set forth in Boyle's many treatises, exhibited an incomplete understanding of the changes in knowledge then underway. Kircher's views on method relied too heavily on esoteric systems of memory derived from such scholars as the thirteenth-century occultist Raymond Lull to approach the call for a new form of inductive logic. Lullism enjoyed great popularity in the seventeenth century.[53] Newton possessed a copy of Johann Heinrich's *Key to the Lullian Art, and True Logic* (1609) and some of the works of Lull, but he did not organize knowledge according to Lullian principles.[54] Instead, his "rules of reasoning" owed a great deal to his youthful

and Jan Golinski, "The Secret Life of an Alchemist," in Fauvel et al., *Let Newton Be*, p. 155.

51 See Kircher, *Oedipus Aegyptiacus*, tome II, vol. III, pp. 427, 433; and *Mundus subterraneus*, 2:232–325 (alchemical dictionary on p. 325). Newton's alchemical dictionary was also created in the 1660s; Golinski, "Secret Life," p. 156. Kircher's attitudes toward alchemy are placed in their confessional context in Martha Baldwin, "Alchemy in the Society of Jesus," in *Alchemy Revisited*, ed. Z. R. W. M. von Martels (Leiden: Brill, 1990), pp. 181–7.

52 McGuire and Tamny, *Certain Philosophical Questions*, pp. 23–4.

53 Frances Yates, *The Art of Memory* (Chicago: University of Chicago Press, 1966); and Paolo Rossi, *Clavis universalis. Arti della memoria e logica combinatoria da Lullo e Leibniz* (Milan and Naples: Ricciardi, 1960).

54 John Harrison, *The Library of Isaac Newton* (Cambridge: Cambridge University Press, 1978), p. 85 (Johann Heinrich, *Clavis artis Lullianae, et verae logices*, Argentorati, 1609), pp. 183–4 (works of Raymond Lull). The Lullian system of memory emphasized symbolic associations among topics, in contrast to the Cartesian emphasize on reason as the supreme tool of knowledge.

infatuation with Descartes, even if he ultimately repudiated the French philosopher's subordination of sense experience to reason.

In establishing experimental knowledge as a crucial element of sound natural philosophy, Newton participated in a more widespread movement to make experimentation the most important test of truth.[55] For him, it was the most secure means of establishing a portrait of nature that approached certainty, when combined with the tools of mathematical analysis. "For since the qualities of bodies are only known to us by experiments, we are to hold for universal all such as universally agree with experiments," he wrote in Rule 3 of the *Principia*. He exhorted natural philosophers "not to relinquish the evidence of experiments for the sake of dreams and vain fictions of our own devising."[56] Only fifteen years earlier, in his paper on optics published in the *Philosophical Transactions* on February 19, 1672, Newton had borrowed Hooke's term, *experimentum crucis*, to define the sort of demonstration that lived up to these expectations. Experimenting was a source of unassailable physical knowledge.

Newton was not the only seventeenth-century philosopher convinced that his experiments were decisive. Many contemporaries did not initially find Newton's prism experiment as decisive as he did; white light was not commonly viewed as heterogeneous until the early eighteenth century. In 1672, the second edition of Kircher's *Great Art of Light and Shadow* was only two years old and enjoying great popularity on the Continent; scholars and learned patrons still built *camera obscura* from instructions in the Jesuit's treatise.[57] Kircher also presented himself as an experimental philosopher whose knowledge was sound because it was tested in public view. "I have introduced nothing, however small, into this book which could not be, so far as lies within my power, personally tested and established," he informed readers of his *Loadstone*.[58] Visitors to his museum enjoyed an endless array of experiments that proved his theories about everything from the impossibility of perpetual motion and the vacuum to the certainty of spontaneous generation and nature's attractive powers. In his *Sub-*

---

55 The fundamental point of departure on this subject for seventeenth-century England is Steven Shapin and Simon Schaffer, *Leviathan and the Air-Pump: Hobbes, Boyle and the Experimental Life* (Princeton: Princeton University Press, 1985).

56 Newton, *Principia*, bk. III, in *Newton's Philosophy of Nature: Selections from His Writings* (New York: Hafner, 1953), pp. 3–4.

57 One example: Pontificia Università Gregoriana, *Kircher*, MS 555 (I), ff. 292–3 (Paris, July 6, 1666). This letter regards the construction of a camera obscura for the duke of Holsten.

58 Kircher, *Magnes*, in R. J. W. Evans, *The Making of the Habsburg Monarchy, 1550–1700: An Interpretation* (Oxford: Clarendon Press, 1985), p. 338.

terranean World, for instance, Kircher described a simple demonstration of electrical attraction as a "truly beautiful experiment, by which spectators all over my Museum usually are entertained."[59]

Such language suggests one of the important differences between Kircher's and Newton's views of experimentation. Newton experimented with virtually no reference to any audience. His experiments were not "crucial" because they were witnessed by nobles and gentleman, but because they spoke God's truth through his mortal messenger.[60] In essence, they were private revelations that he selectively made available to the scholarly community. Accordingly, they were not intended as entertaining spectacles, though by the 1730s scientific popularizers treated Newton's prism experiments as if they were Kircher's magic lantern.[61]

In contrast to Newton's reticence about his experiments, Kircher participated fully in the social world of experimentation familiar to other members of the Royal Society such as Boyle. When challenged by other experimenters, who could not replicate his results, Kircher produced legions of witnesses in Rome and at the imperial court in Vienna who could validate his claims.[62] He, too, belonged to an intellectual community that would support his findings, affirming his ability to produce "experimental certainty."[63] That certainty, however, emanated from his intellectual standing at the center of the Catholic world of knowledge as opposed to the certainty Newton enjoyed as a result of his personal relationship with God.

As the foregoing examples suggest, experimentation included a wide array of practices in the mid-seventeenth century that did not simply mirror the confessional divisions of the day. Although hindsight made it easy to distinguish good work from bad, in the initial enthusiasm about experimenting all possibilities merited serious consideration. For these reasons, Newton's senior colleagues in the Royal Society took a keen interest in Kircher's work, just as they would later do with Newton's. Writing to Benedict de Spinoza in 1665, Henry Oldenburg argued that Kircher deserved more credit than he often

---

59 Kircher, *Mundus subterraneus*, 2:77.
60 In this respect, Newton represents a different sort of experimenter than the one embodied by Boyle in Shapin and Schaffer, *Leviathan and the Air-Pump*; and Steven Shapin, *A Social History of Truth: Civility and Science in Seventeenth-Century England* (Chicago: University of Chicago Press, 1994).
61 Barbara Stafford's *Artful Science: Enlightenment Entertainment and the Eclipse of Visual Education* (Cambridge, Mass.: MIT Press, 1994) offers an interesting perspective on this development in the eighteenth century.
62 See Baldwin, "The Snakestone Experiments"; and Findlen, *Possessing Nature*, pp. 236–7.
63 Gioseffo Petrucci, *Prodomo apologetico alli studi chircheriani* (Amsterdam, 1677), p. 108.

was given. His attempts to utilize the techniques of the new science, Oldenburg felt, set him apart from the other Aristotelian philosophers. "I have turned over part of Kircher's *Subterranean World*," he wrote, "and all his arguments and theories are no credit to his wit, yet the observations and experiments there presented to us speak well for the author's diligence and for his wish to stand high in the opinion of philosophers." Disillusionment followed rather quickly, as members of the Royal Society put Kircher to the test. One month later, Oldenburg had revised his opinion. Attempting to replicate the findings reported by Kircher, he wrote to Boyle that the "very first Experiment singled out by us out of Kircher" had failed, "and yt 'tis likely the next will doe so too."[64] Such problems had also occurred with Boyle's famous air-pump experiments and would later occur with early Continental attempts to replicate Newton's prism experiments. In his failure to convince others that his experiments produced universal truths about nature, Kircher was not alone.

Until tools such as mathematics achieved a more universal status – an event that occurred in no small part due to Newton's work – experimenting remained a qualitative art. As historian Peter Dear demonstrates in his recent study of the intersections between mathematics and experimental culture, the Jesuits played a crucial role in enhancing the status of mathematics as an experimental tool, laying the groundwork for the rejection of the old Aristotelian physics.[65] At the height of his career, Kircher diverged from the path taken by such Jesuit astronomers as Christopher Clavius, Christoph Scheiner, and Giovanni Battista Riccioli, but he began his experimental work practicing a form of the "physicomathematics" they pioneered. In preferring the natural magic tradition to the mathematical sciences, Kircher envisioned experimentation as a demonstrative afterthought to a philosophy already in existence. Experiments played a rhetorical role in his natural philosophy; they were a means of visualizing a philosophy rather than creating one.[66] To the extent that he remained interested in the study of mathematics, Kircher focused primarily on mystical, cabalistic, and numerological traditions, highlighting the "chain of natural things expressed by numbers."[67] For him, mathematics re-

---

64 *The Correspondence of Henry Oldenburg*, ed. and trans. A. Rupert Hall and Marie Boas Hall, 8 vols. (Madison: University of Wisconsin Press, 1966), 2:567 (London, October 12, 1665); 2:615 (London, November 21, 1665).
65 Peter Dear, *Discipline and Experience: The Mathematical Way in the Scientific Revolution* (Chicago: University of Chicago Press, 1995).
66 Ibid., pp. 172–3; Hankins and Silverman, *Instruments and the Imagination*, pp. 14–36, esp. pp. 30, 32, 44.
67 Kircher, *Arithmologia sive de abditis numerorum mysteriis* (Rome, 1665), p. 80.

mained a secret language for initiates. Newton later imagined it to be the one form of communication that could provide satisfactory public expression of nature's laws.[68] Such investigations could not lead to an *experimentum crucis*.

Newton's dissatisfaction with the messy state of natural philosophy in the 1660s was surely one of the reasons that led him to create quantifiable experiments. His optics began as a wholesale reclamation of the subject of light from the realm of Aristotelian physics. "[A]lthough colours may belong to physics, the science of them must nevertheless be considered mathematical," he lectured to Cambridge undergraduates in 1669. It was his hope that "philosophical geometers and geometrical philosophers" would combine forces to produce a new form of knowledge: "instead of the conjectures and probabilities that are being blazoned about everywhere, we shall finally achieve a natural science supported by the greatest evidence."[69] The type of undisciplined philosophical culture that had allowed encyclopedists such as Kircher to flourish was precisely what Newton hoped to reform. Inspired and enraged by Descartes's solution to this problem, Newton systematically reexamined the fundamental tenets of natural philosophy in search of a new truth.

Potentia

Many elements of Newton's reformation of knowledge were novel indeed, and his transformations of optics, mathematics, and physics justly merited contemporary praise of Newton as an immortal among mortals. Yet other aspects of Newton's thought placed him squarely in the messy debates of the seventeenth century, revealing him to be a scholar who did not always have prompt solutions at hand. In order to understand these different aspects of Newton's career, we need to examine the larger goals that determined his intellectual choices. What did Newton hope to gain by finding certainty in nature?

Mathematics seemed to be an especially promising tool with which to achieve a resolution of key philosophical debates, but it was, more importantly, a means of communion with the Deity. In an unpublished corollary to Proposition VI of the *Principia*, Newton wrote: "There exists an infinite and omnipresent spirit in which matter is

---

68 Rivosecchi, *Esotismo*, p. 65. I find it quite striking that Kircher essentially treated mathematics the way Newton treated alchemy – private sciences for a select audience.
69 Newton, *Optical Lectures* (given in 1669; publ. London, 1729), in Fauvel et al., *Let Newton Be*, p. 86. Compare this with Newton's remarks about optics in his 1664–5 Trinity notebook: "Colors arise either from shadows intermixed with light, or stronger and weaker reflections. Or, parts of the body mixed with and carried away by light." McGuire and Tamny, *Certain Philosophical Questions*, p. 389.

moved according to mathematical laws."⁷⁰ More broadly, he argued that nature was a literal scripture. "For there is no way (w^{th}out revelation) to come to ye knowledge of a Deity but by ye frame of nature," he explained in a much cited passage.⁷¹ Such comments surely reflected Newton's frustration in trying to commune with God through other channels that he found less satisfactory. In making such statements, Newton belonged to a growing community of scholars who increasingly viewed nature as the first rather than the second scripture, seeing it as a less corrupt rendition of God's word than biblical and patristic writings.

The religious motivations informing Newton's scientific investigations were not unknown to his contemporaries. John Craig informed one early biographer that "his great application in his inquirys into Nature did not make him unmindfull of the great Author of Nature."⁷² By the 1670s, as historian Richard Westfall notes, Newton initiated his theological studies that were to occupy a great deal of his energies, at times to the exclusion of natural philosophy. They included close examination of biblical chronologies and intense engagement with the writings of the early church fathers.⁷³ In time, Newton's humanistic studies evolved into a wholesale examination of antiquity, all in service of Newton's urgent questions about the introduction of idolatrous practices into pristine Christianity and his reformation of faith.

Questions of faith occupied Newton, initially to the exclusion of concerns about the consequences his views might have on his academic career. As is well known, by 1675 Newton was an avowed Arian. His private heresy – both the denial of the Trinity and the demotion of Christ to the status of a prophet – fueled many of his questions about the search for divine truth. For Newton, truth did not literally emanate from the writings of the ancients but emerged through a careful appraisal of the various images of antiquity that shaped the present state of knowledge. This conviction led him to condemn Egypt as the source of idolatry and the Catholic Church for perpetuating such practices. The imposition of Trinitarian doctrine – "this strange religion of ye west"⁷⁴ – by church fathers such as Saint Athanasius of Alexandria marked a watershed in the history of Chris-

---

70 In Fauvel et al., *Let Newton Be*, p. 172.
71 Yahuda, MS 41, f. 6, in Richard S. Westfall, "The Rise of Science and the Decline of Orthodox Christianity: A Study of Kepler, Descartes, and Newton," in Lindberg and Numbers, *God and Nature*, p. 233; also Dobbs, *Janus Faces*, p. 151.
72 In Frank E. Manuel, *Isaac Newton Historian* (Cambridge, Mass.: Belknap Press, 1963), p. 3.
73 Westfall, "The Rise of Science," p. 229. More generally, see Frank E. Manuel, *The Religion of Isaac Newton* (Oxford: Clarendon Press, 1974).
74 Yahuda, MS 1.4, f. 50, in Westfall, "The Rise of Science," p. 231.

tianity that Newton sought to discredit. Demonizing Kircher's namesake as the originator of the twin evils of the Trinity and monasticism, Newton cast Saint Athanasius as the principal villain in the long corruption of Christianity, devoting an entire treatise to the *Paradoxical Questions Concerning the Morals and Actions of Athanasius and His Followers*.[75] His writings figured prominently in the theological works Newton collected in his library and scoured for clues about the origins of suspect doctrines.[76]

By the time Newton began to shape his unpublished treatise on the origins of gentile theology in the 1680s, his ideas about the early history of Christianity had crystallized. Egyptian theosophy appeared as the first corrupter of true religion, enticing men such as Saint Athanasius to incorporate pagan practices into Christianity. Years later, in *The Chronology of Ancient Kingdoms Emended* (1728), Newton chastised Egyptian theologians for inflating the value of their religion for posterity: "Thus great was ye vanity of the Priests of Egypt in magnifying their antiquities."[77] Instead, Newton identified Noah as an originator of true doctrine and praised the prophets for persuading the Jews to give up their idolatrous practices. Both belonged to an uncluttered world of Christianity that lacked the excessive emphasis on ritual and ceremonial, which Newton found so distasteful in the seventeenth-century Anglican and Catholic churches. "Truth is ever to be found in simplicity, & not in ye multiplicity and confusion of things," he wrote in his unpublished *Rules for interpreting ye words & language in Scripture*.[78] Just as the laws of gravity revealed the uncorrupted state of nature, intense examination of the Bible (which Newton owned in multiple, annotated copies) yielded *prisca theologia*.

Newton's conclusions about the entangled relations between the history of civilizations and the history of religion followed a path taken by many biblical scholars in this period. As historian Frank Manuel observes, "much of his learning and erudition and most of his ideas look backward to the polyhistors and Bible commentators of the mid-seventeenth century rather than forward to the new historiography."[79] Although Newton rejected Jesuit chronologies as religiously suspect, preferring Hebrew chronologies and the works of Renaissance scholars such as Joseph Scaliger, he nonetheless participated in the same project that involved sifting through ancient writings in

---

75 Manuel, *Isaac Newton Historian*, p. 158; Westfall, *Never at Rest*, p. 345; Fauvel et al., *Let Newton Be*, p. 177. *Paradoxical Questions* is Keynes MS 10.
76 Harrison, *Library*, p. 92 (Paris, 1627 edition of St. Athanasius's *Opera*).
77 In Manuel, *Isaac Newton Historian*, p. 91.
78 Newton, "Rules," in Christianson, *In the Presence of the Creator*, p. 261.
79 Manuel, *Isaac Newton Historian*, p. 47.

order to discern fragments of truth in an incomplete and often confusing past. Surely the difficulties of understanding the written record, in contrast to the clarity Newton at times found in nature, helps to explain his decision to suppress *deliberately* those parts of the *Principia* that dealt with the history of the ideas he advocated. Newton was personally convinced of the antiquity of Copernican astronomy and his ideas about gravity, which he dated back to such figures as Plato, Pythagoras, and Hermes Trismegistus. Yet he vacillated about including such information in the published version of his work.[80]

History for Newton was a form of private knowledge rather than public record. Science was surely as corrupted and idolatrous as religion, but the imperfections of the historical record themselves undermined the project of knowing fully where the truth lay. Since Newton only published what was certain, very little of his nonmathematical writings ever reached a broad readership.

Newton's taciturn revelations about the mysteries of Scripture, at least in print, contrasted strongly with Kircher's profusion of commentary on the problem of idolatry. In such works as *Egyptian Oedipus* and the *Tower of Babel*, Kircher shaped a history of Christianity that identified the Israelites, beginning with Moses and culminating especially in Noah's son Cham, as corrupters of Egyptian wisdom. Such prophets and philosophers had obscured the importance of Hermetic wisdom in shaping the doctrines of Christianity and denied the piety of the Egyptians.[81] The folly of Babel had precipitated the demise of a unified Christianity, making idolatry an inevitable result of miscommunication in a linguistically fragmented world. Kircher's search for the principles that unified nature and language belonged to a larger and more urgent project about the reunification of faith. In this baroque vision, the Catholic Church would suppress idolatry by becoming the new Egypt.[82]

At the center of Kircher's quest for unity lay the Trinity. A good disciple of his namesake, this seventeenth-century Athanasius saw evidence of the Trinity across time and space as the most important proof of the universality of true faith. From the Zoroastrian mysteries to the Jewish cabala, from the Egyptian pyramids and Aztec monuments to Confucian doctrine, Kircher discerned the presence of this symbol. While Newton's numerological interests revolved around

---

80 Fauvel et al., *Let Newton Be*, pp. 122–3, 187–93; and Robert Iliffe, " 'Is He Like Other Men?' The Meaning of the *Principia Mathematica*, and the Author as Idol," in *Culture and Society in the Stuart Restoration: Literature, Drama, History* (Cambridge: Cambridge University Press, 1995), pp. 164–8.
81 Godwin, *Athanasius Kircher*, pp. 15, 57; Pastine, *La nascita dell'idolatria*.
82 Cipriani, *Gli obelischi egizi*, p. 104.

such numbers as seven, Kircher made three the mystic symbol of his entire philosophy. He inscribed Osiris, Isis, and Horus in the Pythagorean triangle in order to prove to his readers that mathematics offered further proof of the righteousness of his doctrine. Even the Sun was not exempt from this sort of analysis, since Kircher declared that its trifold properties of growth, light, and warmth made it a perfect exemplar of the Trinitarian powers of the deity.[83]

Writing in support of a theological perspective that found great support among various Catholic rulers, Kircher enjoyed the privileged position of publicizing and elaborating a philosophy that was, in essence, the official position of the baroque church. In the midst of evangelizing to the entire world, the Catholic Church needed constant proof of the means by which one could find traces of its version of *prisca theologia* in all parts of the world. As a result, the church chose to ignore Kircher's heterodoxy on many points of science and belief because it suited its purposes.[84] By contrast, Newton's Arianism fundamentally conflicted with the central tenets of the Church of England, making him enormously circumspect about his means of expressing his ideas. He could know God but never acknowledge the fullness of what such knowledge entailed. Newton belonged more fully to the Pythagorean traditions of silence that were a fundamental part of early modern occult philosophies. Ironically, his work, much more than Kircher's, remained frustratingly incomplete at the end of his lifetime.

In the second half of the seventeenth century many scholars experimented with a variety of techniques by which to resolve the most pressing intellectual problems of their society. As the long and fruitful careers of Kircher and Newton suggest, interpreting nature and defining faith were closely intertwined occupations. Both deeply religious men, Kircher and Newton viewed each other across the chasm of the Reformation. For Kircher, Rome was the citadel of faith and site of a vibrant tradition of humanistic learning – a "unique theater for all hieroglyphic literature."[85] By contrast, Newton's fears of papists in England were compounded by his absolute condemnation of Rome as the center of all idolatry. For him, Rome was the whore of Luther's Babylon: "a woman arrayed in purple & scarlet & decked with

---

83 Ibid., p. 123; Pastine, "Le metamorfosi della trinità," in his *La nascita dell'idolatria*, pp. 191–233; and Rivosecchi, *Esotismo*, p. 66.
84 This interesting detail is noted in Baldwin, "Athanasius Kircher," p. 461. Kircher attributed many of the attractive powers normally associated with Christ to the Holy Spirit.
85 Kircher, *Oedipus Aegyptiacus*, vol. 1, n.p.

gemms, who lives deliciously & sits a Queen upon seven mountains & upon the horned Beast in a spritualy barren wilderness & commits fornication with the kings of the earth."[86] Given such strongly worded sentiments, it is perhaps not surprising that Newton never once purchased a book by the new Athanasius, whose doctrines, monastic profession, and allegiance to the papacy embodied everything he despised.[87]

At the same time, Kircher would have found Newton's relationship to antiquity puzzling and contradictory. A great believer in the art of communication, Kircher surely would have seen Newton as a provincial scholar whose limited linguistic competence – reasonable Latin, some French, and unsatisfactory attempts to learn the languages necessary for biblical scholarship – precluded his ability to judge properly the place of the past in the current shape of knowledge. Ironically the one part of the *Corpus Hermeticum* that Newton copied, translated, and found worthy of commentary, the alchemical *Emerald Tablet*, was precisely the treatise that Kircher singled out as an Arabic forgery. Modern scholars concur with this evaluation.[88] Surely Kircher would have suggested that a fuller knowledge of antiquity led to a different understanding of the past and its documents.

Kircher also would have described Newton's statements about the possibility of knowing the causes of "occult qualities" both philosophically unsound and religiously unorthodox. Certain divine mysteries, he felt, lay permanently in the hands of God, removed from all possibility of human knowledge. Those unseen qualities that philosophers could explain were knowable precisely because a natural force provided the immediate cause, working at the behest of God. Newton's concept of action at a distance and his statements that things currently unknown harbored the potential to be known belonged to a different philosophy of nature that placed greater emphasis on the evidence of the senses. Yet it is not unreasonable to suggest that it was the logical

---

86 Newton, Keynes MS 5, ff. 120–1, as quoted in Manuel, *Isaac Newton Historian*, p. 152.
87 However, I am allowing for the possibility that he may have read some of Kircher's work in the library of Isaac Barrow, or other Cambridge and London libraries, where the Jesuit's books were often found.
88 B. J. T. Dobbs, "Newton's *Commentary* on the *Emerald Tablet* of Hermes Trismegistus: Its Scientific and Theological Significance," in *Hermeticism and the Renaissance*, ed. Ingrid Merkel and Allen G. Debus (Washington, D.C.: Folger Shakespeare Library, 1983), pp. 182–91. On Kircher's views, see Godwin, "Athanasius Kircher and the Occult," in John E. Fletcher, ed., *Athanasius Kircher und seine Beziehungen Zum gelehrten Europa seiner Zeit* (Wiesbaden: Harrassowitz, 1988), p. 27. The original discussion can be found in *Oedipus Aegyptiacus*, vol. 2, pt. 2, pp. 427–8. For the definitive account of the textual status of the *Emerald Tablet*, see Brian P. Copenhaver, ed. and trans., *Hermetica* (Cambridge: Cambridge University Press, 1992), p. xlvi.

evolution of Kircher's own investigations into such hidden forces as magnetism.[89]

Two worlds, only a hair's breadth apart. The dream of unifying all knowledge into a new system motivated many seventeenth-century scholars. Without the fantasy of the Scientific Revolution – its dreams, utopias, and imaginary voyages – it is hard to imagine that the actual science of this period would have ever emerged. Umberto Eco wrote movingly of the sort of personality who ushered in modern science as a man lost in a fantasia of knowledge who ceases to remember how knowledge ought to be and creates something new:

> It's odd how, among the adepts of sapiential studies, eccentric personalities are sometimes found.... I don't mean the usual seekers after transcendental consolation, I don't mean the melancholy spirits, but men of profound knowledge and great intellectual refinement who nevertheless indulge in nocturnal fantasies and lose the ability to distinguish between traditional truth and the archipelago of the prodigious.[90]

That baroque land mass, the archipelago of the prodigius, occupied a unique moment in the history of ideas. It created a space for curious scholars to survey every domain of thought with minimal regard for the conventions restricting knowledge. And in that imaginary world of wonders, Newton's inverse-square law lay contiguous to the quest for Egypt, his reconstruction of Solomon's Temple in London yet another variation of Kircher's meticulous rebuilding of Noah's Ark and the Tower of Babel in Rome.

89 This is certainly Keith Hutchison's perspective. Though he does not mention Kircher in his studies of occult qualities, Kircher is the perfect example of a philosopher who enhanced the natural magic tradition to such an extent that it became useful for the development of experimental philosophy. See Hutchison, "What Happened to Occult Qualities," pp. 250–1.

90 Eco, *Foucault's Pendulum*, p. 282.

12

The Nature of Newton's "Holy Alliance" between Science and Religion: From the Scientific Revolution to Newton (and Back Again)

JAMES E. FORCE

It is time for another look at the Scientific Revolution and Newton. The immediate occasion for further discussion of this perennial issue is Betty Jo Teeter Dobbs's 1994 essay in *Isis* and Richard S. Westfall's reply, both appearing in this volume. These two giants in the modern-day history of science, both now sadly deceased, disagreed about the meaningfulness of the concept of "the" Scientific Revolution – the "big one" that happened between Copernicus and Newton. Dobbs was conspicuously critical of the generalizations of historians who emphasized the notion of the Scientific Revolution at the expense of the particularity and uniqueness of the individuals crushed beneath the weight of this venerable, grand theory. She didn't buy into the concept except with reservations. It was too anachronistic, she claimed, following Cohen.[1] It was also too metaphorical and, therefore, problematic, to talk – along with Whiggish historians such as Butterfield and a host of others[2] – about a revolution that portrays "a change that is *sudden, radical,* and *complete.*"[3] Worst of all, exponents

---

1 Dobb's first footnote is appropriately to I. Bernard Cohen, *Revolutions in Science* (Cambridge, Mass.: Harvard University Press, 1985). See B. J. T. Dobbs, "Newton as Final Cause and First Mover," *Isis* 85 (1994): 633, n. 1, reproduced here as Chapter 2. Cohen's approach – tracing how the term is used by people in the centuries since "it," that is, "the" Scientific Revolution, occurred – is widely echoed and often cited. See, for example, Roy Porter's excellent brief article on the Scientific Revolution in *The Blackwell Companion to the Enlightenment*, ed. John Yolton et al. (Oxford: Blackwell, 1991), s.v. "Scientific Revolution."
2 See, especially, W. E. H. Lecky, *History of the Rise and Influence of the Spirit of Rationalism in Europe*, 2 vols., rev. (New York: D. Appleton, 1880); John W. Draper, *A History of the Conflict between Religion and Science* (New York: D. Appleton, 1875); Andrew D. White, *A History of the Warfare of Science with Theology in Christendom*, 2 vols. (New York: D. Appleton, 1896).
3 See Chapter 2. I have emphasized those aspects of the Scientific Revolution which Dobbs found problematic. For a popular example of one who agrees that the universe changed suddenly, radically, and completely, see the popular book (and, most

of the Scientific Revolution, distorted the highly individualized genius of Newton and appropriated him into their theory as either the heroic "First Mover" of the great change or as the heroic "Final Cause" of the Scientific Revolution.[4]

In his typically gallant manner, Richard S. Westfall politely disagreed and firmly continued to align himself with the "Final Cause" school of the mid to late twentieth century, a position he outlined in his first major book and which he maintained until his death. In a seminal work, first published in 1958, Westfall outlined what he considered "the problem"[5] of integrating the "new intellectual current" of natural philosophy into the framework of the Christian religion. Westfall was well aware of the "complex of forces operating to change Christianity" in the early modern era, but argued that the "achievements" of science raised pressing questions for religion, questions that could "not be ignored." He wrote that:

> There was the possibility that investigation of nature might so absorb a man that he would neglect the worship of God, and investigations were particularly to be feared if the discovery of second causes called the existence or power of the First Cause into doubt. Natural science rested on the concept of natural order, and the line that separated the concepts of natural order and material determinism was not inviolable. The mechanical idea of nature, which accompanied the rise of modern science in the 17th century, contradicted the assertion of miracles and questioned the reality of divine providence. Science, moreover, contained its own criteria of truth, which not only repudiated the primacy of ancient philosophers but also implied doubt as to the Bible's authority and regarded the attitude of faith enjoined by the Christian religion with suspicion. Confronted with such problems, a man was not faced, to be sure, with an exclusive choice – either natural science or Christianity: every one of the problems could be resolved in a variety of ways to reconcile science with religion. But the mere fact of reconciliation meant some change from the pattern of traditional Christianity. With the growing prestige of science – it achieved immense prestige

importantly, the television series) by James Burke, *The Day the Universe Changed* (Boston: Little, Brown, 1985).

4 Dobbs writes of the various interpretations of Newton that "If he was viewed as the First Mover or efficient cause in the nineteenth- and early twentieth-century science, in our accounts of the Scientific Revolution we see Newton emerging as the Aristotelian Final Cause. Have you ever stopped to count the number of books that either begin or end with Isaac Newton?" See Chapter 2.

5 Richard S. Westfall, *Science and Religion in Seventeenth-Century England* (Ann Arbor: University of Michigan Press, 1973), p. 1. "The Problem" is the title of chapter 1.

with the publication of Newton's *Principia* – its reconciliation with Christianity came more and more to mean the adjustment of Christian beliefs to conform to a position of intellectual dominance over Christianity.[6]

From the start of his intellectual career, it was, for Westfall, simply axiomatic that the Scientific Revolution was in fact a relatively quick, very radical moment that commenced with the publication of Copernicus's *De revolutionibus orbium coelestium* in 1543 and was in fact completed with the publication of Newton's great *Philosophia Naturalis Principia Mathematica* in 1687. True to the end to the interpretation that he had worked out in the 1950s, in his article in response to Dobbs in this volume Westfall wrote almost his last word on "the problem" of how revolutionary modern science evicted theology from its position as the "focus of European life and culture."[7] He wrote that:

> Before the Scientific Revolution, theology was queen of all the sciences. As a result of the Scientific Revolution, we have redefined the word "science," and today other disciplines, which once took their lead from Christian doctrine, strive to expand their self-esteem by appropriating the word in its new meaning to themselves. Theology is not even allowed on the premises anymore. Here is the very heart of Butterfield's specific statement. A once Christian culture has become a scientific one. The focus of change, the hinge on which it turned, was the Scientific Revolution of the sixteenth and seventeenth centuries. As you may gather, I am convinced that there has been no more fundamental change in the history of European civilization. Dispense with the concept of the Scientific Revolution? (The capital letters seem ever more important to me.) How can we dream of it?

How indeed? Dobbs, in her article in *Isis* was speaking directly to interpreters such as Westfall about just how to revise our thinking about the Scientific Revolution so as not to do away with it completely while rendering it more historically accurate. She regarded the term as anachronistic, overly metaphorical, and – above all – untrue to the whole Newton. The crux of the issue, as Dobbs realized, is where one starts. Westfall started from his initial, well-worked-out position regarding the Scientific Revolution as the result of "the problem" and moved on to the capstone of his life's work, his monumental study of Newton.[8] His great book is a lengthy reprise of his earlier interpreta-

---

6 Ibid., pp. 2–3.
7 Ibid., p. 2.
8 Richard S. Westfall, *Never at Rest: A Biography of Isaac Newton* (Cambridge: Cambridge University Press, 1980).

tion of Newton as a sort of protodeist who did not realize the paradoxical nature of his own thought.[9]

Dobbs, in her own writing, began with Newton. Her two major books attempt to get Newton right by reassembling the bits and pieces lopped off and ignored or, as with Westfall, simply misinterpreted because of ideological blinders. In 1975, in *The Foundations of Newton's Alchemy or, "The Hunting of the Greene Lyon,"* Dobbs explained how the experimental aspect of Newton's alchemy could be completely ignored by historians who allowed Newton's "successes in mathematics and physical sciences" to color their thoughts, "subtly and deeply," about what constituted the "Newtonian world view." By ignoring the fuller seventeenth-century context in which Newton's thought developed, Westfall made Newton into the prototype of the "modern" physical scientist – ultimately the first or the final cause of the Scientific Revolution: "Thus it became a curious anomaly – and one to be explained away – that Newton's studies in astronomy, optics, and mathematics only occupied a small portion of his time. In fact most of his great powers were poured out upon church history, theology, 'the chronology of ancient kingdoms,' prophecy, and alchemy."[10]

In her subsequent book from 1991, *The Janus Faces of Genius: The Role of Alchemy in Newton's Thought*, she suggested that: "Perhaps because of the remarkable success of the *Principia* itself in restoring true natural philosophy, Newton shifted his focus to *more* study of natural philosophy as the *best* way to restore true religion. He sought the border where natural and divine principles met and fused."[11]

Whereas Westfall worked out his version of the grand theory of "the problem" and how it culminated in the Scientific Revolution through the heroic exertions of Newton, among others, once Dobbs

---

9 Westfall, *Science and Religion*, chap. 8, "Isaac Newton: A Summation."
10 B. J. T. Dobbs, *The Foundations of Newton's Alchemy or, "The Hunting of the Greene Lyon"* (Cambridge: Cambridge University Press, 1975), p. 6.
11 B. J. T. Dobbs, *The Janus Faces of Genius: The Role of Alchemy in Newton's Thought* (Cambridge: Cambridge University Press, 1991), p. 170. It has since been suggested that Newton thought himself specially chosen as an interpreter of millennial prophecies because of his immense successes in natural philosophy. "The 'wisdom' mentioned in Daniel, which is God's gift to the chosen, was interpreted by Newton, above all, as the ability to decode the secrets of nature. Despite the fact that the idea of the chosen sprouted up against the background of Newton's ability to decode prophetic works, his authority for interpreting these texts apparently emanated from his unique attainments in natural philosophy." See Matania Z. Kochavi, "One Prophet Interprets Another: Sir Isaac Newton and Daniel," in *The Books of Nature and Scripture: Recent Essays on Natural Philosophy, Theology, and Biblical Criticism in the Netherlands of Spinoza's Time and the British Isles of Newton's Time*, ed. James E. Force and Richard H. Popkin (Dordrecht: Kluwer, 1994), p. 116.

had reconstituted a more complicated historical Newton (who "is best understood as a product of the late Renaissance, a time when the revival of antiquity had conditioned the thinkers of Western Europe to look backward for the Truth",)[12] she proceeded, in the *Isis* article, to draw out the implications of her work for the Scientific Revolution. As noted earlier, she challenged the usefulness of this term if it is taken to mean a "sudden, radical, and complete" change. What happened was more a part of a continuum than historians such as Westfall maintained. Starting with a Newton who is more fully integrated into the variety of intellectual currents of his day, she naturally questioned the ruthless reduction of Newton to a first or final cause of such a "complete" change. Only by distorting the historical Newton can he be converted into the hero of such a revolution.

When two people of such stature apparently disagreed so starkly, further examination — that is, another volume of essays reappraising the Scientific Revolution[13] — is warranted. Revising the latest revision is perhaps the essence of scholarship. There are many other standard (and not-so-standard), ongoing, revisionistic debates among intellectual historians, after all. Westfall himself cites Butterfield's example of the "Bacon-Shakespeare controversy." Was there ever a decline and fall of the Roman Empire? Did "Puritanism" give a special impetus to early modern science as Merton argued? Was "Athena" black?

Heeding the wise words of yet another great modern master of the history of science, A. Rupert Hall — who once wrote, "I dislike dichotomies: of two propositions, so often neither *a* nor *b* by itself can be wholly true"[14] — it is extremely tempting to stand, in academically balanced moderation, somewhere near the middle. Indeed, I must register my caveats regarding Dobbs's position. She was, in my opinion, wrong that proponents of the "radical and complete" view of the Scientific Revolution necessarily also thought that it was "sudden." Westfall stretched it over 144 years (from the publication of Copernicus's *De revolutionibus orbium coelestium* in 1543 until Newton's *Principia* in 1687). And she may also be wrong (with Cohen) about the actual first usage of the term "revolution" by the first of the early modern scientists themselves such as Boyle.[15]

12 Dobbs, *Janus Faces*, p. 10.
13 See the essay by David C. Lindberg, "Conceptions of the Scientific Revolution from Bacon to Butterfield: A Preliminary Sketch," *Reappraisals of the Scientific Revolution*, ed. David C. Lindberg and Robert S. Westman (Cambridge: Cambridge University Press, 1990), pp. 1–26.
14 A. Rupert Hall, "The Scholar and the Craftsman," in *Critical Problems in the History of Science*, ed. Marshall Clagett (Madison: University of Wisconsin Press, 1959), p. 21.
15 See Chapter 15 in this volume by Margaret C. Jacob. Cf. my note 1.

Even so, I must take my cue in this debate between Dobbs and Westfall from the great medievalist, Lynn Thorndike, who wanted his critics among the scholars of the "Renaissance" to know just how much they had underestimated his "aversion" to the term "Renaissance." Thorndike ridicules the medievalist's definition of the Renaissance as a literal "rebirth":

> Religion may have its resurrections and revivals, but I have even less faith than Nicodemus in rebirths or restorations and revivals of whole periods of human history. I take my stand with the blind writer of Christian hymns, Fanny Crosby, who sang, "But the bird with the broken pinion never soared so high again"; with William Muldoon who said of former heavyweight champions, "They never came back"; with Omar Khayyam who mused;
>
> > The moving finger writes and having writ
> > Moves on; nor all your piety nor wit
> > May lure it back to cancel half a line
> > Nor all your ears wipe out one word of it;
>
> and with a verse from the light opera, Tom Jones,
>
> > Time is not a necromancer;
> > Time's a thief and nothing more.
>
> Legacies from the past? Yes. Inheritances from previous periods? Yes. Survivals? Yes. Resemblances to our forebears? Yes. Reformations? Perhaps. Reactions? Unfortunately. But no rebirths and no restorations![16]

In the reemergence of the Scientific Revolution dispute arising from the pages of Dobbs's article in *Isis* and Westfall's courtly response in this volume, I finally side, Thondike-like, with Dobbs who gets much more right than Westfall does. While she was critical, she did not want to eliminate the term Scientific Revolution as Westfall implies or to claim that nothing significant happened in science in the early modern period. Westfall started with the notion of the Scientific Revolution as a given. Dobbs began by getting Newton right and then went on, in her recent *Isis* article, to rescue Newton from the clutches of the grand theorists of the Scientific Revolution. The crucial problem with Westfall's overly generalized view of the Scientific Revolution is, finally, that it distorts the historical Newton. Dobbs wished to do justice to the uniqueness and singularity that mark Newton's thought and which, she rightly thought, ought to prevent him from being forcibly hijacked by the grand theorists of the Scientific Revolution

---

16 Lynn Thondike, "Renaissance or Prenaissance?" *Journal of the History of Ideas* 4 (1943): 65.

and forced to play the role that they assigned to him, either heroic "first mover" or heroic "final cause" of the Scientific Revolution.

In this chapter, I want to focus, yet again, upon "the problem" described by Westfall, on the attempt by Newton to synthesize science and divine providence, and on the implications of this synthesis for those who would cast Newton in the heroic role of either first mover or final cause in a great historical drama written by nineteenth- and twentieth-century historians. The first issue I reexamine is the relationship between Newton's synthesis of God's general and special providence and the so-called "design argument." Then I briefly trace how early modern thinkers view the relationship between the world of mechanical nature and the providential deity of Christianity. Implicit or explicit in the thought of Wilkins, Charleton, Sprat, Boyle, Whiston, Clarke, and Newton is a reconciliation between a purely mechanistic conception of the cold, mathematically calculable order of nature and natural law and a traditional Christian conception of a specially provident deity active daily in the governance of his creation and powerful enough to set aside the generally provident laws of created nature through acts of special providence such as miracles. After setting out the nature of the "holy alliance" between general (creative) providence and special (miraculous) providence – what might be regarded as one of Newton's most important and distinctive intellectual legacies – I examine Newton's particular synthesis of general and special providence.

Finally, I trace what I believe follows from Newton's synthesis of general and special providence regarding his expectations about the kind of knowledge of nature that the new science would yield in the millennial world he fervently believed would soon come. In this last section, I conclude that we must cease to consider Newton as, in any way, a cause or the final product of the Scientific Revolution and argue that Dobbs was, in large part, right in her astute moderation of the extreme generalities of the grand theorists of the Scientific Revolution.

Providential Design and Providential Intervention

No modern scholar ignores how Newton relies upon the design argument to infer a Lord God of general providence – a divine architect.[17] Modern scholars have increased our appreciation of how keenly Newton feels the presence of the creator/architect of the design argument

---

17 John Herman Randall Jr., "What Isaac Newton Started," in *Newton's Philosophy of Nature*, ed. H. S. Thayer, 2nd ed. (New York: Hafner, 1974), pp. xi–xii. Cf. Gale E. Christianson, *In the Presence of the Creator: Isaac Newton and His Times* (New York: Free Press, 1984), pp. 312–13.

through an analysis of his extensive theological manuscripts. In his unpublished papers, Newton quests widely as he attempts to fathom a whole range of theological questions from an analysis of scripture. Even so, many Newton scholars share an explicit assumption that Newton's characteristic metaphysical theory of the all-powerful nature of the Lord God of dominion, though clearly important to the design argument of natural religion and though perhaps "important" to Newton psychologically, has no necessary logical connection with Newton's science. For them, the design argument is merely a corollary that follows from Newton's articulation of the great machine of nature. It is *not*, in any way, essentially linked to his physical theory.

At the root of Westfall's interpretation of Newton was his long-held view that Newton's thought contained within it contradictory parts, which, when dissolved, terminate in the scientific world that he discerns everywhere in the current era. Dobbs's contrary view (and mine) is radical. Most modern scholars side with Westfall and have similarly discerned a paradoxical tension in the synthesis of both a mechanical frame of nature (obedient to natural laws) and a still active, specially provident, prayer-answering, prophecy-fulfilling God, whose power is sufficient to alter the created natural mechanism whenever God wills. For such writers, there is an uneasy truce between these two sorts of divine providence – the generally provident act of the creator and the specially provident (miraculous) interventions in the natural order by a Lord God who intervenes in his own creation (as described in Scripture) – and the idea of nature that each emphatically entails. The more detailed and defined the picture of the natural and mechanical order of general providence becomes in the early modern period, the less room there appears to them to be for specially provident divine intervention. Within the past thirty years, scholars have analyzed the nature of the "holy alliance"[18] between science and religion in this period and questioned the logical tenability of such a position. Norman Hampson argues that "However orthodox men like Descartes, Locke and Leibniz might be, their Christian orthodoxy was tacked on to systems of thought which were logically viable without it."[19] Westfall, after skillfully tracing how numerous seventeenth-century English thinkers attempt to synthesize science and religion, concludes that "the skepticism of the Enlightenment was already present in embryo among them. To be sure, their piety kept it in check,

---

18 Basil Willey describes the "holy alliance between science and religion" in eighteenth-century England, which scholars such as Westfall have called into serious question. See Basil Willey, *The Eighteenth-Century Background* (Harmondsworth: Penguin, 1972), p. 162.

19 Norman Hampson, *The Enlightenment* (Harmondsworth: Penguin, 1979), p. 28.

but they were unable fully to banish it."[20] Not surprisingly, for Westfall, Newton was a deist.[21] In another essay, which he also wrote toward the end of his life, Westfall showed what must be done to Newton when one is such a strong exponent of the Scientific Revolution. Westfall concluded a brief encyclopedia article about Newton by observing that, "In addition to science, Newton also showed an interest in alchemy, mysticism, and theology. Many pages of his notes and writings particularly from the later years of his career are devoted to these topics. However, historians have found little connection between these interests and Newton's scientific work."[22]

Newton's attempt to synthesize these disparate elements was seen by Westfall as a paradox that contains within it the seeds of a modern, deistic secularism, which provides the terminus of the Scientific Revolution.[23]

Thus, in his pioneering examination of Newton's hitherto little-known theological manuscripts, Manuel likewise insists that Newton achieves, in his private thought, an "intimate union of science and religion."[24] However, in explaining the nature of this union, Manuel

20 Westfall, *Science and Religion*, p. 219.
21 See James E. Force, "The Newtonians and Deism," in *Essays on the Context, Nature and Influence of Isaac Newton's Theology*, ed. James E. Force and Richard H. Popkin (Dordrecht: Kluwer, 1990), pp. 43–74, for a discussion of Westfall's view that Newton is a deist.
22 Richard S. Westfall, *Microsoft Encarta 94 Encyclopedia*. This work is available on a CD-ROM. The article is mounted on the worldwide web by the "Isaac Newton Institute for Mathematical Sciences": <http://www.newton.cam.ac.uk/newtlife.html>. See, also, Richard S. Westfall, "Newton's Theological Manuscripts," in *Contemporary Newtonian Research*, ed. Zev Bechler (Dordrecht: Reidel, 1982), pp. 139–40. Here Westfall also concludes that Newton's theology could not have influenced his science. If it did, then, of course, Newton would not be the first cause of the Scientific Revolution.
23 Westfall, *Science and Religion*, p. 208, also writes that "if Newton's basic contention, that the Golden Rule is dictated by the light of reason, is granted, he succeeded in his effort to place Christianity as he understood it on a rational plane, but what was the Christianity that he understood? Through his words blew the chill wind of death for Christianity, for Newton equated Christ with reason and pruned Christianity of all supernatural elements." The theme of a latent skepticism in Newton's rationalistic biblical interpretation is overshadowed in the magisterial detail of Westfall's magnificent biography, *Never at Rest*. Westfall returns to this theme in "Isaac Newton's *Theologiae Gentilis Origines Philosophicae*," in *The Secular Mind: Essays Presented to Franklin L. Baumer*, ed. Warren Wagar (New York: Holmes and Meier, 1982), pp. 15, 32, where he argues that, while Newton has none of the "essentially negative deistic spirit," his rationalistic conclusions are similar to theirs.
24 Frank E. Manuel, *The Changing of the Gods* (Hanover, N.H.: University Press of New England, 1983), p. 15. Cf. Frank E. Manuel, *The Religion of Isaac Newton* (Oxford: Oxford University Press, 1974), esp. p. 49.

emphasizes that, in his view, Newton's scientific rationality affects his approach to interpreting biblical prophecies. The tests of truth in biblical interpretation, as in scientific demonstration, are "constancy and consistency."[25] Westfall also concludes that Newton brings the rigorous standards of scientific demonstration to the interpretation of biblical prophecies. Both Manuel and Westfall agree that it is primarily Newton's science, or his scientific rationality, that influences his theology and not his theology that influences his science. Westfall explicitly states that it is "more likely to find the flow of influence moving from science, the rising enterprise, toward theology, the old and (as we know from hindsight) fading one."[26]

By limiting Newton's "holy alliance between science and religion" strictly to the design argument – which they consider merely to be a concomitant corollary following from Newton's science – and to a rationalistic approach to Scripture, Westfall tended to prise the main body of the *Principia* apart from the scripturally based theology, which Newton devoted so much time to perfecting in his voluminous unpublished manuscripts. Newton's theology, for various biological or psychological reasons – for example, because of senility or because of an awareness of his own approaching death – is viewed as logically and historically disconnected from Newton's method in natural philosophy. A. Rupert Hall writes that "Perhaps it was inevitable that the imagination which in youth was fertile and original in comprehending the natural world by means of experiment and mathematics should seem so banal, in old age particularly, when facing the imponderables of mortality and immortality."[27]

25 Manuel, *The Changing of the Gods*, p. 23. Several scholars disagree with what I call the "disconnectedness thesis" according to whose proponents, such as Westfall, Newton's science is logically separable – and so separated – from his theology. See William H. Austin, "Isaac Newton on Science and Religion," *Journal of the History of Ideas* 31 (1970): 521–42; James E. Force, "Newton's God of Dominion: The Unity of Newton's Theological, Scientific, and Political Thought," in Force and Popkin, *Essays*, pp. 75–102; Dobbs, *Janus Faces*; and Rob Iliffe, " 'Making a Shew': Apocalyptic Hermeneutics and the Sociology of Christian Idolatry," in Force and Popkin, *The Books of Nature and Scripture*, pp. 55–88.
26 Westfall, "Newton's Theological Manuscripts," in Bechler, *Contemporary Newtonian Research*, pp. 139–40.
27 A. Rupert Hall, *Isaac Newton: Adventurer in Thought* (Oxford: Blackwell, 1992), p. 374. In his *Microsoft Encarta Encyclopedia* article, Westfall states that most of Newton's reflections on "alchemy, mysticism, and theology" come in his "later years." The first influential interpreter to attribute Newton's interest in theology to a mental breakdown owing to age and senility and, hence, to remove his natural philosophy beyond an intellectual "cordon sanitaire" where it was uncontaminated by his theology was Jean-Baptiste Biot. In 1821 Biot penned a brief biography of Newton in the *Biographie Universelle*. Biot relied upon a note by Christiaan Huygens according to which Newton became deranged around 1692–4 when his dog tipped over a candle causing a fire in which all his writings from the preceding twenty years burned.

Finally, however much attention is paid to the presence of the generally provident creator of Newton's mechanical universe or to his hard-nosed rationalistic reading of scripture, Newton's mechanical universe becomes theologically neutered: "Newton's universe, when stripped of metaphysical considerations, as stripped it would be, is an infinite void of which only an infinitesimal part is occupied by unattached material bodies moving freely through its boundless and bottomless abyss, a colossal machine made up of components whose only attributes are position, extension, and mass."[28]

Such writers, in ignoring or downplaying what Newton regarded as its nobler parts, thereby fitting him up for the role of heroic "first mover" or "final cause" in the grand unfolding of the Scientific Revolution, yield an anachronistic and partial Newton. Nor is it valid to assert that "historians have found little connection" between his alchemical and theological works and his science. Newton does not strip his universe of metaphysical considerations simply because in his voluntaristic theory of God's nature God is always supervising nature, whether directly through acts of special providence or indirectly through the creative act of general providence, which is revealed in the design argument. To appreciate and understand the subtlety of Newton's method in science, one must come to terms with its metaphysical substrate: the Lord God whose supreme dominion includes his specially provident power to continue to intervene in his creation through miraculous acts. Newton's universe is not disconnected from God after the divine architect creates it and, consequently, his universe cannot be shorn of metaphysics. Newton's view of divine special providence, willful acts of supreme divine power since the creation recorded in fulfilled scriptural prophecy, has marked consequences for his view of what sort of knowledge we may expect the new science to provide. Newton is apparently immune to whatever logical tension any modern interpreter discerns embedded in his thought. In his own "holy alliance" between science and religion, Newton seeks the nature of God in both the divinely authored books of nature and Scripture and blends general and special divine providence in his conclusion, which is fully, even logically, satisfying to him. We must revise our

> According to Biot, Newton never recovered the full use of his mental faculties, which precipitated a turn from science to theology. Henry Brougham's English translation of this article in 1829 brought it wide circulation in nineteenth-century England. See Richard H. Popkin, "Newton and Fundamentalism, II," in Force and Popkin, *Essays*, p. 165. Not to be overlooked is Richard Yeo, "Genius, Method, and Morality: Images of Newton in Britain, 1760–1860," *Science in Context* 2, no. 2 (1988): 257–84.
>
> 28 This recent statement of the "disconnectedness" thesis is by Christianson, *In the Presence of the Creator*, p. 312. Other versions of it are to be found in Randall, "What Isaac Newton Started," pp. xi–xii, and Edward W. Strong, "Newton and God," *Journal of the History of Ideas* 7 (1952): 167.

understanding of Newton's role in the Scientific Revolution. Dobbs, by beginning with the whole Newton and not with some rigid conception of a cataclysmic, 144-year-long revolution that left the modern world in large part secular has succeeded, in her article in *Isis*, in forcing us to revise our own view not just of Newton but also of the Scientific Revolution. Because "Newton's intellectual development is best understood as a product of the late Renaissance, a time when the revival of antiquity had conditioned the thinkers of Western Europe to look backward for Truth,"[29] we must also, she argues in *Isis*, revise the long-established idea of a radical, 144-year-long shift in world views in which, as Westfall states, "A once Christian culture has become a scientific one." The proper line of interpretation must, as she realized, begin with Newton and then move on to determine what effect an unexpurgated Newton has upon a concept such as the Scientific Revolution.

Newton's Synthesis of General and Special Providence

Newton's Legacy

Many of the Christian virtuosi responsible for founding the Royal Society insist that their scientific inquiry into the operations of the laws of nature is the strongest buttress for religion. In a series of apologetic books and pamphlets these scientist-theologians maintain that a scientific study of nature reveals two specific kinds of divine direction and control of created nature, that is, two types of divine providence. A study of nature illustrates both divine "general providence" and divine "special providence." General providence refers to God's action in the original creation of nature. In the beginning, God created the material frame of nature *and* he structured it to function in obedience to the laws of nature, which he also created. The original creative act of general providence is unique. In contrast, special providence refers to particular acts of direct divine intervention since the one-off act of creation cancels or contravenes the ordinary course of the generally provident operation of nature. The historical record of the Bible reveals, in accomplished miracles and fulfilled prophecies, God's continued beneficent care and governance of the natural world through direct acts of special providence. In striving to retain both a generally provident celestial watchmaker and the specially provident God of revelation fully capable of miraculous intervention in the regular course of nature, the Christian virtuosi of the later seventeenth

---

29 Dobbs, *Janus Faces*, p. 10.

century – such as Wilkins, Charleton, Boyle, and Sprat – and Newtonians of the early eighteenth century, such as Whiston and Clarke, achieve an internally coherent synthesis that satisfied them even if it strikes modern interpreters as incoherent. For all these thinkers, God possesses both sorts of providential power.[30]

Newton and the Lord God of General Providence

By 1703, when Newton was elected president of the Royal Society, the Society's apologetic ideology *generally* reflected Newton's personal view that his own scientific theories provided, through the design argument, a solid foundation for demonstrating a generally provident deity.[31] In the second edition of the *Principia* (1713), Newton himself sets out the creator-architect deity of general providence in his famous General Scholium. After his eloquent description of the solar system and his conclusion that "This most beautiful system of the Sun, planets, and comets, could only proceed from the counsel and dominion of an intelligent and Powerful Being," Newton emphasizes the specially provident nature of a God who sternly governs his generally provident creation: "This Being governs all things . . . as Lord over all; and on account of his dominion he is wont to be called *Lord God . . .* or Universal Ruler; for God has a respect to servants; and *Deity* is a dominion of God . . . over servants."[32]

Long before the General Scholium, however, Newton busily traced the effects of a generally provident deity in the book of Scripture, a book as open to Newton as the heavens. Reading about the nature of God as it is revealed in Scripture or as it is revealed in the Book of Nature does not matter for Newton who experiences no modern anx-

---

30 James E. Force, "The Breakdown of the Newtonian Synthesis of Science and Religion," in Force and Popkin, *Essays,* pp. 144–7.
31 Despite the goal of buttressing religion with science, announced in the Charter of the Royal Society, recent scholarship has shown the impossibility of linking members of the early Royal Society with any particular set of religious beliefs. The early society included adherents to a variety of sects including Puritans ejected from ecclesiastical office, converts to Catholicism, Anglican courtiers, country gentlemen, Presbyterian Royalists, incipient latitudinarians, and even two deists. L. Mulligan, "Anglicanism, Latitudinarianism and Science in Seventeenth Century England," *Annals of Science* 30 (1973): 218. Cf. Charles Webster, *The Great Instauration: Science, Medicine and Reform, 1626–1660* (New York: Holmes and Meier, 1976), p. 496, and Michael Hunter, *Science and Society in Restoration England* (Cambridge: Cambridge University Press, 1981), chap. 7.
32 *Sir Isaac Newton's Mathematical Principles of Natural Philosophy,* translated into English by Andrew Motte in 1729. The translations revised, and supplied with an historical and explanatory appendix, by Florian Cajori, 2 vols. (Berkeley: University of California Press, 1934), 2:543.

iety about maintaining tidy boundaries between reason and faith. Faced with the necessity of entering the Anglican priesthood in order to retain his fellowship at Trinity College, in the early 1670s Newton began an intensive study of theology and of the history of the early church. From the period between 1672 and 1675 there is a sheet summarizing Newton's conclusions about both the nature of Christ *and* the nature of God the Father. Proposition 5, for example, proclaims that "The Son in several places confesseth his dependance on the will of the father."[33] Proposition 10 is even more instructive:

> It is a proper epithete of ye father to be called almighty. For by God almighty we always understand ye Father. Yet this is not to limit the power of ye Son, for he doth what soever he seeth ye Father do; but to acknowledg y^t all power is originally in ye Father & that ye son hath no power in him but w^t derives from ye father for he professes that of himself he can do nothing.[34]

Frank E. Manuel discusses a fragment from Yahuda MS 15.5. Manuel argues that Newton utilizes his concept of the Lord God too many times in too many contexts for it to be merely coincidental to his great dispute with Leibniz. The text from this early manuscript prefigures the "General Scholium" of 1713:

> If the father or son be called *God*, they take the name in a metaphysical sense as if it signified Gods metaphysical perfections of infinite eternal omniscient omnipotent whereas it signifies the same thing with Lord and King, but in a higher degree. As we say my Lord, our Lord, your Lord, the King of Kings, and Lord of Lords, the supreme Lord, so we say my God, our God, your God, the God of Gods, the supreme God, the God of the earth, the servants of God, serve other Gods: but we do not say my infinite, our infinite, your infinite, the infinite of infinities, the infinite of the earth, the servants of the infinite, serve other infinities. When the Apostle told the Gentiles that the Gods which they worshipped were not Gods, he did not meane that they were not infinities, (for the Gentiles did not take them to be such:) but he meant that they had no power and dominion over man. They were fals Gods; not fals infinities, but vanities falsely supposed to have power and dominion over man.[35]

For Newton, only a God of true and supreme dominion is a supreme and true God. Newton relentlessly pursues the logical consequences of this voluntaristic monotheism in a variety of vital doctrinal

---

33 Newton, Yahuda MS 14, at the Jewish National and University Library, Jerusalem.
34 Ibid.
35 Manuel, *The Religion of Isaac Newton*, p. 21.

areas such as Christology, idolatry, predestinarianism (what Newton calls crucial doctrines or "milk for babes"), and the more problematic doctrine of psychopannychism (or "strong meat for elders"). A manuscript note from the early eighteenth century reaffirms the Christological consequences of this metaphysical voluntarism regarding the deity. Newton still, some decades after his intensive theological researches of the early 1670s, views God as substantively one. Jesus, admittedly, derives from God a "unity of dominion, the Son receiving all things from the Father, being subject to him executing his will, sitting in his throne and calling him his God, and so is but one God with the Father as a King and viceroy are but one King. For the word God relates not to the metaphysical nature of God but to dominion."[36]

Ordinary mortals are no less under the dominion of God than Jesus Christ. In an entry to his theological notebook from the 1670s there is a tantalizing hint that Newton is much impressed by Saint Paul's comparison of the relation between God and man to that between a potter and his clay. In an entry entitled simply "Predestinatio," Newton quotes the famous ninth chapter of Romans where Saint Paul places the eternal fate of men under the dominion of their supreme Lord God:

What shall we say then? Is there unrighteousness with God? God forbid. For he saith to Moses, I will have mercy on whom I will have mercy, & I will have compassion on whom I will have compassion. So than it is not of him that willeth, or of him y^t runneth, but of God that sheweth mercy. For ye scripture saith unto Pharoh Even for this same purpose have I raised thee up that I might shew my power in thee, & that my name might be declared throughout all ye Earth. Therefore he hath mercy on whom he will have mercy, & whom he will he hardeneth. Thou wilt say then unto me; why doth he yet find fault? For who hath resisted his will? Nay but O man, who art thou that repliest against God? Shall ye thing formed say to him y^t formed it why hast thou made me thus? Hath not the potter power over the clay of ye same lump to make one vessel unto honour and another unto dishonour?[37]

Most significantly, a God of supreme dominion may not be divided up in terms of his substance. Such a dilution of deity in any form of idolatry is one of Newton's central theological concerns throughout his life. Worshiping anything but the Lord God of true and supreme dominion lessens the absolute nature of God's dominion and consti-

---
36 Newton, Yahuda MS 15.1.
37 Newton, *Commonplace Book*, s.v. "Predestinatio," Keynes MS 2. All of the Keynes collection of Newton manuscripts are located in King's College Library, Cambridge.

tutes idolatry. Any Christian communion adhering to the Trinitarian creed (which lessens the Father's dominion by promoting the coeternality and consubstantiality of the Son) becomes his target beginning with his theological notebook of the early 1670s. Roman Catholicism is particularly abhorrent to the young Puritan. "Never," he writes, "was Pagan Idolatry so bad as the Roman."[38] Newton's manuscript entitled "Paradoxical Questions concerning ye morals & actions of Athanasius & his followers" is devoted to showing how such idolatrous doctrines as the "invocation of saints," the Trinity and the continuous, conscious immortality of the soul were introduced into Catholicism by the "anti-Christian" Athanasius.

Finally, Newton is a soul-sleeping Christian Mortalist, that is, a psychopannychist, for whom the immortal substance of the soul is not, at the death of the body and prior to the resurrection, a conscious entity, even though it does continue to exist in a "sleeping" state. In this heretical doctrine, Newton emphasizes the Lord God's miraculous, specially provident power to awaken souls (and bodies) from the sleep of the dead to meet their maker.[39]

Newton imparts his often heretical constructions of various problematic texts to a trusted few. Newton writes that not everyone will understand his interpretations of prophetic doctrine but:

> If they are never to be understood, to what end did God reveale them? Certainly he did it for ye edification of ye church; and if so, then it is as certain y^t ye church shall at length attain to ye understanding thereof. I mean not all that call themselves Christians but a remnant, a few scattered persons which God hath chosen, such as without being led by interest, education, or humane authorities, can set themselves sincerely & earnestly to search after truth. For as Daniel hath said that ye wise shall

---

38 Newton, Yahuda MS 14, f. 9v. Cited in Westfall, *Never at Rest*, p. 315.
39 For Newton, as for many of the Jewish theologians whom he studied as well as for other radical theologians within this native English tradition, the individual soul is not inherently, eternally conscious and it is only immortal if God acts to reawaken it at the resurrection. The key feature of Newton's answer to the idolatrous Athanasians who distort the literal words of Scripture in order to discover a continuously conscious, substantial soul is a list of the primarily Old Testament texts that support his own pscyhopannychism. The fanciful, idolatrous, and – sadly, in Newton's view – "orthodox" Christian interpretation that the soul is, by its nature, consciously immortal is offered by people who do not literally understand "that ye interval between death & ye resurrection is to them that sleep & perceive it not, a moment." See James E. Force, "Jewish Monotheism, Christian Heresy, and Sir Isaac Newton," in *The Expulsion of the Jews: 1492 and After*, ed. Robert Waddington and Arthur H. Williamson (New York: Garland Press, 1993), pp. 259–80.

understand, so he hath said also that none of ye wicked shall understand.⁴⁰

Newton's theory about the absolute dominion of God is central to all the other aspects of his theology in particular and to the rest of his thought in general. To the question of what ties together Newton's "holy alliance" of science and religion, I answer that it is his particular conception of the one supremely powerful Lord God, a doctrine witnessed by the generally provident handiwork of the heavens and by the supreme scriptural commandment to have no other "gods." The Lord God of supreme dominion glues Newton's religion and science together in a synthesis that he regards as indissoluble. A God of such supreme dominion exhibits his power in both the generally provident act of creation and the specially provident, miraculous interventions in that creation which have followed and are recorded in scripture.

Newton's Providential Lord God

Newton, like his predecessors and contemporaries, always attempts to preserve both the generally provident architect of the design argument and the specially provident God who daily exercises dominion over his created servants down to the last sparrow's last flight. In the complete system of theism worked out by some of Newton's followers, God usually displays his specially provident power of dominion through fulfilled biblical prophecies, which record the long history of divine involvement with his creation, and, especially in the past before miraculous gifts ceased in the apostate churches, through miracles.

Newton (and such Newtonians as Clarke and Whiston) believe in "miracles," a term they use in two senses. First, God's continuous act of preserving the natural world is itself an instance of special providence. Second, when pressed by Leibniz about their reduction of God to the status of an inferior clock repairman or when considering scriptural reports of phenomena for which there is no apparent (or conceivable) mechanical explanation, they accept, however reluctantly, standard miraculous accounts and view them as proof of God's directly interactive special providence.

---

40 Newton's large manuscript entitled a *Treatise on the Apocalypse* is in the possession of the Jewish National and University Library and comprises "8 bundles" which contain Sotheby Lot 227 and parts of 228. It is cataloged as "A large *Treatise on the Apocalypse*, incomplete at the end, (various versions) (similar to Keynes MS 5 in King's College, Cambridge, and to Yah MS 9 but materially different from both)." Yahuda MS 1 dates from quite early in Newton's career, probably from the 1660s.

Because of God's sovereign nature as "Lord God," the laws of nature are, in an important sense, *both* necessary and contingent. They are necessary *only* while God, who created them, maintains them in operation. Newton is no enthusiast and he labors mightily to separate the few cases of genuine historical (often catastrophic) miracles from the many cases of idolatrous and false ones.[41] Newton claims that "a continual miracle is needed to prevent the sun and fixed stars from rushing together through gravity."[42] Newton also observes, with Whiston, that miracles in the traditional sense are often simply misunderstandings on the part of the vulgar. Miracles, writes Newton, "are not so called because they are the works of God, but because they happen seldom and for that reason excite wonder."[43] Even so, he accepts the record of past miracles, especially those recorded in miraculously fulfilled prophecies. In his letter to Bentley congratulating him upon his use of the design argument in his inaugural Boyle Lectures, Newton cryptically refers to another argument in addition to the design argument: "there is yet another argument for a Deity wch I take to be a very strong one, but till ye principles on wch tis grounded be better received I think it more advisable to let it sleep."[44]

Locke,[45] Newton,[46] and Whiston use the many instances of apparently fulfilled prophetic predictions recorded in the Bible to supplement the design argument of natural religion, to round it out, and to draw from the total package the special, as well as the general, providence of God. Newton's own theology is always the design argument in conjunction with the argument from prophecy. Together, both illustrate the comprehensiveness of God's providential Lordship over nature. Newton, for example, claims that "giving ear to the Prophets is

---

41 Newton, "Paradoxical questions concerning ye morals & actions of Athanasius & his followers," William Andrews Clark Memorial Library MS, University of California, Los Angeles.

42 *The Correspondence of Isaac Newton*, 7 vols., ed. H. W. Turnbull, J. F. Scott, A. R. Hall, and L. Tilling (Cambridge: Cambridge University Press, 1959–77), 3:336.

43 *Sir Isaac Newton: Theological Manuscripts*, ed. H. McLachlan (Liverpool: University of Liverpool Press, 1950), p. 17.

44 Newton to Bentley, December 10, 1692, in *Correspondence*, 3:233. See James E. Force, "Newton's 'Sleeping Argument' and the Newtonian Synthesis of Science and Religion," in *Standing on the Shoulders of Giants: A Longer View of Newton and Halley*, ed. Norman J. W. Thrower (Berkeley: University of California Press, 1990), pp. 113–19.

45 Locke writes that the miracles of Jesus are significant only because they are predicted in biblical prophecy. See his work *The Reasonableness of Christianity as Delivered in the Scriptures* (London, 1695), p. 55.

46 Newton agrees that "the events of things predicted many ages before will then be a convincing argument that the world is governed by *providence*." *Observations upon the Prophecies of Daniel and the Apocalypse of St. John* (London, 1733), p. 252.

a fundamental character of the true Church. For God has so ordered the Prophecies, that in the latter days *the wise may understand, but the wicked shall do wickedly, and none of the wicked shall understand.* Dan. xii. 8–10."[47]

Newton's Specially Provident Millennium: Part of the Scientific Revolution or a Problem with Westfall's "Problem"?

Interpreting scripturally based doctrines and understanding prophecies correctly is essential to Newton's complete theology, and the proper method of biblical interpretation constitutes a large portion of Newton's unpublished theological manuscripts. Because Newton's criteria of meaning are rational, in contrast to modern interpreters such as Westfall, who approach this aspect of Newton's thought armed with the view that Newton was the most important "scientist" during the era when theology was evicted from the premises, they have mistaken his method for his message and emphasized his rational approach to Scripture as the essence of his "holy alliance" between science and religion rather than the doctrines which that analysis reveals. In his "Rules for Methodizing the Apocalypse," Newton describes Rule 9 for interpreting millennial and apocalyptic language in biblical prophecies. The exegete must

> choose those constructions which, without straining, reduce things to the greatest simplicity. . . . Truth is ever to be found in simplicity, and not in the multiplicity and confusion of things. As the world, which to the naked eye exhibits the greatest variety of objects, appears very simple in its internal constitution when surveyed by a philosophic understanding, and so much the simpler, the better it is understood, so it is in these visions. It is the perfection of all God's works that they are done with the greatest simplicity.[48]

Yet the substantive doctrinal points that this method yields are, by far, the most important aspect of Newton's interpretative method. In addition to yielding heretical doctrines about, for example, the generally provident nature of the one Lord God, his interpretations of the millennial kingdom and the circumstances of the Day of Judgment show his expectations regarding the Lord God's continuous, miracu-

47 Ibid., p. 14.
48 Newton, Yahuda MS 1.1, reprinted in Manuel, *The Religion of Isaac Newton*, p. 120. Newton's Rule 9 in his "Rules for Methodizing Scripture" clearly echoes the famous "principle of simplicity" with regard to physical phenomena stated in Rule 1 of the second edition of the *Principia*.

lous, specially provident involvement in his creation right up to the apocalypse (and beyond).[49]

Newton is a premillennialist who believes that the second coming of Christ will precede his thousand-year reign on earth.[50] In Newton's view, following the ruin of the wicked nations, following the "end of weeping and of all troubles," and following the "return of the Jews," then Christ will return for his millennial reign.[51] For Newton, the millennial New Jerusalem is the inheritance of the mortals then alive *and* individually resurrected saints specifically chosen by God to help Christ rule during the millennium. During Christ's initial thousand years of dominion over this earth and the New Jerusalem, the earth will be at peace in a manner unknown since the primeval paradise of Eden. Newton goes on to describe the interaction of the immortal "children of the resurrection" with the mortals who share their abode during the millennium in most striking terms:

> we are not to conceive that Christ and the Children of the resurrection shall reign over [mortals] the nations after the manner of mortal Kings or convers with mortals as mortals do with one another; but rather as Christ after his resurrection continued for some time on earth invisible to mortals unless [when] upon certain occasions when he thought fit to appear to [mortals] his disciples; so it is to be conceived that at his second coming he and the children of the resurrection shall reign invisibly unless they shall think fit upon any extraordinary occasions to appear. And as Christ after some stay in or neare the regions of this earth ascended into heaven so after the resurrection of the dead

---

49 Westfall writes that "Since Newton explicitly spurned the notion that prophecies can enable man to foretell the future, he did not use them to predict any millennium or utopia. By comparing the prophecies with recorded history, which of course still lay in the future as far as the authors of the prophecies were concerned, he merely sought to demonstrate God's governance of the world through His plan for human society." *Science and Religion in Seventeenth-Century England*, pp. 216–17. Newton definitely prohibits predictions about when future, as yet unfulfilled prophecies (such as those regarding the millennium), will come to pass. But a strong, inductively grounded, rational expectation based upon the record of what he takes to be the specially provident intervention of the deity in natural and human history in historically fulfilled prophecies, is not the same thing as a "prediction." Newton expects a millennial utopia but he does not predict when or how absolutely. He feigns no such hypotheses absolutely but he does nevertheless have good grounds for expecting a particular scenario, which he describes with ardor.

50 See James E. Force, "Millennialism," in *The Blackwell Companion to the Enlightenment*, ed. John Yolton (Oxford: Blackwell, 1991), pp. 331–2.

51 Newton, *The Synchronisms of the Three Parts of the Prophetick Interpretation*, "Of the [world to come,] Day of Judgment and World to come," published as appendix B in Manuel, *The Religion of Isaac Newton*, pp. 126–8.

it may be in their power [also] to leave this earth at pleasure and accompany him into any part of the heavens, that no region in the whole Univers may want its inhabitants.[52]

At the end of this thousand year reign of Christ (with his specially chosen, individually resurrected assistants, the children of the resurrection), Newton believes that Satan will rise up from his pit and make apocalyptic war. Satan's final defeat will be followed by the general resurrection of martyrs and saints not resurrected for the millennium, that is, those who "sleep in the dust."[53] Newton is ultimately convinced that, in all likelihood, the millennium is nearly at hand and that those with eyes to see and ears to hear – the true Christian "remnant" – will understand the truth. Newton's constant reference to the following passage of Daniel is crucial to understanding his silence *and* his view of himself as "scientific" biblical exegete: "*In the time of the end the wise shall understand, but none of the wicked shall understand. Blessed is he that readeth, and they that hear the words of this Prophecy, and keep those things which are written therein* [Dan. xi. 4, 10; Apoc. 1.3]."[54]

What, one may now legitimately ask, do Newton's intensely private musings about the Lord God who will, at some point, providentially set in train the Revelation scenario, have to do with Newton's method in natural philosophy or with his expectations of what that method can achieve? Are not the scientific portions of the *Principia* still scientifically valid even if Newton embeds them in his fervent millennial expectations regarding the "future state" of the "children of the resurrection? Cannot the *Principia* itself, if not Newton, be considered as the "final cause" of the Scientific Revolution?

Given how Newton understands the absolute power and total dominion of the specially and generally provident Lord God, Newton's metaphysical theology necessarily interacts with his scientific epistemology. He accepts the reality of direct divine intervention in nature through miraculous ("specially provident") acts of will (which are simultaneously supreme acts of power) that interrupt the ordinary coursing of nature and nature's generally provident laws, and his reading of prophecy leads him to expect a "new heaven and a new earth" when the laws and principles of the current system may no longer apply. For Newton, the primacy of God's power results in a distinctive contingency in the natural order even while Newton ac-

---

52 Ibid., pp. 135–6.
53 James E. Force, "The God of Abraham and Isaac (Newton)," in Force and Popkin, *The Books of Nature and Scripture*, pp. 179–200.
54 For Newton's view of the centrality of this text, see his *Observations upon the Prophecies of Daniel, and the Apocalypse of St. John. In Two Parts* (London, 1733), 2:251.

knowledges the virtual necessity of that order in its ordinary ("generally provident") current operation and provides a rational method for studying its operation. For Newton, the whole of creation is "subordinate to [God], and subservient to his Will."[55] This is the theological and metaphysical background to Newton's most famous methodological statement in his fourth rule of reasoning:

> *In experimental philosophy we are to look upon propositions inferred by general induction from phenomena as accurately or very nearly true, notwithstanding any contrary hypotheses that may be imagined, till such time as other phenomena occur, by which they may either be made more accurate, or liable to exceptions.*
>
> This rule we must follow, that the argument of induction may not be evaded by hypotheses.[56]

Modern interpreters have discerned in this epistemological injunction to proceed empirically, while eschewing the goal of absolute certainty, good reason to disconnect Newton's epistemology from his theology. They have seen in this injunction the triumph of modern, mitigated skepticism in natural philosophy, which some detect, at least embryonically, in the "holy alliance" between general and special providence. But Newton's universe is *not*, and for Newton can *never* be, stripped of "metaphysical considerations" because its creator, owner, and operator is the Lord God of Israel. Fifty years prior to Hume, Newton, from a vastly different metaphysical and theological starting point, argues emphatically that the future need not resemble the past simply because of the Lord God's absolute power to change natural law and that, consequently, we must mark all the consequences of this fact in regulating our expectations about what sort of human knowledge scientific empiricism can provide. As Newton understands them, natural laws and mechanisms work, in general for now, but, in the millennium and beyond, the "children of the resurrection" will live in a "new heaven" and a "new earth" where the former natural laws need not apply if the Lord God so chooses. Because of the power of the Lord God to change natural law at will – and because of his prophetic promise to do so soon in unfulfilled prophecies – the study of this natural order must be cautiously provisional. When Newton refers above to "such time as other phenomena occur" that alter our inductively grounded propositions, it is not unreasonable to suppose that he may be referring to the radical changes in the natural order that he expects during the millennium.

55 Isaac Newton, *Opticks or a Treatise of the Reflections, Refractions Inflections and Colours of Light*, based on the 4th ed. (London, 1730), p. 403.
56 *Sir Isaac Newton's Mathematical Principles of Natural Philosophy*, 2:400. This rule is not added to the *Principia* until the second edition of 1713.

When Dobbs questioned the Whiggish notion of Newton as either "first mover" or "final cause" of the Scientific Revolution," she also questioned the entire Whiggish attempt to sever Newton's particular alliance between general and special divine providence and, also, between divine providence – as a whole – and science. The *Principia* may indeed be, as Hampson might prefer to say, "logically viable" without the theological baggage of divine special providence. For us, as for Laplace who famously remarks to Napoleon that his own particular mechanical "system of the world" has no need of the hypothesis of God, Newton's theology need not necessarily be related to his science.[57] But, because God's Absolute Dominion profoundly affects Newton's natural philosophy just as it affects his particular theological and political doctrines, the keystone of Westfall's entire structure of the Scientific Revolution – when a "once Christian culture has become a scientific one," presupposing their watertight separation – is removed.

Of Pythagoras, F. M. Cornford has written that "The vision of philosophic genius is a unitary vision. Such a man does not keep his thought in two separate compartments, one for weekdays, the other for Sundays."[58] What is true of Pythagoras is true of Newton, and it alters how we must think of him and how we must be critical of "the" early modern revolution in science. The latest and by far the best and most detailed recent work coming out about Newton from such outstanding young historians as Robert C. Iliffe and Stephen David Snobelen[59] illustrates further the necessity of starting with the project of getting Newton right rather than with any a priori hypothesis regarding an insoluble "problem" of reconciling religion and science, which

---

57 See Alexandre Koyré, *From the Closed World to the Infinite Universe* (Baltimore: Johns Hopkins University Press, 1957), p. 276.
58 F. M. Cornford, *Before and after Socrates* (Cambridge: Cambridge University Press, 1968), p. 66.
59 Iliffe, " 'Making a Shew,' " pp. 55–88; " 'Is He Like Other Men?': Newton, the *Principia*, and the Author as Idol," in *Culture and Society in the Stuart Restoration: Literature, Drama, History*, ed. Gerald Maclean (Cambridge: Cambridge University Press, 1995), pp. 159–78; "Isaac Newton: Lucatello Professor of Mathematics," in *Science Incarnate: Historical Embodiments of Natural Knowledge*, ed. C. Lawrence and S. Shapin (Chicago: University of Chicago Press, 1998), pp. 121–55; "Those Whose Business It Is to Cavill: Newton's Anti-Catholicism," in *Newton and Religion: Context, Nature, and Influence*, ed. James E. Force and Richard H. Popkin (Dordrecht: Kluwer, 1999), pp. 97–119. Stephen David Snobelen, "Newton as Heretic: The Strategies of a Nicodemite" (M.Phil. diss. Cambridge, 1997); "Caution, Conscience and the Newtonian Reformation: The Public and Private Heresies of Newton, Clarke and Whiston," *Enlightenment and Dissent* 16 (1997): 151–84; "On Reading Isaac Newton's Principia in the 18th Century," *Endeavour* 22 (December 1998): 159–63.

ultimately culminated in "the triumphant advance of secularization on all fronts"[60] but particularly in the revolutionizing of natural philosophy and the concomitant demotion of religion. Dobbs was one of the leaders in the vanguard of this revisionistic movement. She was correct that Newton cannot to be regarded as either the "first mover" or the "final cause" of the Scientific Revolution. He is, rather, a far more complex thinker for whom the Lord God of supreme dominion constitutes the key to understanding the nature of his particular "holy alliance" between science and religion. For Newton, "the problem," as defined by Westfall, did not exist.

60 Westfall, *Science and Religion in Seventeenth-Century England*, pp. 1–2.

13

The Fate of the Date: The Theology of Newton's *Principia* Revisited

J. E. McGUIRE

> And I, spoke eternity, how shall I appear before mankind so that they will not hold me in terror? Thus spoke the lord: I will grant to mankind one moment to grasp you. And he created love.[1]

Betty Jo Teeter Dobbs argues that Newton's *De gravitatione* was composed in the year 1684 or early 1684/5.[2] Most scholars (myself included) have fixed its date of composition no later than 1673. Nevertheless, Dobbs's claim is persuasive if not decisive. I shall assume her dating in reassessing the theological outlook that underlies the structure of Newton's *Principia*. My concern is how Newton's religious and natural philosophical beliefs interrelate within the intellectual context of their making. If Dobbs is right, *De gravitatione* is part of a creative process (begun in the mid-1680s) that includes the composition of the *Principia*. This is a significant point. Seen from this perspective, *De gravitatione* expresses the theological world view that animates the *Principia* itself. To elaborate this theme, I consider "Tempus et locus," a manuscript which can be dated to the early 1690s.[3] This important manuscript, I argued, is part of Newton's revisions for the first edition of the *Principia* (1687), begun directly after that treatise was published. In both *De gravitatione* and "Tempus et locus" he considers the infinity of spatial extension. In *De gravitatione* he articulates its geometrical structure in relation to the being and activity of God.[4] Once these

---

1 Arthur Schnitzler, *Beziehungen und Einsamkeiten: Aphorismen* (Frankfurt: Fischer Taschenbuch Verlag, 1987), p. 34. I thank Peter Long for calling my attention to this aphorism and providing a more nuanced translation than I could provide.
2 Betty Jo Teeter Dobbs, *The Janus Faces of Genius: The Role of Alchemy in Newton's Thought* (Cambridge: Cambridge University Press, 1991), pp. 143–6.
3 J. E. McGuire, "Newton on Place, Time, and God: An Unpublished Source," *British Journal for the History of Science* 2 (1978): 114–29.
4 Isaac Newton, De Gravitatione et Aequipondio Fluidorum, University Library Cambridge (ULC) MS Add. 4003, in Rupert Hall and Marie Boas Hall, ed. and trans., *Unpublished Scientific Papers of Isaac Newton: A Selection from the Portsmouth Collection*

connections are evident, the preface to the first edition of the *Principia* (1687) – which outlines relations between mathematics and nature – throws significant light on the structure of Newton's masterwork.[5]

Let me begin by relating my project to Dobbs's and Westfall's views on the Scientific Revolution. As will become apparent, Dobbs's strictures against the "Whig" historical style are germane to my attempt at delineating the setting of Newton's mathematical and theological views in the *Principia*. That said, I cannot agree with her that the category of the Scientific Revolution needs to be undermined. Westfall is surely right in maintaining that a profound transformation in human thinking occurred during the seventeenth century, the consequences of which still shape the worlds in which we live and flourish. Call them revolutions or what you will, great historical transformations can neither be thought away nor go unnoticed in our historiographies. Dobbs is, of course, right to warn against the willful projection of later values on earlier historical movements: it leads us to see them as inevitable grounds for history's culminating moments. We must try to get the context of historical actors right. Only then can we establish a nuanced historical appreciation of their achievements – and the consequences both intended and unintended. It is in just such a context that I want to place my analysis of Newton's *Principia*.

The Fate of the Date

Dobbs uses a host of arguments to justify her claim that *De gravitatione* was composed in 1684. She first provides some essential background. She rightly conjectures that Newton was astonished on realizing that Kepler's area law (which Newton derived in 1684 from his inertial/centripetal model of planetary motion) agreed too closely with the best established observational data.[6] To Newton this indicated that the heavens may not be filled with a material medium that can retard the passage of the planets in the manner in which terrestrial projectiles are deviated from their predicted paths under the action of the earth's atmosphere. Dobbs argues that Newton's recognition of this fact is recorded in his *De motu corporium in gyrum* written in November 1684. There he first defines "centripetal force" as "that by which a body is

---

in the University Library, Cambridge (Cambridge: Cambridge University Press, 1962), pp. 90–156.

5 Isaac Newton, *Mathematical Principles of Natural Philosophy and System of the World*, trans. Andrew Motte (1729) and rev. Florian Cajori, 2 vols. (Berkeley: University of California Press, 1962), 1:xvii–xviii.

6 Dobbs, *Janus Faces*, p. 131.

impelled or attracted towards some point regarded as its centre."[7] This constituted a conceptual breakthrough: prior to this Newton still appealed to vortical components and to centrifugal force in conceptualizing astronomical motions. In Dobbs's view, 1684 signals a deep transformation in his thinking. It marks the end of his commitment to material and mechanical aethers and to the use of vortical devices in dealing with astronomical motions.[8]

Dobbs also notes that in "De aere et aethere" (ca. 1684) Newton attempts to find evidence for the existence of the aether.[9] There he conjectures that the dampening of a pendulum bob's motion in a vacuum might be potential evidence for the aether's existence. The dampening indicated to Newton that the pendulum's motion "ought not to cease unless, when the air is exhausted, there remains in the glass something much more subtle which damps the motion of the bob."[10] Dobbs observes that this experiment is designed to separate from one another the individual effects of the air and the aether as they act on bodies. Was there an experiment, however, that could detect a property of the aether "by its varying reactions under experimentally varied conditions?"[11] For Newton the answer was in the affirmative.

Given that it permeates the inner pores of terrestrial bodies, the aether ought to interact with their internal parts. Assuming the air can act only on the surface of a pendulum bob, whereas the penetrative aether interacts with its inner parts, if the retardation due to the air is held constant, and if the quantity of matter within the bob is varied, one should detect an increase in the retarding action of the aether commensurate with the increased quantity of matter in the bob. Just such an experiment is reported in the first edition of the *Principia*. Newton fills an empty box, acting as a pendulum bob with various and increasingly heavy metals. The increase expected in retardation due to the increased quantity of matter made very little if any differ-

---

7 Ibid., p. 130; D. T. Whiteside, ed., *The Mathematical Papers of Isaac Newton*, with the assistance of M. A. Hoskin, 8 vols. (Cambridge: Cambridge University Press, 1967–80), 6:30–1.
8 Dobbs reviews carefully all the evidence that indicates Newton's adherence to many aspects of Cartesianism well into the 1680s (*Janus Faces*, pp. 122–9). I have argued that Newton learned many of the terms used in his epistemological discourse from his early reading of Descartes. See J. E. McGuire, *Tradition and Innovation: Newton's Metaphysics of Nature* (Dordrecht: Kluwer, 1995), chap. 4.
9 Dobbs, *Janus Faces*, pp. 134–5.
10 Isaac Newton, "De Aere et Aethere," ULC MS. Add. 3970, ff. 652–3, in Hall and Hall, *Unpublished Scientific Papers*, pp. 227–8.
11 Dobbs, *Janus Faces*, p. 136.

ence: the total resistance of the empty box was to the total resistance of the full box in the proportion of 77 to 78.[12] Newton had failed to detect any clear evidence that the aether existed. Dobbs dryly notes that "if he (Newton) wrote 'De aere et aethere' and performed the new pendulum experiments shortly before he wrote the first draft of *De motu* in 1684, as seems likely, one has there a full and sufficient explanation for Newton's use of 'impelled or attracted' in his first definition of centripetal force."[13] And in his *De motu sphaericorum in fluidis* (December 1684), Newton claims that the aether's resistance is nonexistent, but not the aether itself.[14]

Dobbs argues that Newton's metaphysics of empty space in *De gravitatione* relates to his belief that if the aether possesses any resistance at all, it is nil or very small. At this point he seemed to believe that "by far the largest part of the aetherial space is void, scattered between the aetherial particles."[15] This plausibly explains why Newton aborted his plan in *De gravitatione* to mathematize a "nongravitating fluid."[16] If the aether fails to retard the passage of the heavenly bodies, it need not be brought under mathematical law. Accordingly, in Dobbs's view, the pendulum experiments helped to motivate Newton's metaphysics of nonresisting spaces. As I have noted, she supports her claim by arguments dating the treatise to the period of these experiments and to the composition of the *Principia*.[17] In the larger perspective, Dobbs's interpretation is strengthened by the fact that the conceptual structure of *De gravitatione* (as well as the pendulum experiments) is foreign to the framework of Newton's earlier aethereal systems. These systems, the "Hypothesis" of 1675, the letter to Boyle of 1678/9, and "De aere et aethere" (ca. 1684) employ ontologies that are inconsistent with Newton's emerging ontological commitments in the immediate pre-*Principia* period. As such, these aethers represent a style of thinking that Newton was forced to abandon in the mid-1680s.

Let me now consider Dobbs's main arguments for dating *De gravitatione* in late 1684 or early 1684/5: (1) That dating easily explains the title of a section in a *De motu* draft entitled *De motu corporum in spatiis non resistentibus* written in winter or early spring of 1684/5. In a slightly earlier version, the relevant section is entitled *De motu corporum in mediis non resistentibus*. For Dobbs, the shift from "non-resisting

---

12 Newton, *Mathematical Principles of Natural Philosophy*, pp. 325–6.
13 Dobbs, *Janus Faces*, p. 137.
14 Whiteside, *Mathematical Papers*, 6:261, 285–6.
15 Newton, *De Gravitatione*, in Hall and Hall, *Unpublished Scientific Papers*, p. 147.
16 Ibid., pp. 121, 151–2.
17 Dobbs, *Janus Faces*, p. 143.

media" to "non-resisting spaces" is significant; it reflects Newton's belief, caused by the pendulum experiments, that the aether is mostly constituted by "empty spaces."[18] (2) Newton's reference to force as an "internal principle" conserving a body's motion or rest so that it "endeavors to continue its state and opposes resistance" fits into the conceptual world of the *Principia* in which resisting media are no longer present.[19] (3) The explanation for Newton's violent attack on Descartes in *De gravitatione* becomes evident. Descartes equated body with extension and extension with the material aether. If aetherial media, including Descartes's, had to be rejected (a result that pendulum experiments indicated), space and body must differ. That is, if the resistance of a corporeal aether, even one composed of the finest particles, was negligible or nil, body "cannot be equated with extension."[20] Therefore, if there are aethereal particles, they must scatter widely in the *empty* extension of space, an extension generically different and ontologically separate from bodies: "it is manifest that all this force [i.e., that of retardation] can be removed from space only if space and body differ from one another."[21] Against this background, one is reminded of Newton's devastating critique of vortical mechanisms in book II of the *Principia*.

The High God or the Dwarf God?

Newton's "theological turn" is clear in "Tempus et locus." This manuscript articulates the theological position Newton had reached in the 1690s. His God is a biblical God of dominion who decrees and enacts the laws that govern created things. Providence expresses divine omnipresence so that God, in virtue of being actually present in space and time, is "able to act in all times & places for creating and governing the Universe."[22] In "Tempus et locus" Newton contrasts eight opposing opinions on divine nature. In each case, he supports the second opinion of each pair. They are worth quoting at length:

> Let them therefore consider whether it is more agreeable to reason that God's eternity should be all at once (*totum simul*) or that his duration is more correctly designated by the names Jehovah and "He that was and is and is to come"; (1) that the

18 Ibid., p. 144.
19 Hall and Hall, *Unpublished Scientific Papers*, p. 146.
20 Dobbs, *Janus Faces*, p. 145.
21 Newton, *De Gravitatione*, in Hall and Hall, *Unpublished Scientific Papers*, p. 147.
22 Alexandre Koyré and I. Bernard Cohen, "Newton and the Leibniz-Clarke Correspondence with Notes on Newton, Conti, & Des Maizeaux," *Archives Internationales d'Histoire des Sciences*, nos. 58–9 (1962): 101.

substance of God is not present in all places, or that the Jews more correctly call God Place, that is the substance essential to all places in which we are placed and (as the Apostle says) in which we live and move and have our being; (2) that God is everywhere as regards power and nowhere as regards substance, or that God's power should subsist everywhere in the divine substance as its proper substrate, and exist nowhere separately, and have no medium by which it be propagated from its proper substance into external places; and (3) that place itself and thus the omnipresence of God was created in finite time, or that God was everywhere from eternity; (4) that all the properties of created things argue imperfections to the extent that they are absolutely removed from God, or that creatures share so far as possible the attributes of God as fruit the nature of the tree, and an image the likeness of man, and by sharing tend towards perfection, and to that extent God be discerned in the more perfect creatures as in a mirror; (5) that the more perfect God is he who produces the more imperfect and fewer creatures, or that produces more perfect and countless ones; (6) that the creator's power is infinite, and the possibility of creating only finite, or that the power of God in no wise extends to that what is impossible; (7) that a dwarf-god should fill only a tiny part of infinite space with this visible world created by him, or that the best and the greatest God willed everywhere what was good and did everywhere what he willed.[23]

The passage presents a definite picture of God's nature and existence: God is necessarily in all places and at all times; God's existence is neither supratemporal nor "frozen" in a timeless moment (the *nunc stans* and *totum simul* of eternity); God's existence is substantial and omnitemporal; God acts directly in creation and not through intermediaries; God's presence is not that of a "dwarf-god" lost in the infinities of space, but the direct presence of a "universal ruler" (*pantokrator*) who is actually omnipresent. Newton's God is an artificer who creates in accordance with the principle of creative likeness. Consequently, it is a divinity worthy of worship to whom one should pray.

I have argued that "Tempus et locus" relates closely to topics in the Scholium on space and time of the *Principia* (1687), that its arguments resemble those of *De gravitatione*, and that it anticipates the theological ethos later expressed in the General Scholium to the second edition of the *Principia* (1713).[24] Here I focus on Newton's use of the phrase

23 McGuire, "Newton on Place, Time, and God," pp. 121–2.
24 Ibid., pp. 124–8.

"dwarf-god." He contrasts this conception with the notion of divine omnipresence, pointing to the insignificance of such a god in relation to the vastness of space. But Newton probably has a more important contrast in mind: the contrast between the high God who is actively present in creation and a god similar to the Arian Christ, a separate and distinct god, who enacts the will of the high God by making and governing the cosmos. This contrast highlights a persistent tension in Newton's thinking to which I shall return. This will also allow me to assess the character of Newton's rejection of orthodox Trinitarianism.

Dobbs argues that Arianism made an impact on Newton's thought after he converted to it in 1673.[25] In strict Arian theology Christ is denied equality and coeternity with God the Father. There was a time, Arius argues, when Christ was not. Only the transcendent God existed who would not be named father until after the creation of the son. The Arian Christ, accordingly, shares qualities with other creatures. It is a being first among creatures, the mediator between God and humankind, but not of the same essence as the high God. Arius lost his battle with his essentialist, Trinitarian opponents. In the Nicene Creed of 325, the son is defined as "being of one substance with the Father," a conception that became canonical in Christianity. Although he comes into being, the Arian Christ nevertheless has cosmogonic nobility. Existing before all the ages, the son is the framer of the cosmos, the conqueror of chaos, and the primal agent of God's creativity. Subordinate to God, the Arian Christ is nevertheless central to the operations of divine volition since the high God's creative will is channeled through the agency of the cosmic son.[26]

In his alchemical writings and elsewhere Newton homologizes the Arian Christ to the creative logos of Genesis, the crafting demiurgos of Plato's *Timaeus*, the active Hermes of alchemy, and the divine magus of the Hermetic *Pimander*. In Newton's view, "God and his Son cannot be called one God upon account of their being consubstantial" but only in virtue of a "unity of dominion." By this Newton understands the son as "receiving all things from the Father, being subject to Him, executing His will, sitting in His throne and calling Him his God, and so is but one God with the Father as a king and his viceroy are but one king."[27] Accordingly, acting as the high God's viceroy, the

---

25 Dobbs, *Janus Faces*, pp. 81–2.
26 For discussions of Arianism, especially in relation to seventeenth-century thought, see Dobbs, *Janus Faces*, pp. 80–7, 106–10, 243–9; and James E. Force, *William Whiston, Honest Newtonian* (Cambridge: Cambridge University Press, 1985), pp. 105–13. For a general discussion, see Robert C. Gregg and Dennis E. Groh, *Early Arianism: A View of Salvation* (Philadelphia: Fortress Press, 1981).
27 Yahuda MS Var. I. Newton MS 15.7, f. 1545, quoted in Dobbs, *Janus Faces*, p. 82.

cosmic Christ is the messenger of divinity in the world. From this religiocosmic perspective the high God is transcendent and radically "other" than his creation, but he is still the ultimate cause of all that exists. Consequently, whenever God acts in the world, it is through the offices of a mediating agent. Newton unequivocally states this conception of the mediating agent in the early 1680s: "That God the ffather is an infinite, eternal, omniscient, immortal & invisible spirit whom no eye hath seen nor can see . . . & God does nothing by himself wch he can do by another."[28]

It is the Arian Christ, therefore, who prepares and forms the cosmos and governs it according to God's will. Dobbs argues that the term "protoplast," which Newton uses in "An Hypothesis Explaining the Properties of Light . . ." (1675), reflects his recent conversion to Arianism. In one sense the term means the models, the exemplars, the originals, according to which all things are divinely created. Dobbs observes that Newton also uses it to mean first agent: "Newton's shift from the common meaning of 'protoplast' as first prototype or exemplar of created belongs to the rare second meaning of 'protoplast' as first agent in the creative process can only be a reflection of his recent shift from orthodox Trinitarianism to the Arian heresy."[29] Essential to Dobbs's interpretation is her claim that the protoplast that "has fashioned the copies for nature to imitate must not be interpreted as the supreme Deity but rather as the Christ acting in his capacity as cosmological agent."[30]

But there is an entirely different conception of how the high God's creative power relates to nature in Newton's "physicotheology" from 1690 to 1710. This conception is distinct from his Arian Christology and is prominent in the "classical scholia" manuscripts, material that Newton intended to incorporate into the opening propositions of book III of the *Principia*.[31] These propositions argue for the universality of gravitational action, and the scholia that Newton intended for them are part of a systematic overhaul of the first edition that he began in the early 1690s. In the scholia the high God is conceived as the direct source of activity in nature and the immediate cause of gravitation. In sum: God, being literally present in the limitless expanses of space and time, activates matter directly through his absolute power without the instrumentality of intermediary agents.

In his mature years, however, Newton continues to entertain two

28 Yahuda MS Var. I, Newton MS 15.4, f. 67r, quoted in Dobbs, *Janus Faces*, p. 36.
29 Dobbs, *Janus Faces*, p. 108.
30 Ibid., pp. 109–10.
31 J. E. McGuire and P. M. Rattansi, "Newton and the 'Pipes of Pan,' " *Notes and Records of the Royal Society* 21 (1966): 112–21.

solutions to the cause of gravity: either the supreme and omnipresent Deity directly subsumes gravity providentially, or an immaterial agent activates gravitation without impeding the passage of the heavenly bodies. Newton vacillates from 1690 to 1710 – indeed to the end of his life – as to which solution is compatible with his ongoing attempt to recover the uncorrupted sources of ancient religion and natural philosophy. One thing is certain, however. After 1684, the dense, material and mechanical aethers of the 1660s and 1670s never appear again in Newton's work as serious solutions to problems of activity in nature. My immediate aim, in what follows, is to discuss Newton's view of God's direct action in nature and the conception of "Divine geometry" that this implies. I shall raise doubts about the cosmological implications of his Arianism and suggest that this may not best characterize Newton's position. Certainly, in his attempt to recover primitive Christianity, Newton arrived at a religious position not unlike Arianism. It does not follow from this that his religious belief in a cosmic Christ necessarily shaped his cosmological speculations as Dobbs assumes.

De gravitatione: The Structure of Space and God's Sacred Field

I turn now to Newton's metaphysical account of the creation of bodies in *De gravitatione*. Recall: I am assuming with Dobbs that *De gravitatione* is a foundation document for the composition of the *Principia* of 1687. It is noteworthy and striking that it is a sustained attack on the basic principles of Descartes's philosophy of nature. This indicates that Newton may not have broken decisively with Cartesianism until the eve of composing his own *Principia*.[32]

Mounting a frenzy of criticisms, Newton resolutely dismisses Descartes's theory of motion, especially the paradoxical consequences of its relationalism.[33] Using versions of arguments later to appear in the Scholium on space and time of the *Principia* (including the first version of the bucket experiment), he lays out the ontology of infinite space: (1) Space exists because God exists and is coeternal with God's omnipresent being; it is neither a substance nor an accident but, in combination with absolute time, space possesses a mode of being that is presupposed by all other forms of being; (2) space is eternal and uncreated, and is the receptacle within which all created things come into being; (3) lastly, infinite space posseses an inherent structure of geometrical solids, limitless in number, and differing in size, together

32 See note 26.
33 Newton, *De Gravitatione*, in Hall and Hall, *Unpublished Scientific Papers*, pp. 123–31.

constituting all variety of shapes. These figures are juxtaposed throughout the tridimensional expanses of space without the presence of intervening gaps.[34]

Newton's account resembles the creation myth of Plato's *Timaeus* in which the demiurgos brings order out of chaos by fixing its impermanent elemental traces with geometrical form. Moreover, it is related to the long-standing commentary tradition on the *Timaeus*, which includes figures such as Proclus, Chalcidius, Philoponus, and Patrizi.[35] As I have argued elsewhere, Newton is also in the company of writers like Pierre Gassendi, Walter Charleton, Ralph Cudworth, and Blaise Pascal. These seventeenth-century writers, like the Stoics, the middle Platonists, Philoponus, and the sixteenth-century writer Patrizi, hold that space and time are prior ontologically to the way of being of other created things.[36]

By mobilizing his ontology of uncreated space, Newton explains the genesis and nature of bodies. Things that have bodily natures, unlike space, do not "exist necessarily but by divine will."[37] In regard to divine will, Newton distinguishes three things: divine will itself, the effects of divine will, and space itself that receives these effects. Divine will acts on particular regions in space. By this act it endows the geometrical solids in space with those "features" of bodies present in conditions under which they are perceived in sensory experience. Thus, instead of thinking in terms of "bodies," Newton thinks of space as a system of regions into which material features are ascribed by the divine will. Consequently, space's invisible figures are made manifest and become impenetrable physical realities. Newton conceives these "material features" as active powers able to bring about representations of bodies in the mind of the external presence of bodies.[38]

Accordingly, the concrete effects of God's actions on space "will either be bodies or like bodies. If they are bodies, then we can define bodies as determined quantities of extension which omnipresent God endows with certain conditions."[39] These determined quantities are

34 Ibid., pp. 129–40; J. E. McGuire, "Space, Infinity, and Indivisibility: Newton on the Creation of Matter," in *Contemporary Newtonian Research*, ed. Zev Bechler (Dordrecht: Reidel, 1982), pp. 152–3.
35 Edward Grant, *Much Ado about Nothing: Theories of Space and the Vacuum from the Middle Ages to the Scientific Revolution* (Cambridge: Cambridge University Press, 1981); Richard Sorabji, *Philoponus and the Rejection of Aristotelian Science* (London: Duckworth, 1987), pp. 18–23; McGuire, "Space, Infinity, and Indivisibility," p. 162.
36 McGuire, *Tradition and Innovation*, pp. 2–16; McGuire, "Space, Infinity, and Indivisibility," pp. 162–3.
37 Newton, *De Gravitatione*, in Hall and Hall, *Unpublished Scientific Papers*, p. 138.
38 Ibid., pp. 139–40.
39 Ibid., p. 140.

movable; that is, they are not numerically identical with space but are definite "physical" quantities "such as may be transferred from space to space," are regularized by laws of motion, and are able to excite "various perceptions of the senses and of the representative faculty in created minds."[40] In this manner, corporeal active properties are superadded to space via the direct action of divine will. In short, Newton conceives body as determinate sets of features localizable in space, "things with motion itself constituted by those features pervading now one and now another of space's geometrically structured regions."

This is one vocabulary Newton uses to characterize the ontological nature of what God creates in space. In this vocabulary, bodies are geometrical solids endowed with active powers enabling them to affect other "physical" things and to excite perceptions in the mind. But he adduces another vocabulary and another conception:

> for the existence of these entities there is no necessity that we suppose some unintelligible substance to exist in which as subject there may be an inherent substantial form: extension and the act of Divine will are sufficient. Extension, in which the form of body comes about and is conserved by Divine will, substitutes for the substantial subject; and the effect of the Divine will is the form or the reason of the body denominating all dimensions of space in which body is to be produced.[41]

There are a number of interesting points here. Newton uses the notion of formal causation, stating that God informs the geometrical figures in space with that which makes bodies what they are: in other words, God actualizes a formal nature that exists eternally in the divine mind. Thus, in place of material substance, conceived as a composite of matter and form, Newton substitutes infinite extension as the receptacle that receives the form of matter by God's direct action. In the spirit of the *Timaeus*, uncreated space is a matrix into which divine will impresses the form of matter. Reasoning in this way, Newton rejects the scholasticized version of Aristotelian first matter, namely, the notion of pure potentiality out of which all informed actualities come into being.[42] In its place he articulates the positive structure of spatial extension whose prior reality grounds the genesis of bodies by the direct action of God. Thus the existence of bodies, that is, clusters of properties ascribed to space, depends directly on God such that bodies are no longer conceived to be coextensive with extension in the manner of Cartesian ontology.

40 Ibid.
41 Ibid., pp. 106–7 (my translation).
42 Ibid., p. 140.

Notice that the identity of a determinate quantity of matter over time is grounded in God's action since God instantiates a formal nature in a particular geometrical figure at different times. The criteria identifying the nature of matter are embodied in the forms of creation. Here Newton stands the Aristotelian notion of individuation on its head. Instead of matter as that which individuates a particular instance of a form, it is formal natures themselves that qualitatively individuate the particular figures in space. Furthermore, God does not recreate similar conditions in successive regions of space; he maintains the same formal reality in different parts of space through a succession of times. In this way the continuity of motion is the real effect of God's motion.

Newton's conception of how God stands to creation goes beyond the notion that spatial solids are rendered corporeal by divine will. It is important to note that *De gravitatione* is a story of how the material cosmos comes into being in sacred space. If divine action is directly embodied in the creative act, its effects are sacred as they manifest themselves in the created world. Since divine will inscribes effects in space by making geometrical solids into bodies, *De gravitatione* depicts a sacred cosmology – in fact, a cosmology of sacred places. Newton's spatial solids, now constituted as sacred spaces infused with divine power, are breaks in the homogeneity of spatial extension. God's act of "informing" space constitutes an hierophany; that is, an act that infuses divinity into the limitless expanses of space. The term "hierophany" combines the Greek term ἱερός (meaning that which is filled with or manifests divine power) and φαίνω (to bring to light, to reveal or make manifest). Thus we have the conception that created things, *qua* created things, are holy and sacred.[43] Understood in this way, Newton's *De gravitatione* depicts a sacred cosmogony. Central to its metaphysics is the idea that the divine act makes physical reality in space an act that embodies the ontological founding of the material cosmos. In short, basic to this conception is the vision that infinite and uncreated space *is* the sacred field of God.

The Arian demiurgos is not present in *De gravitatione*. Nevertheless, its account of God's creative activity can be homologized to the divine craftsman image of the *Timaeus*, namely, the maker of artifacts according to the archetypes of eternity. But there is an important difference. Plato's demiurgos, in bringing forth permanence in the physical cosmos, faces the "nurse of becoming" and the recalcitrance of chaos. But the God of *De gravitatione* freely makes bodies as artifacts within the

---

43 Mircea Eliade, *The Sacred and the Profane: The Nature of Religion* (New York: Harcourt Brace, 1987), pp. 29–62.

eternal moulds of space. In so doing, divine volition breaks the undifferentiated homogeneity of space.

"In Him Are All Things Contained and Moved": Sacred Wisdom and the Ground of Being

It seems that these religious sensibilities drive Newton's thinking in the period from 1684 to 1710. Moreover, they largely underwrite the theology of the 1687 *Principia*. The phrase quoted earlier is from the General Scholium to the 1713 edition of the *Principia*.[44] In his validatory note to the phrase, Newton refers to Aratus's *Phaenomena* (in fact its original home) and to Saint Paul in Acts 17.27-8, where the Apostle's famous "In him we live and move and have our being" may have been suggested to him by the Stoic writer. As I mentioned earlier, after the demise of his mechanical and material aethers, particularly as explanations for the operation of gravity, two types of solution remained open to Newton as explainers of cosmic changes. His first solution is my chief concern here, namely, the idea that the omnipresent god subsumes the action of gravity directly.

As a way into this, I reconsider some of the material discussed by Rattansi and me in 1966 and more recently by Dobbs.[45] While busying himself with revising the *Principia* in the early 1690s, Newton constructed a distinctive fusion of religious and natural philosophical ideas.

Let me summarize the main elements of this creative fusion.

1. In his important *Theologiae gentilis origines philosphicae* (ca. early 1680s), Newton argues that the ancients knew the true religion, the original monotheism, before corrupting themselves and becoming idolatrous by worshiping the heavenly bodies as gods.[46] Parts of the *Theologiae* are outcrops in Query 31 of the 1717 *Opticks* and also appear in the opening paragraphs of "The Treatise Concerning the System of the World."[47] According to Newton, Noah and his sons worshiped one true God, the sole creator of the cosmos. This view later became corrupted by ancient civilizations who invented other gods and deified the heavenly bodies. Noah's descendents had worshiped God in temples, such as Solomon's, constituting in their structure a prytaneum or sacred fire in the middle, which, according to the ancient Egyptian religion, symbolically represented the sun of the heliocentric

---

44 Newton, *Mathematical Principles of Natural Philosophy*, p. 545.
45 McGuire and Rattansi, "Newton and the 'Pipes of Pan' "; Dobbs, *Janus Faces*.
46 See Richard S. Westfall, "Newton's Theological Manuscripts," in Bechler, *Contemporary Newtonian Research*, pp. 136-9.
47 Newton, *Mathematical Principles of Natural Philosophy*, pp. 549-50.

universe. Newton believed that the ancient Jews and alchemists were less corrupted than other ancient peoples. Moreover, he believed that the more ancient sources, including those of alchemy, were less likely to be corrupted. For these reasons they were probably closer to the original wisdom in religion and natural philosophy. Accordingly, Newton looked to antiquities as sources of Divine Wisdom for clues pertaining to the structure of the cosmos, and for insights into the hieroglyphic meaning of symbols that might express the original heliocentrism. For example, he studied the proportions and spatial orientations of Solomon's temple to the extent that they provided clues to the structure of the cosmos.[48] Likewise he studied correspondences between the seven traditional metals of alchemy and the seven planets symbolized as encircling the central sun. In this scheme, the Sun is represented by the central fire of the prytaneum and by the active fire of alchemy, an agent that the alchemists believed to reside at the heart of matter.[49]

2. Dobbs argues that Newton's orientation toward ancient wisdom led him to ancient natural philosophy. In these sources he sought clues to God's role as the cause of gravity and to the question whether divine omnipresence subsumes the mathematical laws of universal gravity. In the 1690s we find Newton culling the sacred writers for images of a God-filled space: among these is the notion that all things are contained in an omnipresent God, who is the "ground of all being." Although God is the Place (*Makom*) of the world, the world is not God's Place: "For God is alike in all places. He is substantially omnipresent, and as much present in the lowest Hell as in the highest heaven."[50] Newton's use of the Hebrew term *Makom* is meant to convey the idea that God dwells *in* space, not that space itself is a property of God. This view is conveyed in another passage from the "classical scholia." Newton copied the following sentence from Macrobius: "The entire universe was rightly designated the Temple of God." He comments: "This one God they [the Ancients] would have it dwelt in all bodies whatsoever as in his temple, and hence they shaped ancient temples in the manner of the heavens."[51]

3. Newton's use of these expressions does not rule out an Arian Christ as the high God's agent in creation. In a manuscript intended for the unimplemented edition of the *Principia* in the 1690s, however,

48 Isaac Newton, *The Chronology of the Ancient Kingdoms Amended* (London, 1728), pp. 332–46.
49 Dobbs, *Janus Faces*, pp. 157–66.
50 Frank E. Manuel, *The Religion of Isaac Newton* (Oxford: Clarendon Press, 1974), p. 101.
51 ULC MS Add. 3970. f. 292v, as quoted by McGuire and Rattansi, "Newton and the 'Pipes of Pan,'" p. 108.

Newton is definite as to the cause of gravity: "Those ancients who more rightly held unimpaired the mystical philosophy as Thales and the Stoics, taught that a certain infinite spirit pervades all space into infinity, and contains and vivifies the entire world. And this spirit was their supreme divinity, according to the Poet cited by the Apostle. In him we live and move and have our being."[52] The poet is Aratus and the apostle is Paul. I shall comment shortly on the significance of Newton's reference to the Stoics. But note that here Newton's language is not Arian; indeed, the high God is conceived as "diffused" throughout all space and thus by his presence constitutes the cosmos as an hierophany. So conceived, uncreated space becomes an infinite nexus in which the sacred and the real are vividly intermingled.

4. But the ancient sources also provided Newton with clues regarding the atomic structure of matter and the mathematical laws according to which gravity operates upon the planets. We have Fatio de Duillier's and David Gregory's testimony that Newton believed the ancients knew both the cause and the manner of the operation of gravity:

> The ancient philosophers who held Atoms and Vacuum attributed gravity to atoms without telling us the means unless in figures: as by calling God Harmony representing him & matter by the God Pan and his Pipe, or by calling The Sun the prison of Jupiter because he keeps the Planets in their Orbs. Whence it seems to have been an ancient opinion that matter depends up on a deity for its laws of motion as well as for its existence.[53]

Regarding the laws of motion Newton observes, in a scholium intended for Proposition IX of book III of the *Principia*, that the ancient philosophers were alluding to them when they called "God harmony and signified his actuating matter harmonically by the God Pan's playing upon a pipe and, attributing musick to the spheres, made the distance and motions of the heavenly bodies to be harmonical, and represented the Planets by the seven strings of Apollo's Harp."[54] Here Newton invokes the Pythagorean-Platonic conception of world harmonies, a view later absorbed into Christianity through Augustine's *De musica*, and culturally transmitted in biblical sources such as the Wisdom of Solomon, with its view of creation as a divine enactment according to weight, number, and measure. Newton knew these and related sources intimately.

52 ULC MS Add. 3965.12, f. 269, as quoted by McGuire and Rattansi, "Newton and the 'Pipes of Pan,' " p. 120.
53 ULC MS Add. 3970, f. 619r, as quoted in McGuire and Rattansi, "Newton and the 'Pipes of Pan,' " p. 118.
54 Ibid.

"The Sensorium Dei": The Divine Mind That Reigns According to the Laws of Motion

Dobbs argues that Justin Martyr – who Christianized the Pythagorean tradition in the second century A.D. – reveals two pictures of God important to Newton: God is an all-pervasive spirit diffused throughout the universe but also artificer who orders things harmoniously.[55] I want to explore these pictures, especially their connection with Newton's reading in Stoic sources. From the Cicero of *De natura deorum* and from Virgil's *Georgics* he would encounter the idea that the high God is an infinite mind who freely penetrates all things. The idea that God is a penetrative and spiritual presence in the material world is found in Newton's *Quaestiones quaedam philosophicae* as early as 1664.[56] It is also present in the work of his Cambridge contemporary Henry More who held that God is an incorporeal yet tridimensional being, the presence of whom fills the world.[57] Newton knew More's writings well.

Dobbs suggests that Newton's reading of Philo Judaeus's *Allegorical Interpretation of Genesis,* and possibly Justus Lipsius's *Physiologiae stoicorum,* made an important difference to his understanding of these ideas.[58] Philo Platonized the Stoic notions of active and passive principles and "spiritualized" their conception of the material pneuma. His views emerge in the writings of Justin Martyr, Clement of Alexandria, Eusebius, Ambrose, Origen, Saint John Damascene, and Saint Augustine. The "Platonized" Stoicism that flows through this patristic current is augmented by Lipsius during the later Renaissance. He turned the Stoic notions of active and passive principles into a dualism transformed by his vision of an incorporeal deity who actively binds the material world into a harmonious whole.

This shift from Stoic materialism to Platonic dualism reconceptualized the nature of the Stoic deity for Newton. The Stoic God is conceived by Lipsius as a tensional force – an omnipresent *tonos* – able to penetrate, mingle with, and bind together all parts of the Cosmos. I shall return to the notion of a spiritualized divine *tonos* in connection with the notion of world harmony. Here it is important to note that the significance of this reinterpretation for Newton lies in the notion of an immaterial God who animates nature. In Stoic thought itself, the

---

55 Dobbs, *Janus Faces*, pp. 195–6.
56 J. E. McGuire and Martin Tamny, *Certain Philosophical Questions: Newton's Trinity Notebook* (Cambridge: Cambridge University Press, 1983), pp. 408–9.
57 McGuire, *Tradition and Innovation*, pp. 6–9, 14–15.
58 Dobbs, *Janus Faces*, pp. 200–5.

ever present and all-penetrating Deity is synonymous with the all-penetrating pneuma.[59] For the Stoics these terms are interdefinable, if not identical. In the holograph manuscript, "Out of Cudworth," in which he made notes on ancient theology from Cudworth's *True Intellectual System of the Universe* (1678), Newton rejects Stoic materialism and chides Cudworth for thinking that the earliest of the ancients were atheists. He agrees with Cudworth, however, that most of the ancients viewed the divine mind as a force that penetrates the workings of nature.[60] Indeed, in the *True Intellectual System* (on a page from which Newton took notes), Cudworth says that the Platonizing ancients hold the corporeal world to proceed from a *Rational* and *Intellectual* Principle, diffusing itself through all."[61] It is tantalizing to speculate that the spiritualized *tonos* may be the other "Argument for a Deity" that Newton mentioned to Bentley during their correspondence on the theological implications of the *Principia*. On that occasion, Newton remained silent "till the Principles on which it is grounded are better received."[62]

Newton's knowledge of this reconceptualized Stoicism prompted two important ideas in his writings. In a draft fragment related to "Tempus et locus," Newton asks approvingly "whether the Prophets more correctly say that God is present absolutely in all places, and constantly sets in motion the bodies contained in them according to mathematical laws, except where it is good to violate those laws."[63] This was written in the early 1690s, and is related to a sentence in a draft version of the Scholium on space and time of the *Principia*. Newton writes in the published Scholium: "It is indeed a matter of great difficulty to discover, and effectively to distinguish, the true motions of particular bodies from the apparent; because the parts of that immovable space in which these motions are performed, do by no means come under the observation of our senses."[64] In the draft version, the sentence that follows begins: "Solus enim Deus, qui singulis immobiliter et insensibiliter..." The sentence is unfinished, but

---

59 S. Sambursky, *Physics of the Stoics* (New York: Macmillan, 1959), pp. 36–7.
60 "Appendix: 'Newton's *Out of Cudworth*,'" in James E. Force and Richard H. Popkin, eds., *Essays on the Context, Nature, and Influence of Isaac Newton's Theology* (Dordrecht: Kluwer, 1990), pp. 207–13.
61 Ralph Cudworth, *The True Intellectual System of the Universe*, 2 vols. (London, 1678), 1:344.
62 Newton to Bentley, December 10, 1692, in *Four Letters from Sir Isaac Newton to Doctor Bentley, Containing Some Arguments in Proof of a Deity*, in *Isaac Newton's Papers and Letters on Natural Philosophy*, ed. I. Bernard Cohen (Cambridge: Cambridge University Press, 1958), p. 290.
63 ULC MS Add. 3965, sec. 13, f. 542r.
64 Newton, *Mathematical Principles of Natural Philosophy*, p. 12.

Cohen completes it as follows: "For God alone, who (gives motion to) individual (bodies) without moving and without being perceived (can truly distinguish true motions from apparent)."[65]

These drafts highlight Newton's conception of the spiritual *tonos*, the structuring intensity that is God's omnipresence in the geometry of space. For Newton, God subsumes created things and the laws of their operations. As all things "are contained" in God, so too are the laws that bind them into a cosmos. In God's sacred field, it is absolute change and its laws that are truly present in immovable space as embodiments of the divine mind. These statements strengthen Dobbs's conjecture that the Philonian and Stoic conception of the divine mind as an active principle informs Newton's notion of space as the *sensorium dei*.[66] Interestingly enough, the sensorium image occurs in "Tempus et locus," a document that is itself a part of the revision of the *Principia* in the 1690s. Referring to God's omnipresence and being "live and making live," Newton says that he contains "all other substances in Him as their underlying principle and place; a substance which by his own presence discerns and rules all things, just as the cognitive part of man perceives the forms of things brought into the brain and thereby governs his own body."[67] This remarkable image recurs in Newton's later writings, especially in the 1706 *Optice*, where its presence annoyed Leibniz.[68] It has often been misunderstood, however. There is no reason to think that Newton held space to be literally God's sensorium. By using that figure, Newton was attempting, however ineffectively, to understand the infinity of the divine mind on analogy with finite minds. In fact, the sensorium image fits well with the idea that the cosmos lives, moves, and has its being in God; and with the idea that the divine *tonos* binds the cosmos together in infinite space. Indeed, Newton's picture of the divine mind as that which "discerns and rules all things" fits the hierologic topos that guides his religious thought.

God's Sacred Field: Geometry and Natural Motion

This brings me back to *De gravitatione* and the first edition of the *Principia*, especially its preface. If *De gravitatione* depicts a sacred cos-

---

65 I. Bernard Cohen, "Isaac Newton's *Principia*, the Scriptures, and the Divine Providence," in *Philosophy, Science, and Method: Essays in Honor of Ernest Nagel*, ed. Sidney Morgenbesser, Patrick Suppes, and Morton White (New York: St. Martin's Press, 1969), p. 528.
66 Dobbs, *Janus Faces*, p. 209.
67 McGuire, "Newton on Place, Time, and God," p. 123.
68 Alexandre Koyré and I. Bernard Cohen, "The Case of the Missing *Tanquam*: Newton, Leibniz, and Clarke," *Isis* 52 (1961): 555–66.

mogony that is the result of the divine will informing space with bodies, Newton's views on geometry in the preface are in a similar vein. That is, he speaks of natural geometry based on natural motions and implies they are "constructed" by God and contained in God. In *De gravitatione*, in his account of how God materially "informs" figures inherent in space, Newton conceives "bodies" resulting from divine action as mobile in themselves and as moving "in accord with certain laws."[69] On the assumption that *De gravitatione* was composed around 1684, and is thus a foundation document for the *Principia* itself, the *Principia*'s preface addresses the status of these laws and the natural motions they regularize. Although God is not invoked in the preface, there is no doubt who "constructs" natural motions and the geometry based upon them. In short, they too are among the things "contained and moved" in God.

In the preface, Newton considers the magnitudes of the motions generated and configured by natural powers. The difference in the operations of "rational" and "practical" geometry is that the latter is a manual art where "the errors are not in the art, but in the artificers." Moreover, the figures upon which practical geometry purchases are "founded in mechanical practice."[70] Traditionally, the manual powers generate motions, and geometry measures their magnitudes. If the ancients cultivated the manual powers, Newton cultivates natural powers: "those things which relate to gravity, levity, elastic force, the resistance of fluids, and the like forces, whether attractive or impulsive ... for the whole burden of philosophy seems to consist in this – from the phenomena of motions to investigate the forces of nature, and then from these forces to demonstrate the other phenomena."[71]

There is a constructivist orientation here. Newton's focus is on the technician who constructs the postulates of geometry. There are three candidates: God, nature, or the human technician. Newton's perspective on geometrical foundations rests on the content of its postulates and on the technician who in fact constructs them. In this regard, God is the perfect artificer. Newton's concern is with any force that is "required to produce any motion."[72] If the manual powers produce the motions of contrivances, the magnitudes of which geometry measures, the "natural powers" generate natural motions that the science of mechanics measures and expresses in the language of geometry. Hence, Newton bases geometrical mechanics – his mathematical principles of philosophy – on natural motions generated by natural pow-

69 Newton, *De Gravitatione*, in Hall and Hall, *Unpublished Scientific Papers*, p. 140.
70 Newton, *Mathematical Principles of Natural Philosophy*, p. xvii.
71 Ibid., pp. xvii–xviii.
72 Ibid., p. xvii.

ers. In short, Newton's geometry has a definite content: it is about the motions of physical objects generated by real forces in absolute space and time.

Physical motions therefore constitute more than a possible interpretation of geometry: they are its postulates as constructed by God. If we know geometry by constructing it, it is because we made it: in contrast, the "natural" geometry of the cosmos is made by God, yet open to our apprehension if we succeed in decoding the mathematics of the Book of Nature. Newton holds, in voluntarist fashion, that the postulates of "natural" geometry are not open to ultimate demonstration. They are produced by God: this means that although the human mind is unable to grasp the full nature of the infinite mind in its act of structuring the geometry of space, it can nevertheless decipher the effects of that action through the language of geometry.

The standard view that the *Principia* is an exercise in positivism is wide of the mark. Newton never doubts the possibility of revealing the geometrical structure of nature and of establishing working criteria by which to define the difference between relative and absolute motions. This is indeed the whole thrust of his argument in the scholium on space and time. More to the point, therefore, is his religious vision of natural geometry. Consider his ontology again, beginning with *De gravitatione*. In that manuscript Newton speaks of the informing form that conditions spatial figures to make bodies what they are. He says that the "proper affections that denominate substance" are "actions," recognized by an ability to "act upon things." Indeed, he speaks of force as "an internal principle by which existing motion or rest is conserved in a body, and by which any being endeavours to continue in its state and opposes resistance."[73] Newton conceives of bodies as active substances that causally underwrite their natural changes, a view captured in the *Principia* by the use of the phrase "the *vis insita* of bodies."[74]

Newton did not later abandon the view put forth in *De gravitatione* concerning the manner in which God creates bodies in the uncreated expanses of space. In a footnote to the third edition of his French translation of Locke's *Essay Concerning Human Understanding*, Pierre Coste reports a conversation with Newton sometime around 1710. Coste noticed that Locke speculates in the *Essay* whether, if we made the effort, we might conceive how God had created matter. As Locke gives no details, his speculation remained a mystery to Coste. When

---

73 Newton, *De Gravitatione*, in Hall and Hall, *Unpublished Scientific Papers*, p. 148.
74 J. E. McGuire, "Natural Motion and Its Causes: Newton on the 'Vis Insita' of Bodies," in *Self-Motion from Aristotle to Newton*, ed. Mary Louise Gill and James G. Lennox (Princeton: Princeton University Press, 1994), pp. 305–29.

he mentioned this to Newton, Newton replied that he had discussed this very question with Locke. Coste quotes the content of Newton's reply, which is a summary of the metaphysics of *De gravitatione* concerning the creation of matter.[75] We have no reason to reject the validity of Coste's report. It is clear, therefore, that long after the publication of the *Principia* Newton still adhered to the metaphysics of *De gravitatione* and saw the *Principia* in its terms, a view he surely held when he first composed it.

In the *Principia* Newton conceives of gravity as a nonmechanical force that acts instantaneously upon any group of bodies by penetrating their quantity of matter. Moreover, bodies are bound together dynamically into ever changing actions and reactions according to the third law; and their action on one another is mutual and opposite with respect to the strict proportionality of mass and inertia and operates inversely as the square of the distances between them. Most significantly, this drama of cosmic change is enacted in absolute space and time under the subsuming presence of the supreme God. These principles, seen through the metaphysics of *De gravitatione*, constitute an hierophany: God activates space's preexisting parts to make "bodies," each one manifesting in turn an individual *vis insita* or an inner principle of change.

Accordingly, the *Principia*'s God can be seen as the spiritual *tonos* that makes "bodies" in space and geometrically configures their laws of motion. A rich picture emerges. The term *tonos* carries a number of interrelated connotations: it indicates the notion of being stretched or strained; it invokes the idea of measure and proportion; and, lastly, it conveys the idea of force, energy, and intensity. In Newton's mind, consequently, the spiritualized *tonos*, assimilated and transformed from Stoic thought, evokes the dynamical harmony that activates the geometry of space. That is, Newton combines the conception of the protoplast, as the archetype of creation, with the notion of God's divinity understood as *tonos* or tensional force existing in the world. In short, God is a "structuring structure," the archetypal being that imprints structures on creation. This means that the penetrative force of gravity is a modulation of God's vivifying presence, and the externalization of divine volition in creation. From this perspective, divine geometry, the measure and proportion of the divine *tonos*, inscribes the action of gravity throughout the expanses of space.

This is the vision that lies behind Newton's reference to God as "an universal life" who attends the laws of motion, the very laws to which

---

75 Alexandre Koyré, "Newton and Descartes," in *Newtonian Studies* (Cambridge, Mass.: Harvard University Press, 1965), p. 92.

"the ancient Philosophers seem to have alluded when they called God Harmony and signified his actuating matter harmonically by the God Pan's playing upon a pipe."[76] In discussions with Newton recorded in 1705, David Gregory states that "The plain truth is, that he believes God to be omnipresent in the literal sense . . . for he supposes that as God is present in space where there is no body, he is present in space where a body is also present . . . *What cause did the ancients assign to gravity?* He believes that they reckoned God the cause of it, nothing else." [77] Gravity has a special status in Newton's natural philosophy, especially in reference to his commitment to an ontology of forces. The law of its operation does not flow from the nature of matter as do the laws of motion that arise from the *vis insita* of individual bodies. Gravity is an interactive force that depends on simultaneous interrelations among many bodies, and reflects the contingent structure that the cosmos in fact possesses. It is not surprising, therefore, that it should be linked in Newton's mind with the direct operation of divine volition.

The scholium on space and time reveals a theological ethos similar to what we see in the documents mentioned earlier. We are told that the terms time, space, place, and motion are commonly taken in their relative measures, unlike mathematically understood measured quantities: "Upon which account, they do strain to the Sacred Writings [*sacris literis*], who there interpret those words for the measured quantities. Nor do those less defile the purity of Mathematical and Philosophical Truths, who confound real quantities themselves with their relations and vulgar measures." Though Scripture is written in plain language, its interpretation is not straightforward. In natural philosophy the situation is also serious: careless interpretation risks confounding the mathematically revealed structure of God's cosmos with the relative measures of sensory experience. Levels of meaning must be kept distinct. Scriptural understanding is "strained" if it is interpreted through inappropriate categories: so likewise is philosophy if absolute quantities are mixed with relative terms. God alone "can truly distinguish true motions from apparent." Cajori deleted the phrase "sacred writings" from his redaction of Motte's translation of Newton's *Principia*. It apparently had no place on a treatise on mathematical physics![78]

It is clear, then, that Newton's commitment to the idea of God as an omnipresent being was profound. For him the Divinity that directly

76 McGuire and Rattansi, "Newton and the 'Pipes of Pan,' " p. 118.
77 W. G. Hiscock, ed., *David Gregory, Isaac Newton and Their Circle* (Oxford, 1937), p. 29.
78 Cohen, "Isaac Newton's *Principia*," p. 528.

subsumes the operations of *gravitation and activates the motion of matter is in nature but not of nature*. Indeed, in *De gravitatione* Newton dismisses the possibility that God created a world soul from which the rest of creation "necessarily emanates." He tells us, "But I do not see why God does not directly inform space with bodies; so long as we distinguish between the formal reason of bodies and the act of divine will." Accordingly, bodies are the externalized effects of divine volition, namely, the effects that "that act produces in space."[79]

This brings me to the vexed question of Newton's Arianism. Dobbs makes much of draft materials written sometime between 1710 and 1720 in which Newton expresses views that appear to be Arian.[80] However we characterize his religious outlook, Newton clearly opposes the canonical doctrine of the Trinity, and Dobbs is right to emphasize that. But this in no way puts Newton squarely in the Arian tradition. Like everything else about the man, his religious convictions were far from straightforward. And, as Westfall has pointed out, his theology also bears the marks of innovation and novelty.[81] Newton shows interest in what Cudworth says about "The Egyptian Trinity."[82] Cudworth discusses how many ancient writers in the Platonic and Neoplatonic traditions express a "Trinitarianism" of the gods, especially in the context of their emanationist metaphysics.[83] Newton's consideration of Cudworth's views on the role of the higher and lower gods in emanationist metaphysics puts him in neither the Arian nor the Trinitarian camp. But the religious sensibility that both Cudworth and Newton evince indicates that the Arian and Trinitarian ontologies are varied and complicated and that these labels need to be used with care. In fact, Newton's "Arianism" is probably as much indebted to emanationist Trinitarianism as it is to the high and low god of Arius!

Dobbs's assumption that the ontology of the Arian Christ guides Newton in his attempt to establish other intermediary agents is open to question. Certainly Newton considered aetherial media, the action of light, and the diffused effects of an electric spirit as possible sources for activity in the material world. Moreover, he continued to the end of his life to gather evidence for intermediate agents, which act as second causes, in explaining gravity and other phenomena. It is not clear, however, that Newton's search was driven by Arian convictions.

79 Newton, *De Gravitatione*, in Hall and Hall, *Unpublished Scientific Papers*, p. 142.
80 Dobbs, *Janus Faces*, pp. 223–49.
81 Richard S. Westfall, *Never at Rest: A Biography of Isaac Newton* (Cambridge: Cambridge University Press, 1980), pp. 314–30.
82 "Appendix: 'Newton's *Out of Cudworth*,' " in Force and Popkin, *Essays*, p. 211.
83 Cudworth, *True Intellectual System*, 1:340–50.

Nor is it clear that his "Arianism," as Dobbs maintains, precluded the possibility that God's direct action in nature could be Newton's final solution to the cause of gravitational action.

This said, it seems that Newton never fully resolved two distinct pictures of how God relates to the world: the Arian, or something similar to it, in which the high God is transcendent and works in nature through an intermediary, the cosmic Christ; or the God of dominion of the "classical scholia" who is directly present and active in creation. Newton struggled not only to uncover the secondary causes of change, but also to ease the theological tensions in his views regarding the nature of the ultimate cause. In my view, it is wrong to think that his religious "Arianism," which he could not easily set aside, compromised his picture of God's activating present in the limitless expanses of space. Rather, it is Newton's "Arian" impulses that conflict with what is surely his preferred picture of how God relates to the world: the divine *tonos* who in virtue of its concrete omnipresence acts harmoniously in creation.

Conclusion

There is clearly a creative fusion of religious, theological, and philosophical vocabularies in Newton's discourse on natural philosophy. Indeed, they function correlatively and none is privileged over the other. Consider what I. Bernard Cohen has called the "Newtonian style." According to Cohen, the methodology of the *Principia* is not an exercise in the method of analysis and synthesis: it is a unique use of mathematical models designed specifically to assimilate, transform, and extend physical data.[84] Newton assumes physical entities and conditions that are far simpler than anything in nature. He first reduces planetary motion to a one-body system, for example, a single body moving within a central-force framework. Next he considers a mass point, rather than a physical body, and treats its motion in reference to mathematical space and time. Because Newton's mathematics maps onto his idealized physical system, he can move from one to the other. Consequently, by moving between mathematics and idealized physical systems, by introducing successively more entities, conditions, and data, and by extending his mathematical techniques, Newton generalizes his approach to the world of finite bodies and the forces that act between them.[85]

Cohen has caught the gist of Newton's procedure. But his analysis

---

84 I. Bernard Cohen, *The Newtonian Revolution, with Illustrations of the Transformation of Scientific Ideas* (Cambridge: Cambridge University Press, 1980), pp. 52-3.
85 Ibid., pp. 61-5.

treats the *Principia* as an exercise in mathematical positivism. If my interpretation is correct, Cohen's perspective finds its true focus in Newton's view of divine geometry. If God is the ground of all being – the spiritual *tonos* and "structuring structure" of the cosmos – the *Principia* becomes the conduit through which that structure is disclosed. That is, the geometry of the *Principia* is the interpretative framework that reveals the harmonies of nature. As Cohen claims, Newton's discourse does weave a complex tapestry combining physical idealizations, mathematical modeling, and successive appropriations of physical data. But for Newton more than that is at stake. In his eyes, the *Principia* is the framework that allows the geometry of motion to become uncovered within the sacred space of the "System of the World." Viewed in this way, it is the vehicle of aletheia – it is that which allows the true structure of creation to manifest itself. Nature's harmonies, thus disclosed, stand forth manifesting their roots in the omnipresence of God. Certainly, Newton, like Galileo before him, sees the Book of Nature as a text written by God in mathematical characters. Moreover, the key to this book is mathematics, the language that enables the human mathematician to "decode" the effects of what God has inscribed in nature.[86]

We can expect nothing less from the man who sought the disclosure of truth in the discourses of history, biblical chronology, Scripture, ancient philosophy, and alchemy. These disparate elements Newton strove to combine into a mosaic that would express the primal unity of God's truth. It is this sensibility that informs his statement to Richard Bentley in 1692: "When I wrote my Treatise about our System, I had an eye upon such Principles as might work with considering Men, for the belief of a Deity."[87]

86 On Galileo, see James J. Bono, *The Language of God and the Languages of Man: Interpreting Nature in Early Modern Science and Medicine* (Madison: University of Wisconsin Press, 1995), pp. 193–8.
87 Newton to Bentley, December 10, 1692, in Cohen, *Newton's Letters and Papers*, p. 280.

14

Newton and Spinoza and the Bible Scholarship of the Day

RICHARD H. POPKIN

We now think that any serious religious person would be dismayed by the Bible scholarship of Spinoza and Father Richard Simon. But, though they shocked many of the theologians of the late seventeenth century and afterward, they also carried further historical and critical inquiries about the biblical text that had been developing from the late Middle Ages and through the Reformation and Counter-Reformation.[1] Isaac Newton, who devoted about sixty years of his life to studies about the Bible seems to have been affected positively by points made by Spinoza and Simon. And this is curious in that two hundred years later Newton's own exegesis of the books of Daniel and Revelation was republished by the former head of the British Medical Association, Professor William Whitla of Queens College, Belfast, as an answer to the higher criticism of the Bible that had been developed from the ideas and methods of Spinoza and Simon by German scholars in the nineteenth century.[2]

First, did Newton actually know the texts of either of these radical biblical exegetes? We know that he knew of Simon's work, since he owned copies of five of his books, and cited them occasionally in his own work.[3] We know he had ample opportunity to know of Spinoza's

---

I should like to thank Ms. Anna Suranyi of UCLA for her able assistance in editing this chapter.

1 Cf. Richard H. Popkin, "Spinoza and Bible Scholarship," in *The Books of Nature and Scripture: Recent Essays on Natural Philosophy, Theology, and Biblical Criticism in the Netherlands of Spinoza's Time and the British Isles of Newton's Time*, ed. J. E. Force and R. H. Popkin (Dordrecht: Kluwer, 1994), pp. 1–20. See also the comments by Amos Funkenstein which follow this essay, pp. 21–3.

2 William Whitla, *Sir Isaac Newton's Daniel and the Apocalypse with an Introductory Study of the Nature and Cause of Unbelief of Miracles and Prophecy* (London: John Murray, 1922).

3 John Harrison, *The Library of Isaac Newton* (Cambridge: Cambridge University Press, 1978), entries 1513–17, p. 239. Newton owned Simon's *Critical History of the Old Testament*, and *Critical History of the New Testament* (both in the English editions), his critical inquiries into the various editions of the Bible, and his comparison of the

writings, especially the work that deals extensively with the Bible, the *Tractatus Politico Theologicus*. It was in the library of Isaac Barrow that Newton had cataloged. The work was known to his colleagues at Cambridge, Ralph Cudworth and Henry More, both of whom were exercised about Spinoza's views and attacked them in print.[4] And we know that Newton discussed his biblical interpretations with More.[5] Recent studies show that knowledge of Spinoza was far more widespread in England than had previously been indicated by studies such as those of Rosalie Colie.[6] Besides the texts, which appear in many personal libraries of the time, some of Spinoza's personal friends, like Charles de Saint-Evremond and Dr. Henri Morelli were lively and active figures on the London intellectual scene. Spinoza's work was translated into English in 1689, probably by Charles Blount. Newton's good friend, John Locke, with whom he discussed the meaning of the prophecies in Revelation, knew about Spinoza, and was accused at the time of being a secret Spinozist. Father Simon's writings also appeared in English, and were discussed by many theologians and writers on the Bible.

Newton was concerned about knowing all that could be known about the history and the meaning of biblical texts and followed in the footsteps of many of the leading English biblical scholars of the time. He, like them, knew an enormous amount about ancient Jewish history, about the political, social, and cultural situations stressing when the books of the Bible came into existence, about the ways in which the texts were transmitted, about the varying manuscripts and printings, and so on.[7] A recent dissertation by Matt Goldish done at Hebrew University goes into great detail about Newton's knowledge of Jewish sources.[8] And, although Newton hardly knew Hebrew, he read and owned an enormous amount of Latin Judaica, that is, Latin translations and commentaries of Jewish writings from Josephus and

ceremonies of the Jews with the church. He also owned some of the anti-Simon writings of the time.

4 Cf Frank E. Manuel, *The Religion of Isaac Newton* (Oxford: Clarendon Press, 1974), p. 84; and R. H. Popkin, "Newton as a Bible Scholar," in *Essays on the Context, Nature and Influence of Isaac Newton's Theology*, ed. J. E. Force and R. H. Popkin (Dordrecht: Kluwer, 1990), p. 104.

5 See Henry More's letter to Dr. John Sharp, later archbishop of York, August 16, 1680, published in *The Conway Letters*, ed. Marjorie Nicolson, rev. Sarah Hutton (Oxford: Clarendon Press, 1992), pp. 478–9.

6 Rosalie Colie, *Light and Enlightenment: A Study of the Cambridge Platonists and the Dutch Arminians* (Cambridge: Cambridge University Press, 1957).

7 How some of these scholars dealt with the same biblical passages that Spinoza stressed as indicating conflicts, or different authors of Scripture, are treated in Popkin, "Spinoza and Bible Scholarship," pp. 1–20.

8 Matt Goldish, *Judaism in the Theology of Sir Isaac Newton* (Dordrecht: Reidel, 1998).

the Mishnah and Talmud through the Middle Ages and even seventeenth-century Jewish writings. Newton also knew an amazing amount about ancient history, Jewish, pagan, and Christian, which he used adroitly in his *Chronology of Ancient Kingdoms Amended*.

The importance of Latin Judaica is just beginning to be recognized. All through the sixteenth and seventeenth centuries Latin editions of postbiblical Hebrew writings appeared with commentaries by modern biblical scholars. They were trying to glean what they could of the Jewish understanding of biblical texts, and of the context of the Bible. In the seventeenth century when there was a widespread learned belief that the millennial kingdom would be established as a continuation of the ancient Kingdom of Israel, when England saw itself as the New Israel, it was essential to learn as much as possible about the ancient Hebrew world through whatever sources one could find. The very popular work by the Dutch theologian, Petrus Cunaeus, *De republica Hebraerorum*, presented information about the ancient Hebrew republic as very related to what was going on in Europe in the seventeenth century.[9] Many of the Jewish sources from antiquity to the present were translated into Latin for scholarly use. The Marsh Library in Dublin, which contains the library of Newton's contemporaries, Edward Stillingfleet, bishop of Worcester, and Bishop Marsh of Dublin, has a truly amazing amount of this Latin Judaica.[10] Goldish has examined quite a bit of this material and has shown how it was used by Newton and others for both scholarly and political purposes (political in the sense that one could evaluate late-seventeenth-century English politics in terms of whether it conformed to the laws of ancient Judea).

Given Newton's interest in what was being written about the Bible, about ancient Jewish history, and about interpreting the Bible, it would not be at all surprising if he looked into what Spinoza was suggesting, and into the erudite histories of the Old and New Testaments by Father Richard Simon. In the main passage we have in which Newton dealt with the history of the Bible texts, Newton's own explanation of the apparent contradictions in the Scriptures seems very close to the kinds of accounts given by Spinoza and Simon, and in fact appears to be based on what they say. But Newton's explanation of how and why we can accept the texts of Daniel and Revelation with

---

9 This work, first published in 1617 in Leiden, was often reprinted, and was translated into English in 1653, as well as French and Dutch.

10 A collective volume of papers on the treasures of the Marsh Library in Dublin can be found in Allison P. Coudert, Sarah Hutton, Richard H. Popkin, and Gordon M. Weiner, eds., *Judaeo-Christian Intellectual Culture in the Seventeenth Century: A Celebration of the Library of Narcissus Marsh (1638–1713)* (Dordrecht: Kluwer, 1999).

certainty, while realizing that other texts that we possess are the result of all sorts of accidents and distortions in the course of human history, is quite different from Spinoza's.

In two previous essays I have shown the relationship of Newton's views on the Bible to those of fundamentalists of the past two centuries, and have shown how interested they have been in Newton's views on various biblical questions, and how they regard him as one of their most revered forebears.[11] Although the Westminster Confession of 1643 had declared that, "The Old Testament in Hebrew (which was the Native Language of the People of GOD of old), and the New Testament in Greek . . . being immediately inspired by God and by his singular care and Providence kept pure in all Ages, are therefore Authentical,"[12] Newton, like Spinoza and Simon, took seriously the problems that had arisen in the collection, editing, and transmission of Scripture that made it difficult if not impossible to find the pure original text. Newton, unlike the fundamentalists of the past century and a half, was not committed to claiming the inerrancy of the biblical text, but was committed to finding its message for mankind.

Newton had been working on a commentary on Daniel and Revelation for much of his adult life. Manuscripts exist of many versions, and notes for it are in the vast collections of Newton's still unpublished manuscripts.[13] The work was finally published in 1733 by the Royal Society after Newton's death, *Observations upon Daniel and the Apocalypse of St. John*. In part I, on Daniel, Newton began with a theme of great interest to Spinoza and Simon, concerning the history, accuracy, and reliability of the text of the Old Testament. Similarly, in the letters he sent to John Locke arguing that the doctrine of the Trinity does not appear in the New Testament, Newton gave some of his views about the history, accuracy, and reliability of the New Testament. These letters were originally intended for publication in Jean

---

11 R. H. Popkin, "Newton and the Origins of Fundamentalism," in *The Scientific Enterprise*, ed. E. Ullmann-Margalit (Dordrecht: Kluwer, 1992), pp. 241–59; and "Newton and Fundamentalism II," in Force and Popkin, *Essays*, pp. 165–80.

A fundamentalist who is a researcher for Hal Lindsey, Paul N. Moore, has recently published *The End of History: Messiah Conspiracy* (Atlanta: Conspiracy Press, 1996), in which Newton's unpublished views are made central to immediate concerns about the Second Coming and the millennium. Moore was led to study Newton's manuscripts because of what I had said about them to an Israeli reporter in 1985. Moore's book is over 1,200 pages long.

12 *The Confession of Faith . . . composed by the Reverend Assembly of Divines sitting at Westminster* (London, 1658), chap. 1, p. 6.

13 See Popkin, "Newton as a Bible Scholar," p. 103. The larger collection of unpublished materials is the Yahuda collection in the National Library of Israel, which contains many drafts of various parts of Newton's commentaries on Daniel and Revelation.

LeClerc's *Bibliothèque Universelle*. Newton at the last moment withdrew them, probably for fear of the reaction if he publicly stated his anti-Trinitarianism. Two of the letters were published in the mid-eighteenth century,[14] and the last one only in the mid-twentieth century with the publication of all of the then known letters of Newton.[15]

In the first chapter of the *Observations* Newton gave his views on who wrote the Old Testament books, and about how reliable the present text might be. Newton covered the same ground that Spinoza did in *Tractatus* chapter 8 and Simon in *Critical History*, book I. The text as presented by Newton is quite close to Simon's in showing that Moses could not be the author of the whole Pentateuch, in showing that some of the books had to be written by later authors than the authors to whom they are ascribed, and in arguing that many of the books are compilations from earlier writings, now lost.

Spinoza and Simon went into far more detail than Newton did. Spinoza went over some of the opinions of earlier commentators, especially Aben Ezra about the order of composition of the books, and their dates.[16] Simon, much more erudite than Spinoza, went into great detail about the opinions of the Caraites, the views of various rabbis, the commentaries of Abarbanel, the opinions of various modern editors of the Hebrew texts, the different opinions of Spanish Jewish scholars from those in France and Germany, among other topics.[17] Newton seemed to have just culled the essence of Simon's erudite findings and digested them into a fairly simple story of how the texts developed, by whom the various parts were put together, and when the text got stabilized into its present form. Newton's reasons for claiming Moses was not the author of the entire Pentateuch as it now exists, that Joshua and others were not the authors of all of the books that have their names, that various later scribes added and changed the texts are just simplified versions of what Father Simon worked out from his vast erudite knowledge of texts and commentaries.[18]

There is no evidence that Newton attempted the kind of research

14 Newton, *Two Letters of Sir Isaac Newton to Mr. Le Clerc* (London, 1754).
15 In *The Correspondence of Isaac Newton*, ed. H. W. Turnbull (Cambridge: Cambridge University Press, 1961), 3:83–146.
16 Benedictus de Spinoza, *Tractatus Theologico-Polticus*, in *The Chief Works of Spinoza*, ed. R. H. M. Elwes, 2 vols. (New York: Dover, 1955), vol. 1, chap. 7.
17 Richard Simon, *A Critical History of the Old Testament written originally in French by Father Simon and since translated into English by a person of quality* (London, 1682), first chapters.
18 A comparison of Newton's first chapter with the first eight chapters of Simon's *Critical History* shows that Newton had pretty much accepted Simon's picture of how the text developed, who might have been the author of various parts, and so on.

that either Spinoza or Simon had done. He was not a Hebraist and seems to have found his way around the critical literature from Simon's account. In fact, a careful textual analysis of the *Observations*, chapter 1, and Simon's book I would suggest that for the most part Newton was just summarizing what Simon had figured out, namely that the biblical texts developed over time, and that we do not possess the original texts. This much was enough to lead Spinoza to his most radical conclusion, that the Bible is just a compilation of ancient Hebrew writings by human authors having only human significance.

Simon, who said that he agreed with Spinoza's way of examining Scripture, but not his conclusion that the Bible had no supernatural message, offered some reasons for accepting certain portions of the text as part of a genuine revelation that might be recaptured by careful reconstruction of what might have been the original message. Newton, while admitting that much of the text contained errors, additions, and such, went further and declared that we could not recapture what was in the text before its final redaction. He wrote that "such marginal notes or other corruptions, as by the errors of the transcribers, before this [final] Edition was made, had crept into the text, are now scarce to be corrected."[19] The "official" text, Spinoza, Simon, and Newton agreed, was put together after the Romans captured Judea, and the Jews sought to preserve their traditions. "For preserving their scriptures, [they] agreed upon an Edition, and pointed it,[20] and counted the letters of every sort in every book."[21] But, nonetheless, Newton did not develop either Spinoza's complete religious skepticism, or Simon's skepticism about any particular version of the divine message. Instead he insisted that "The authority of the Prophets is divine, and comprehends the sum of religion."[22] Newton adopted the most extreme suggestions of Simon about the multiple authorship of the book of Daniel, with only a small portion being by the prophet himself.[23] "The book of Daniel is a collection of papers written at several times. Only the last six chapters contain prophecies written at several times by Daniel himself."[24] Nonetheless, Newton claimed that the prophetic writings "contain the covenant between God and his people, with instructions for keeping the covenant; instances of God's judgments upon them that break it; and predictions of things to

---

19 Isaac Newton, *Observations upon the Prophecies of Daniel, and the Apocalypse of St. John* (London, 1733), p. 12.
20 Meaning that they put in the markers for the Hebrew vowels.
21 Newton, *Observations*, p. 11.
22 Ibid., p. 14.
23 See Simon, *Critical History*, book I, chaps. 8 and 9.
24 Newton, *Observations*, p. 10.

come."²⁵ "The predictions of things to come relate to the state of the Church in all ages."²⁶ And, in spite of the historical critical comments, the prophecies of Daniel are the most distinct in order of time, and the easiest to understand, "and therefore, in those things which relate to the last times, ... must be made the key to the rest."²⁷

Daniel 12 is one of Newton's favorite texts, foretelling that, as we approach the end, the wise may understand and none of the wicked shall understand.²⁸ Newton, as one of the wise, was able to see that the prophecies in Daniel provided the key for understanding what is to happen as the millennium was approached, and the key for the rest of the prophecies.

In chapter 2 of the *Observations*, "Of the Prophetic Language," Newton offered his own view of how to understand the prophecies, in relation to scientific understanding of nature. He took the Bible to be one more artifact to be analyzed in terms of when it was composed and by whom as the key to understanding scriptural predictions. In this discussion Newton was going beyond and away from Spinoza or Simon. Spinoza claimed that the way to understand Scripture was like the way to understand nature; in fact, Spinoza said, the method was almost the same. For Spinoza this meant that the Bible was composed, and we could determine why it was composed.²⁹ It would not reveal a special message, but just historical literary information.

Simon, on the other hand, was concerned with the best understanding of the biblical text, without worrying about understanding nature. He was not, like Newton, a scientist and a Bible scholar. He was the latter par excellence.

Newton saw that one had to recognize special usages of language in the prophecies, because in them statements about the natural world were metaphorical or even mystical, not to be read literally. Newton in his early manuscript of how to read Scripture had advised against literal reading, except when it was not completely clear.³⁰ Prophetic language was that case, and so needed a special way of dealing with it and interpreting it. Here Newton went off in his own direction, away from both Spinoza and Simon, and claimed to have found the special messages in the prophecies of Daniel.

25 Ibid., p. 14.
26 Ibid., p. 15.
27 Ibid.
28 This is cited in ibid., chap. 1, p. 14.
29 Spinoza, *Tractatus Theologico-Politicus*, chap. 7.
30 Newton, Yahuda MSS 1, f. 12r, 34 "To prefer those interpretations which are most according to the litterall meaning of the scriptures unless where the tenour and circumstances of the place plainly require an Allegory." This fragment of a manuscript has been published in Manuel, *The Religion of Isaac Newton*, p. 118.

With regard to the New Testament, Newton was much more critical about the text because much more was known about the conditions under which it was put together and accepted. He held that the text that we have had been deliberately altered by some of the early church fathers, especially Saint Athanasius, in order to buttress the false doctrine of the Trinity.[31] It is interesting that Newton made no claims of deliberate alteration of texts of the Old Testament, which he explained as the result of human frailty, historical changes, and so on.

The text of the New Testament Newton saw as developing as the early Christian movement emerged, and it reflected various stages of the change from the views of a Jewish sect to a Christian Church. The earliest part of the New Testament, Newton believed, was the book of Revelation, which Newton claimed to have been written by someone who lived before the destruction of the Temple in A.D. 70 and the expulsion of the Jews from Jerusalem. Revelation contains more Hebraisms than other New Testament texts and was therefore written by someone who was part of the Palestinian Jewish world. Newton contended that the author was John, the Evangelist, who "was newly come out of Judea, where he had been used to the Syriac tongue, and that he did not write his Gospel, till by long converse with the Asiatick Greeks he had left off most of the Hebraisms."[32]

Newton assumed that the Gospel of John and Revelation were written by the same person and, contrary to much of Christian tradition and scholarship, he argued that Revelation was an early work, revealed directly by Jesus to the author. From the second or third century Christian authorities had said that Revelation was a late work, having no connection with Jesus' life on earth. Thus it did not have to be taken seriously, and its powerful apocalyptic message could be ignored except as a reflection of some excited mind of ancient times.

On the contrary Newton contended that various historical details pointed to an early date of composition. The Syriac text claimed that Revelation was written at the time of Nero. In the text we have there are references to the fact that the Temple and the Holy City were still standing. Newton was a great scholar in his own right of the history of early Christianity and could feel justified in offering authoritative claims about his view of the early composition of Revelation. In early

---

31 There are three unpublished manuscripts in which Newton set forth his charges against Saint Athanasius and his followers, "Paradoxical Questions Concerning ye morals & actions of Athanasius and His Followers," William Andrews Clark Library, University of California at Los Angeles; Keynes 10, King's College Library, Cambridge; and Yahuda 14, Yahuda Collection, Jewish National Library and University Jerusalem.

32 Newton, *Observations*, part II, chap. 1, p. 238.

Christian history there were many spurious books about the Apocalypse. Some were attributed to Peter, Paul, Thomas, Stephen, Elias, and Cerinthus. These spurious works constituted evidence that there must have been a genuine Apostolic work about the Apocalypse that was being imitated by these others. Because one has some idea when these spurious texts were composed, one knows approximately when the Revelation of Saint John was written.[33]

The question was, How does one tell if the text, or any text, has a divine message? Calvin had raised the question in the *Institutes* of how one ascertains that the Bible has a divine message and that Titus Livy's *Histories* do not. Calvin admitted that as texts they were on the same epistemological level. However, by the action of the Holy Spirit on the reader, the saved recognized the message of the Bible and saw that Livy's text was just a human message. On the other hand, Spinoza had a deaf ear to the actions of the Holy Spirit upon him, and so all texts were epistemologically on the same footing: they were writings by humans, to be studied in terms of their context, authorship, intent, and so on. Simon however accepted the fact that the biblical texts now extant have a human history, a human pedigree, but insisted that this did not preclude another dimension, namely the original revelatory message if only we could find it above and beyond all of the historical trappings.

Newton seems to have found it without any trouble, in terms of the prophetic message that shone through all of the additions and overlays of human involvement with the texts, and the assurance that God, employing Jesus as an agent, guaranteed the text of Saint John. Possibly the most extended statement of Newton's justification for the way he was reading and interpreting Scripture is in manuscript fragments from a *Treatise on Revelation*.[34] At the outset Newton explained that he felt bound to let the world know what he had discovered in Scripture, "For I am perswaded that this will prove of great benefit to those who think it not enough for a sincere Christian to sit down contented with the principles of the doctrin of Christ" because something of the greatest importance was involved in understanding prophecies.[35]

Newton then said that he was not discouraged by the lack of success of other people's attempts to explain the prophecies, for "it was revealed to Daniel that the prophecies concerning the last times should be closed up and sealed until the time of the end: but then the wise

---

33 Ibid., book II, chap. 1, pp. 238–9.
34 Published in Manuel, *The Religion of Isaac Newton*, pp. 107–25. This, of course, is only a tiny portion of Yahuda MS 1 in the National Library of Israel, which is 550 pages long.
35 Ibid., p. 107.

should understand, and knowledge should be increased."³⁶ In other words since there was hope that the time was at hand in which the prophecies are to be made manifest, then they no longer had to remain obscure. And, as Newton asked "If they are never to be understood, to what end did God reveal them?"³⁷ They are to be understood by a remnant, "as few scattered persons which God hath chosen, such as without being [blinded] led by interest, education, or humane authorities, can set themselves sincerely and earnestly to search after truth. For as Daniel hath said that the wise shall understand, so he hath said also that none of the wicked shall understand."³⁸ Here Newton had certainly parted company from Spinoza and Simon. For them the objective scholar, not the chosen adept, could read the texts, the available historical data, just as with any other piece of literature, and could draw the best hypothesis about the text.

Newton immediately went on to underscore the difference, urging people not to trust "to the opinion of any man concerning these things" because of the likelihood of being deceived. One should search the scriptures by oneself and pray that God would enlighten one. When this happens it would give "steddy satisfaction to thy mind which he onely can know how to estimate who shall experience it."³⁹

This realization, which Newton described not in Calvinist terms as the action of the Holy Spirit but just as a self-revealing experience, was of the greatest importance because of what it would help one understand about the forthcoming fulfillment of prophecies, which if one did not know about, would be of great danger. "Wherefore it is thy duty to learn the signes of the times that thou mayst know how to watch, and be able to discern what times are coming on the earth by the things that are already past."⁴⁰ By casting the issue of understanding Scripture in terms of the immediacy of millenarian expectation, Newton completely parted company from the preceding Bible scholars. And Newton emphasized the dangers of not understanding – pointing out what happened to the Jews in antiquity for not understanding the prophecies about the coming of the Messiah. They had been severely punished. We would be even more so if we were to understand nothing of the Second Coming and turn the whole description of it into allegories.⁴¹

In reply to the question of why the Apocalypse was ordained in the

36 Ibid.
37 Ibid.
38 Ibid., p. 108.
39 Ibid.
40 Ibid., p. 110.
41 Ibid., f. 3v.

first place, Newton claimed it must have been in order to guide and direct the early church in the right way. And this was also the purpose of all prophetic Scripture. "If there was no need of it, or if it cannot be understood, then why did God give it? ... Does he trifle?"[42] If the early church needed this kind of guidance, then how can we neglect it? Newton claimed that what led to the Reformation was the recovering of the prophecies, "therefore we have reason to believe that God foreseeing how much the Church would want a guide in these latter ages designed this Prophecy for this end."[43]

Another reason for taking the book of Revelation seriously, Newton said, was the blessing that is promised in the book, and he asked, "does God ever annex his blessings to trifles or things of indifferency?" Obviously no, so "search into these Scriptures which God hath given to be a guide in these latter times, and be not discouraged by the gainsaying which these things will meet with in the world."[44]

Newton realized how one would be considered if one took his advice about studying Revelation. "They will call thee ... a hotheaded fellow, a Bigot, a Fanatique, a Heretique etc." They will tell you that the interpretation of the prophecies is uncertain. But the Jews were punished for not accepting prophecies "of more difficult interpretation."[45]

In view of the importance for Newton of the prophetic message in Revelation, how does one tell that we have the correct text, and how do we find the true or the truest reading? The first question I have only found addressed in an unpublished page or two that is in private hands. Here Newton said that God was so concerned that John present the text correctly that Christ sent his messenger to keep an eye on John as he wrote down the prophecies.[46]

On the second question, "therefore the great concernment of these scriptures and dangers of erring in their interpretation," Newton offered his rules of interpretation to "inable him to know when an interpretation is genuine and of two interpretations which is the best."[47]

Newton's rules are somewhat close to the modern fundamentalist view. His fourth states "To (prefer) chose those interpretations which are most according to the litterall meaning of the scriptures unles

---

42 Ibid., p. 111, f. 3r.
43 Ibid., p. 111, f. 3v.
44 Ibid., p. 111, f. 5r.
45 Ibid., pp. 111–12, f. 5r.
46 Newton wrote out his explanation around a letter he received in his capacity as director of the mint.
47 Newton, *Fragments from a Treatise on Revelation*, p. 115, f. 10r.

where the tenour and circumstances of the place plainly require an Allegory."[48] Newton preferred literalism except where the context indicated some other reading. In this he combined the contextualism of Spinoza and Simon with literalism. And Newton worked out an elaborate way of reading symbols such as "Sun," "Moon," and the like that went far beyond just the literal reading.

Another interesting point raised in Newton's rules is his disavowal of any interpretation of the text of Revelation in terms of contemporary events. Rule 10 states, "In construing the Apocalyps to have little or no regard to arguments drawn from events of things; ... Becaus there can scarce be any certainty in historicall interpretations unless the construction be first determined."[49] Considering how much Bible reading and interpretation from contemporary events went on then as now, Newton was breaking with those who were seeing late seventeenth-century politics in terms of the scriptural text. In the *Observations* he asserted that "The folly of interpreters hath been, to foretell times and things by prophecy, as if God designed to make them prophets. By this rashness they have not only exposed themselves, but brought prophecy also into contempt."[50] Newton was a contemporary of the prophetess Jane Lead and the Philadelphians, and of the French Prophets who were making such predictions all of the time. Newton's disciple and associate, Fatio de Duillier, was a member of the latter group and apparently aroused Newton's interest in its activities, but Newton was also apparently worried about people, including the French Prophets, who were making explicit predictions about when the events in Revelation would take place in human history.

Conclusion

Spinoza and Simon had a great influence on the course of biblical studies from then to now, being the direct and acknowledged ancestors of the higher critics of the Bible. Did Newton, half fundamentalist about accepting divine message as certain and unquestionable, half modern in accepting higher criticism of surviving text, have much influence? Most of Newton's writings on religious topics remain unpublished even today. Frank Manuel, one of the first to look at the huge amount of Newton manuscript materials housed in the National Library of Israel, declared that "Newton's printed religious views have exerted no profound influence on mankind."[51] In two articles, I

48 Ibid., p. 118, f. 12v.
49 Ibid., p. 120, f. 14r.
50 Newton, *Observations*, p. 251.
51 Manuel, *The Religion of Isaac Newton*, p. 1.

have tried to show that, contrary to Manuel's view, Newton had a sizable influence on religious writers from the later eighteenth century down to the present, mainly on the protofundamentalists and the nineteenth- and twentieth-century fundamentalists.[52] Newton was cited favorably as an authority by David Hartley, Joseph Priestley, Bishop Thomas Newton, Charles Crawford, and Bishop Samuel Horsley, among others writing on religious topics in the eighteenth century. The *Observations* appeared in German, and Newton was cited by J. G. Hamann and by J. A. Bengel.

In the nineteenth century the early British fundamentalists almost all cited him favorably. Newton provided important historical information, ways of interpreting texts, and ways of accounting for the failure of prophetic predictions. After the failure of "The Great Expectation" of 1843, followers of William Miller sought refuge in Newton's texts. They found a slight difference in calendric calculations which allowed them to prolong their expectations another year. Then, when the bitter disappointment set in, they appealed to Newton's statement, note 51, that God did make us into prophets.

Without going over the material in the articles, I think it is fair to say that Newton was seen as a counterweight to the so-called scientific Bible scholarship that dominated late nineteenth- and early twentieth-century interpretations. One of the greatest scientists of all time took the prophecies in Daniel and Revelation seriously, and did not denigrate the Bible because of the findings of modern science.

Without just repeating what I said in the two articles, let me just mention three indications of Newton's influence. In 1922 Sir William Whitla, professor emeritus of Queen's University, Belfast, the former head of the British Medical Association, put out a reprint of Newton's *Observations*. It was the first such reprint since 1785. Whitla dedicated it to the leader of the Salvation Army, General Booth, who had done so much to hasten the coming of the Kingdom of God.[53] Whitla introduced the text with a series of lectures that he gave to his church refuting German higher criticism, and tried to make Newton's way of interpreting the prophetic books, Daniel and Revelation, a method of vital concern to twentieth-century readers. Whitla stressed what was to become all-important to many twentieth-century fundamentalists, the interpretation of prophecies about the Jews returning to the Holy Land, as clear indications that the end of history was at hand, and

---

52 Popkin, "Newton and the Origins of Fundamentalism," and "Newton and Fundamentalism II," pp. 165–80.

53 "Who with his illustrious father has accomplished so much among the nations of the world towards the hastening of the coming of the kingdom predicted in the book of Daniel." Whitla, *Sir Isaac Newton's Daniel and the Apocalypse*, p. v.

showed that Newton had pointed the way to this Christian Zionistic reading.⁵⁴

The scholar who collected the huge number of Newton's theological manuscripts that are in Jerusalem, Abraham Shalom Yahuda, published works in the 1930s arguing the claim that parts of the Bible were accurate as historical narrative based on eyewitness accounts. Yahuda contended that Exodus was written by someone whose native language was ancient Egyptian. After Yahuda purchased his collection of Newton manuscripts at the Sotheby auction of 1936, he found Newton making a similar claim about Revelation, namely that it was written by someone whose native language was Aramaic. In Yahuda's papers there is an unpublished essay on Newton's religious views, in which Yahuda used Newton to counter claims of the German higher critics.⁵⁵

These may not be world-shaking indications of Newton's influence as a Bible scholar, but they show some important influence. To close, let me mention something that may show a much greater influence. At the Seventh Day Adventist headquarters in Maryland there is a picture, *The March of the Reformers*.⁵⁶ It shows Daniel at the far left, followed by John, Hippolytus, Joachim de Fiore, Wycliffe, Luther, Knox and then Newton, who is shown handing the torch to Wesley who will pass it on to a contemporary believer, presumably a Seventh Day Adventist. Newton is portrayed as the pivotal figure in this development of reformers. What more influence could a person have in the fundamentalist tradition?

The Seventh Day Adventists own a manuscript in Newton's hand on "Prophecies Concerning Christ's 2nd Coming," forty pages of citations of Biblical sources.⁵⁷ And Isaac Newton is given a most serious and approving chapter in the history of prophecy written by one of their eminent historians, Leroy E. Froom.⁵⁸

Having said this, I think I have shown that Newton is a complex

---

54 Popkin, "Newton and Fundamentalism II," pp. 172–4.
55 Ibid., pp. 174–5.
56 Reproduced in Force and Popkin, *Essays*, opposite p. 1.
57 This manuscript is in the Adventist Source Collection at Barrien Springs, Michigan. There is some mystery about the provenance of this document. The Adventists acquired it around 1942. It is previously listed as belonging to the Library of Congress. Then in 1942 it is listed in manuscript sales for the year. The Library of Congress has told me emphatically that they never sell any manuscripts in their collection. The Adventists have told me that they no longer have any records concerning the acquisition of the document.
58 Leroy E. Froom, *The Prophetic Faith of Our Fathers*, vol. 2 (Washington, D.C.: Review and Herald Press, 1948), pp. 658–69. Froom also notes throughout volumes 2 and 3 the references to Newton in other writers on prophecy.

figure in the history of Bible studies. He more or less accepted the historical claims of Spinoza and Richard Simon, accepted their views about the corruption of the text, and their contextualism. Only two biblical texts survived this kind of analysis for Newton, the prophetic portion of Daniel and Revelation. The meaning of these texts was being progressively revealed to the wise who would understand, like himself. His interpretative views about these texts have had a continuing influence on those who are concerned with understanding the prophecies and with relating them to the ongoing course of human history. Fundamentalists seem to have ignored or put aside Newton's views about the history of biblical texts, his denial of the doctrine of the Trinity, and have concentrated instead on his ideas about prophecy. And in this he still remains a vital thinker. If his unpublished writings are ever made more available, then one can expect much attention being paid to Newton's views on the Bible.[59] And it is not just because he, the great scientist, wrote on these matters, but that he as a major interpreter of prophecy paved the way for various kinds of fundamentalist views up to the late twentieth century.[60]

59 Most of them can be read on microfilm at present put together by Chadwick-Healey. James Force, David Katz, and I are working on making them available in printed or CD-ROM form.
60 A fundamentalist writer whom I have talked to over the last several years, Paul N. Moore (see n. 11), tells me he is preparing a text comparing Revelation with Newton's commentary thereon, as a guide for present-day believers.

PART IV

The Canon Constructed

15

The Truth of Newton's Science and the Truth of Science's History: Heroic Science at Its Eighteenth-Century Formulation

MARGARET C. JACOB

Lecturing on Newtonian mechanics and dynamics around 1800, the natural philosopher John Dalton employed all the standard demonstrations in what had become by then a well-established genre of scientific education. On his tabletop he used oscillating devices, pendulums, balls made of various substances, levers, pulleys, inclined planes, cylinders of wood, lead in water, and pieces of iron on mercury to illustrate phenomena as diverse as gravitation, the "3 laws of motion of Newton," impulse or the "great law of percussion," force and inertia, specific gravity, attraction and magnetism.[1] There was nothing extraordinary in what Dalton was doing, first in his Quaker school then at New College in Manchester. The genre of British lecturing focused on Newtonian mechanics had begun in the second decade of the eighteenth century with the travels and publications of Francis Hauksbee, Jean Desaguliers, and Willem s'Gravesande who lectured in the Dutch Republic.[2] Dalton was deeply indebted to their legacy. His terse manuscript notes on his lectures – charred from a fire in 1940 – tell us that in one lecture he used a "machine with mercury, water-cork," and it was intended to illustrate, of all things, the effect on the planets of the "Cartesian Vortices."

In talking about the Cartesian vortices, and in explaining how wrongheaded they had been as a conceptual device for understanding

---

1 The manuscript lecture notes made by Dalton survive at the John Rylands University Library of Manchester, John Dalton Papers, no. 83, and appear to be dated randomly from 1796 to 1818. They have been partially damaged by fire. I wish to thank National Science Foundation (NSF) grant no. 9310699, which made possible this research and Dale Bowling for his work in the Manchester archives. On Dalton's lectures and his debt to popular Newtonianism, see Arnold Thackray, *John Dalton: Critical Assessments of His Life and Science* (Cambridge, Mass.: Harvard University Press, 1972), pp. 47–8, 66–7.
2 See *A Course of Mechanical, Optical, Hydrostatical, and Pneumatical Experiments. To be perform'd by Francis Hauksbee ... lectures read by William Whiston, M.A.* (London, [1713]); J. T. Desaguliers, *A Course of Experimental Philosophy*, 2 vols. (London, 1745).

planetary motion, Dalton was repeating an old Newtonian trope. In the process he was flogging a truly dead horse. Indeed, even in French colleges after the 1750s Descartes's horse had survived only in a few places and then by artificial resuscitation.[3] By midcentury the pressure to teach Newtonian science had become all but overwhelming. In Britain the vortices had been passé by the 1720s.[4] By then Desaguliers had dismissed Newton's great predecessor with this historical aside: "When M. DesCartes's philosophical Romance ... had overthrown the Aristotelian Physicks, the World receiv'd but little Advantage by the Change."[5] The demise of the vortices among the literate and scientifically curious had been insured by Newtonian lecturers like Desaguliers who never missed an opportunity to attack the fundamentals of Cartesian science. His colleague on the Newtonian lecture circuit, Benjamin Martin, went further and denounced Decartes because "he adopted the old atheistical Tenets of Lucretius."[6] In illustrating the vortices as late as the 1790s, Dalton was perhaps being a bit lazy; he was just following a tried and true lecture format passed along over two generations. He was also – unwittingly – teaching the heroic history of the new science that had been put in place during Newton's lifetime. The science, and the history told about it, were of a piece. But Dalton was a natural philosopher, soon to become famous and join the pantheon of heroes who laid the foundations of modern chemistry. Like his Newtonian predecessors, Dalton was not, nor did he aspire to be, a historian.

The contributors to this volume do aspire to being historians. Indeed, we have written on this occasion to honor a master of the historical genre, the late Betty Jo Teeter Dobbs. To illustrate her importance we have chosen to deal with one of the most fundamental

---

3 L. W. Brockliss, *French Higher Education in the Seventeenth and Eighteenth Centuries* (Oxford: Clarendon Press, 1987), pp. 353–8, 376–80, 366. There was still, however, a strong emphasis on mathematical skills in French university courses. The French colleges are the nearest equivalent to the dissenting academies. In the year XI, the first *Bulletin de la société pour l'industrie nationale* (Paris), p. 179, still complained that "on s'est peu occupé en France de technologie, et jamais cette étude n'a fait partie de l'instruction publique." This reference was supplied by Jeff Horn.

4 On the rise of Newtonian science in the universities, see Gordon Donaldson, *Four Centuries. Edinburgh University Life, 1583–1983* (Edinburgh: University of Edinburgh Press, 1983), p. 34; and John Gascoigne, *Cambridge in the Age of the Enlightenment: Science, Religion and Politics from the Restoration to the French Revolution* (Cambridge: Cambridge University Press, 1989).

5 Preface to Desaguliers, *A Course of Experimental Philosophy*.

6 B. Martin, *A Panegyrick on the Newtonian Philosophy, shewing the Nature and Dignity of the Science; and its absolute Necessity to the perfection of Human Nature; the Improvement of the Arts and Sciences, the Promotion of True Religion, the Increase of Wealth and Honour, and the Completion of Human Felicity* (London, 1754), p. 6.

questions raised by her work, and explicitly by her lecture that opens this collection. Was there a Scientific Revolution in the period from 1543 to 1687? Is it even appropriate to use the term "revolution"? Some historians of science think it was borrowed only in the eighteenth century, largely from political events, and used by Enlightenment polemicists intent upon distancing modern Western culture from its religious foundations. Dobbs argued that the term was an anachronism. She followed I. B. Cohen in believing that none of the major participants in the so-called revolution of the seventeenth century ever used the term to describe what had happened. But Cohen had overlooked evidence of Boyle's having used it precisely to describe the transformation in intellectual life he experienced in midcentury England. In addition, Dobbs believed that "the word 'revolution' hardly began to acquire its modern meaning until the eighteenth century." We now know that, just as we should have expected, the term was being applied to political events by the late 1650s.[7] It was evolving just then to signal the occurrence of irrevocable, dramatic change. In addition, this modern usage in late seventeenth-century political affairs, which then took hold in the eighteenth century to describe the rise of the new science, was far more widespread than simply what can be found in the polemics of the French *philosophes*. Indeed, as I am about to argue, the revolutionary character of science was one of the central premises at work among the post-1700 British and European disseminators of the new science. They described it as revolutionary almost as an afterthought, as a way of introducing their science. The linkage of their heroic and revolutionary account of the history of science with the explication of Newtonian science gave extraordinary staying power to the first long after the second had been transformed. But saying it was revolutionary and heroic – a dramatic and irrevocable change effected by a few individuals – need not make it so.

Old or new, contemporaneous or posthumous, the notion of there having been a Scientific Revolution might still just be quite simply wrongheaded. Or alternatively its eighteenth-century version might be obviously, incontrovertibly, true. In his essay Richard Westfall argues (also posthumously) in contrast to Dobbs's position, that because science is everywhere a part of our present-day lives, there must have been, metaphorically speaking, a revolutionary moment. Unashamedly, Westfall embraces an essentially eighteenth-century narrative.

---

7 Ilan Rachum, "The Meaning of 'Revolution' in the English Revolution (1648–1660)," *Journal of the History of Ideas* 56 (1995): 195–215. Cohen did not have this article available but he also thought that between 1640 and 1660 no changes of lasting value occurred. See I. Bernard Cohen, *Revolution in Science* (Cambridge, Mass: Harvard University Press, 1985), p. 67.

He dates the "moment" according to the publications of its heroes, as commencing with Copernicus in 1543, continued by Galileo, Kepler, and Descartes, completed by Boyle and Newton. The revolution was all over with the *Principia* (1687). Indeed, Westfall would have the Scientific Revolution be more important than the Renaissance or the Reformation, without for a second considering that it would never have happened had those other two epochs not preceded it. So, too, he effortlessly attributes historical causation. Both the Enlightenment and the Industrial Revolution were directly the product of the earlier revolution in science. Westfall never tells us how Newton could have been deeply pious, obsessed with both God's work and word, and an enlightened deist – in Westfall's account almost a secret *philosophe*. Nor does he provide a narrative of how the linkages were established between the new scientific culture of the eighteenth century and the technological innovation that comes at its end. Late in the twentieth century the Scientific Revolution remains a subject fraught with assertions.

Given such massive disagreements among the masters, predictably many of our contributors have implicitly steered away from the meta-issue of – or if – the Scientific Revolution. Instead they have wisely reiterated the importance of religion and "magic" to nearly every one of the major players in the revolution that may never have happened, or they have pointed to how very little about it was self-evident to contemporaries. To them Kircher might have been another Newton. Who in northwestern Europe around 1650 thought that alchemy had at best a generation left to its vitality? Previous historians have seen Boyle and his contemporaries fighting over correct science; since the late 1970s with the work of James R. Jacob we have seen them awash in the revolutionary turmoil of mid-seventeenth-century England.[8] Where once we read only Boyle's *The Sceptical Chymist*, now every tick and dot in his manuscripts have been analyzed. It is not so much that works have been decanonized. Rather the canon of what is deemed relevant to the rise of the mechanical philosophy has been vastly expanded now to include alchemy, theology, religious convictions, and political interests. All the once-designated "heroes" are in fact still very much central to the story this volume addresses, only now their complexity has served to make them seem more accessible and certainly less "scientific." In this volume we are even prepared to take seriously a comparison of Newton with the anti-Copernican Jesuit, Athanasius Kircher. If Protestant heroes could rise up from their graves, I would not want to be Paula Findlen.

8 James R. Jacob, *Robert Boyle and the English Revolution* (New York: Burt Franklin, 1977).

Historical judgments about canonicity and stature depend upon a lengthy process of shifting and weighing, of accumulation, of the new being assessed and assimilated, of values as well as creativity being defined, recognized, and then enshrined. The would-be king of Denmark got to be immortalized as *Hamlet* and the lad from Lincolnshire got to be one of the heroes of modern science as a result of a historical process. Heroes are not born, they are made. However brilliant Kepler or Newton had been, their innovative contributions would have remained relatively unknown, or esoteric, possibly even banished, without a larger transformation in the way literate westerners conceived both nature and history. My task here is to examine the historicity of scientific hero formation complete with its revolutionary implications. I seek not to relativize so as to dismiss the revolutionary character of the new understanding of nature, or to deny the obvious genius of the leading figures whose work helped bring it about. Rather my point is to get beyond the hyperbole and to examine the values – what was at stake – to those who took up the new learning and enshrined its revolutionary and heroic mystique. If this exercise in historicizing the history of science is successful, then it may occur to readers (would that my friend Betty Jo could be one of them) that the stakes were then, and are now, very high. If we make the move to diminish, or dismiss, the depth and breadth of the intellectual transformation in the Western understanding of nature (bracketed between roughly 1540 and 1750), we want to be sure about what it is that we are giving up.

As we broaden the canon and explore the context, we need to understand that although Newton and Boyle – and who knows how many others among their contemporaries – were, in their different ways, practicing alchemists, this fact does not alter the profound character of the intellectual transformation described by the somewhat misleading, shorthand phrase, the Scientific Revolution. That intellectual transformation can be dismissed only if we believe that two elements alone define it: the heroes had to be pure, simply great "scientists," and they alone made it happen. In effect, if the older historiography erred on the side of simplicity, then the history it sought to convey can be dismissed. The logic of the dismissal is flawed. Finally, and not least, if the history of science as a discipline abandons a central problematic, that of explaining how and why Westerners moved to mechanize the world picture, then it would lose, not enhance, one of its most important raisons d'être. In the West, at least, revolutions tend to keep their readerships.

Let me now return to the end of the history making about deification and rapid change in the Western approach to nature, to 1800 and

John Dalton, to the latter stages of a process that had been underway for at least three generations. In one set of lectures Dalton set the stakes and proclaimed the importance of the science he was about to teach. "It will be universally allowed that the cultivation of Mechanical Science, in the present state of society especially, is an object of primary importance. It is true, the arts and manufactures, are all interested in the science." Such interest in mechanics for manufacturing had not always prevailed. The ancients, in contrast to the moderns, knew "little of Mechanics as a science." Both duty and self-interest require that people now employed in mechanical occupations, or in their supervision, need to know mechanical principles in detail. Dalton believed that the very success of British ingenuity in manufacturing, transportation, and instrumentation hung upon a knowledge of Newtonian mechanics. As they contemplated French competition, Dalton's audience – he trusted – would have been the first to acknowledge the importance of national prestige and wealth, and hence the importance of mechanical science. By the late eighteenth century natural philosophers added to the heroic story of science's achievements a message about its essential role in progress and prosperity. It was a global and competitive vision with which we in the late twentieth century are still familiar.

The modern science of mechanics, as well as mathematics, had its unique heroes, its founders and originators. Dalton laid them out in a simple historical narrative: Galileo "who lived about two centuries ago," followed by Newton, Leibniz, and the Bernouillis. There have been many other contributors but only a few, Dalton implies, are worth naming. As had his predecessors, Dalton aimed his lectures at the industrious part of the nation. He told them that only one eighteenth-century person stood worthy to be ranked with the Galileos and the Newtons: John Smeaton, civil engineer, canal builder, harnesser of water power in the service of mechanized industry.[9] So coupled, engineers like Smeaton, as well as their friends and employers, could imagine themselves to be scientists.

The Smeatons and the Watts could even try to become amateur historians, collecting letters and memorabilia of their great predecessors. Writing to Smeaton's and Dalton's contemporary, James Watt, a friend sought to know about letters between Newton and Fatio de

9 Michael Adas, *Machines as the Measure of Men: Science, Technology, and Ideologies of Western Dominance* (Ithaca, N.Y.: Cornell University Press, 1989), chap. 2. On the early history of Cartesian lecturing in France, see J. L. Heilbron, *Electricity in the 17th and 18th Centuries: A Study of Early Modern Physics* (Berkeley: University of California Press, 1979), pp. 146–59; by the 1740s women were more numerous than men at the Parisian lectures (p. 163).

Duillier. Watt was supposed to have acquired them and "I would give anything to have a Scrap, however insignificant of his writing."[10] During his life and long after, Newton enjoyed a large crowd of worshipers. If in those letters Watt had discovered Newton's millenarianism, he would have found it quaint, but hardly a reason to stop his sons from mastering mechanics or dim his own admiration for either science or Newton.

The inspiring history Dalton told, and someone like Watt believed, claimed in passing that science required genius and promoted mechanized industry and progress. The stakes raised by both science and its history could hardly have been higher. Within the culture of practical science that emerged in eighteenth-century Britain participants as diverse as itinerant lecturers, craftsmen in metal and steam, engineers as well as entrepreneurs could imagine themselves – however unoriginal their daily contributions – as standing on the shoulders of giants. Only piety and Godliness were missing from Dalton's remaining lecture notes. But there was no shortage of preachers and teachers willing to make the godly connection.

Dalton and his Newtonian predecessors added to the true science of Newton what could be presumed by his audience to be an equally true history of science. All the history one needed to know was there: Newton had been right about the vacuum and Descartes had been wrong about the vortices. The development of science was a progressive story of great discoveries punctuated by the occasional misdirected theory; such had been the swirling vortices. In the eighteenth century the history of the heroes of true science framed the presentation of natural knowledge. Their history got told with the science, almost as the filler, the enticement to spark interest among a restless, not scientifically literate, audience. Within this history lay the key to human enlightenment, to a new intellectual freedom. Watt and his industrial friends actually believed that true science and philosophy promised to destroy "the very foundation of Enthusiasm, Superstition, and all Kinds of Imposture.... What glaring instances of this Truth has this last century produced? Where are now the Wizards and Necromancers, the Pseudo-Prophets, the Demoniacs, the Wonder-working Relicts, and the Group of Omnipotent Priests that formerly swarm'd in this Island?"[11] The articulation of English liberty, religious freedom for Protestants, the demise of quackery, the progress of industry, and the history of science were seen to be of a piece.

Little wonder that the heroic narrative of socially isolated geniuses,

---

10 See Eric Robinson and D. McKie, eds., *Partners in Science: Letters of James Watt and Joseph Black* (London: Constable, 1970), pp. 272–3, dated 1797.

11 Ibid., p. 50.

first articulated in the eighteenth century, survived well into the twentieth. Decoupling it from the truthfulness of scientific laws has taken a generation or more of historical scholarship. It has been difficult and controversial to create a textured, nuanced, and historically informed understanding of modern science at its origins. The rise of postwar and now contemporary scholarship about science has required the dismantling of assumptions about isolation, about disinterest, about purity as the key to brilliance – in short, about the way history and human beings work – which were once taken to be as certain as the very science learned in lecture hall, pulpit, and school. The search for a richer, more textured history should not, however, lead us to turn away from some of the earliest historical associations of the new science. We cannot imagine that the revolutionary legacy of science has nothing to do with us, that in effect we have never been modern. As both modern historians with our methods, and as citizens with our expectations, we are in part science's beneficiaries.

Our legacy received its most widespread, European dissemination in the revolutionary decade of the 1790s. What Dalton taught to his paying audiences, teachers inculcated into their captive ones. The youngest son of James Watt, Gregory, being of Scottish origin was suitably trained at Glasgow College. There in the 1790s the education for the whole man rested on philosophy and science. Drawn to the radical voices of the decade and deeply interested in reform, the young Watt scribbled in his student notebooks his belief that natural knowledge and virtue bear a relation "more intimate than one can imagine." The "tree of science never flourishes where good dispositions have not prepared the soil." Wealth and power without virtuous industry and learning produce "a crowd of servile sycophants." Even Francis Bacon got his office by his own exertions, and Bacon was, of course, a true son of the Reformation. In rather florid terms Watt was being taught – rightly in my view – that Bacon could not have written as he did had he not been a Protestant. History, he was being told with hyperbole, teaches that when the Catholic Church dominated the "blossoms of science continued to droop." No progress was made until Bacon like the "rising sun illumined the learned world ... [and] The system of Aristotle sunk beneath him."[12] At Glasgow, science and its history revealed the revolutionary progress of Protestants. On their odyssey away from superstition toward science and good government, piety and Godliness ensured a providentially guided progress.

In the 1790s one did not have to be a radical or an industrialist to

---

12 Gregory Watt's exercise book, 1793; James Watt MSS, C4/C18A, Birmingham City Library.

learn the meaning of science and its history. One did not even have to be a man. Teaching science to young women outside of London at precisely the same moment as Dalton lectured and Watt studied, Margaret Bryan told a very similar historical and scientific tale. Her textbook in mechanics, *Lectures on Natural Philosophy: The Result of Many Years' Practical Experience of the Facts Elucidated* (London, 1806), grew out of her many years as headmistress of a girl's school. Bryan dedicated it to Princess Charlotte of Wales and the naturalist, Charles Hutton, who encouraged the project. According to Bryan, piety was the reward promised by the mastery of both science and its history. She proposed to arm young women and all readers "with a perpetual talisman," which will "guard your religious and moral principles against all innovations." Bryan presents herself as "merely a reflector of the intrinsic light of superior genius and erudition" who is translating and moderating knowledge for readers without "profound mathematical energies." Male writers and lecturers often said similarly modest and enticing things.

Bryan confesses to being a follower of William Paley's version of natural theology. In effect, she revered a century-long tradition of Newtonian preaching that became fashionable with the 1705–6 Boyle Lectures of Samuel Clarke. Physicotheology based upon mechanical science taught that order and harmony in the universe sanctioned tranquillity and hierarchy in society as well as the rule of law. A century and more of political stability, born out of revolution, only confirmed the theology.

Like Desaguliers (a mere youth in 1688–9), Margaret Bryan begins with Newtonian definitions of matter and gravity and in the process introduces students to the history of the new science beginning with Galileo and leading to brief historical discussions of Boyle, Guericke, through to Newton. In now tried-and-true fashion, the textbook of 1806 simultaneously turns to levers, weights, and pulleys to illustrate Newtonian mechanics. Air pumps, atmospheric pressure, pneumatics in general, hydrostatics, hydraulics, magnetism, electricity, optics, astronomy (on which she had written another whole book) are all illustrated by experimental demonstrations. Bryan concludes with "Of Man as a Machine," which despite its materialist sounding title, attributes the wonderful mechanism of the human body to divine artifice. She caps off the scientific instruction with a preachy lecture on stoicism, obedience, cheerfulness, affection, and duty. True to her general conservatism, piety, science, and history are then placed in the service of politeness.

British men and women of the eighteenth century, and beyond, found in Newtonian science and its edifying history of genius upon genius, beginning with Galileo and culminating in Newton, one source of immense national pride and accomplishment. Theirs was a nationalist history of science. As Francis Hauksbee told his audiences as early as the reign of Anne, "after many ages had pass'd, with little or no Progress in the True Knowledge of the nature of things, greater advances have been made within the compass of a small number of years, than was easily to be imagined." Boyle and Newton have been principally responsible for the extraordinary, recent progress of science.[13] A few decades later, in the age of Whig oligarchy, the nationalist theme was still going strong, and James Ferguson explained how back in the 1650s, inspired by Francis Bacon, Boyle came on the scene. He eradicated "the old notions of the schools so strongly possessed [in] people's minds at that time." Then in the 1660s came the Royal Society and "true philosophy began to be the reigning taste of the age, and continues so to this day." The British were the first to lay the foundations of physics; the followers of Newton, now named by Ferguson one by one, along with their Continental followers from s'Gravesande to Nollet in France, "have...also acquired just applause."[14] From Newton's lifetime to his second generation of followers, scientific achievement inspired national pride. English science complemented an edifying history created, it was believed, by freeborn English people. It also justified an imperial expansion that supposedly brought light into a darkness bereft of scientific knowledge.

The history of science we inherited was not simply a self-serving British invention. On the European Continent, particularly in the French academies, teachers and eulogists charged with reporting on "an extraordinary revolution in science" had been extolling the virtues of the scientific heroes for much of the eighteenth century.[15] In addition, late-seventeenth-century Cartesians had pioneered the genre of the public lecture aimed at the literate and affluent. During the reign of Louis XIV, even the most cautious among the French lecturers were prepared to proclaim science as vaguely "progressive," if not shackled by the antique and the dogmatic.[16] Here too history fitted implicitly into the story. Where we can find lecture notes comparable with those left by Dalton – for example, Samuel Koenig's vastly more detailed

13 F. Hauksbee, *Physico-Mechanical Experiments On Various Subjects*... (London, 1709), pref.
14 *Ferguson's Lectures on Select Subjects*..., ed. David Brewster (Philadelphia, 1806), preface by Ferguson (d. 1776), pp. xiii–xv.
15 Charles B. Paul, *Science and Immortality. The Éloges of the Paris Academy of Sciences (1699–1791)* (Berkleley: University of California Press, 1980), p. 20; the phrase belongs to Charles B. Paul.
16 Jacques Rouhault, *Physica. Latine reddidit*... *S. Clarke* (London, 1697), author's preface.

lectures of 1751 – we can see the integration of science with its history, all now in the service of an enlightenment for educated elites.[17] By the 1730s the scientific culture of the French, particularly as disseminated by Newtonians like Voltaire, became Western.

Trained by the foremost Newtonian explicator, Madame du Châtelet, Koenig supplemented his income by lecturing, in this instance in The Hague. Of course, by the time of his lectures Dutch audiences had been hearing about Newton and his heroic feats for at least two generations.[18] In the early 1700s students at Leiden learned from Boerhaave that "Newton is the miracle of our time." Perhaps only Francis Bacon could be described as being in some sense his intellectual equal.[19] But by 1750 The Hague enjoyed a rich, international clientele. The city housed embassies and the court of the newly restored Dutch stadholder and his entourage. Among them were Charles and William Bentinck, connoisseurs of the moderate Enlightenment and patrons of various reform-minded intellectuals; both Diderot and Rousseau would become their guests.

Koenig's audience to whom he spoke in French were "un certain nombre de personnes de goût . . . de distinction et de mérite." Here at a provincial outpost of the Enlightenment Koenig aimed to show how the logic of nature, properly understood, could deliver men from their prejudices, "nous armons contre la superstition qui entend la tirannie sur la surface de la terre." Koenig's task was to explain the movement and force of bodies as well as the construction and power of new and old machines. In fact, he said little about machines to these men and women of the court, but much about the history of the science they were meant to illustrate.

The new science, he explained, is distinctively European; America, Africa, and all of Asia are ignorant of things "about which we speak." Astronomy lit the way as Europe came out of the dark ages led by Regiomantus. In between the ancients and the moderns lay the barbarian invasions and Scholasticism. First emerged Copernicus. A great battle ensued over his views, but aided by Tycho Brahe and Kepler –

---

17 MS X B I, "Lecons de physique de Mr. le Prof. Koenig qu'il a donne a la Haye, 1751–52," University Library, Amsterdam. Jacques Rohault was giving lectures in his home in Paris in the mid-1650s. See his *Traité de physique* (1671); by 1730 there were ten separate publications of it. Samuel Clarke did a Latin translation in 1697, expanded upon in 1710, and he added Newtonian footnotes that refuted the vortices among other aspects of Cartesian science.

18 See the lectures of Daniel G. Fahrenheit, University Library, Leiden, MS BPL 772 (from the collection owned by van Swinden), "Natuurkundige Lessen van Daniel Gabriel Fahrenheit . . . 1718," ff. 6–11, on Descartes and Newton, ff. 88ff. on Boyle.

19 E. Kegel-Brinkgreve and A. M. Luyendijk-Elshout, eds., *Boerhaave's Orations* (Leiden: Brill and the University of Leiden Press, 1983), p. 160, and for Bacon, pp. 175ff.

whose laws Koenig explicates – heliocentrism triumphed. Galileo followed in their footsteps and in 1633 he had the "great courage to defend the Copernican system against the Inquisition."[20] But the "great genius" of the period was Descartes, and his system in the hands of Leibniz (who had influenced Koenig as he had Madame du Châtelet) became "le Antichambre de la veritable philosophie."

Koenig's tastes in natural philosophy were catholic and he saw merit in the ideas of Descartes, Leibniz, and Newton. The latter Koenig labeled "the second Archimedes" who had the good fortune of many disciples. In an oblique reference to the Leibniz-Clarke imbroglio, Koenig claims that Newton's disciples did, however, reintroduce occult qualities after Descartes and Leibniz had banished them. Koenig blames a few unnamed Newtonians for hypothesizing too much. Newton, however, comes out blameless and much time is spent explaining the law of universal gravitation. Leibniz was, of course, another "Hercule de science." Yet given "the disputes among the various schools, the true physicist should be neither Newtonian, Cartesian nor Leibnizian." All this history, interspersed with science and mathematics, had been packed into just one lecture.[21]

Subsequent lectures explained that science and mathematics enhance trade and commerce just as reason and evidence drawn from nature promote belief in God's existence. Koenig attacks skeptics and "les mille beaux esprits." He also accuses Spinoza and Locke of trying to establish that miracles had never occurred. Within the context of 1751 and the recent revolution in the Dutch Republic, his attack seems focused on its radical wing, agitators like Rousset de Missy who had sought to use Locke to argue for deeper reform than William IV would endorse and who were known to be pantheists.[22] Science and its history had many uses. In the aftermath of a revolution that restored the stadholderate and threatened more radical reforms, the circle of acceptable philosophy is being drawn very tightly; even Newtonians were suspected of insufficient piety. All these ideological signals are being flashed amid lessons in natural philosophy and science: first, "the preliminaries," that is, definitions of extension and the divisibility of bodies; then, the applications of natural philosophy from density to porosity and specific gravities, from Boyle's law to the microscope.[23]

20 Ibid., ff. 1–16, here summarized.
21 Ibid., f. 31; second lecture is on ff. 32–79. These lectures appear to be verbatim as they were delivered. In the lecture covered on ff. 234–6, Koenig returns to the early history.
22 See Margaret C. Jacob, *The Radical Enlightenment: Pantheists, Freemasons and Republicans* (New York: Harper Collins, 1981).
23 Ibid., ff. 130–91.

Dynamics, hydrostatics, mechanics, optics, explications of Newtonian gravitation with fairly complex mathematics – all appear in Koenig's lectures that were in fact more technically sophisticated than what was being routinely given across the Channel by British lecturers.

English, French, and Dutch language explications of the new science throughout the eighteenth century relied upon the story of heroic science to promote specific and often different, but not necessarily incompatible, agendas. Progress, industry, piety, moderation could all be supported by the same history that reformers, even revolutionaries, enlisted. The progressive nature of scientific knowledge and the rapidity with which its expansion had occurred in the seventeenth century impressed all observers. Most notably in Protestant lands, from as early as the 1660s in England, some clergy took up mechanical explanations and integrated them with traditional theological positions. Also liberal Dutch pastors such as Bernard Nieuwentyt offered their own version of physicotheology, and it in turn was translated into English.[24] Gradually, in select Protestant circles throughout the Euro-American world, God's work came to be more prominent than the reading of his word.

Physicotheology had been intended to shore up the church, to create a stable status quo, and to arm the pious against the radical freethinkers who emerged as early as the 1690s. But storytelling in the interest of order and hierarchy can come back to haunt its promoters. Late in the eighteenth century the revolutionary nature of science, its almost miraculous ability to reveal order in nature, fueled a different set of passions. It fed into a growing discontent found among the literate from Amsterdam to Paris, in both Protestant and Catholic Europe. In the 1780s in the Low Countries and France – each with markedly different sets of issues – educated people grew increasingly critical of the disorder perpetrated by courts and their minions.

Dutch manufacturers who sided with the Revolution of 1787 collected books by Descartes, Locke, and Voltaire, one noting in the margin "superstition is the opposite of religion just as astrology is the opposite of astronomy."[25] Indeed in the generation after the stadholderate of William IV (d. 1751), many *patrioten* and reformers with industrial interests turned revolutionary. For them, like their French counterparts, the heroes of science became beacons in the night; only exceptional men could follow their example. "Newton gave us a wonderful theory of how the heavenly bodies worked but a century passed before men could use the wonderful theory in navigation,"

24 B. Nieuwentyt, *The Religious Philosopher* (London, 1720).
25 Cited in C. Elderink, *Een Twentsch Fabriqueur van de achttiende eeuw* (Hengelo: Broekhuis, 1977), p. 29.

said the Amsterdam patriot and distinguished scientist, J. H. van Swinden.[26] What we need now are men of both theory and practice – like Franklin – who will seize the day and institute reforms in industry and government. For reformers like van Swinden, Newton became an exemplar of the meritorious and the extraordinary, to be imitated in whatever task was necessitated by one's calling.[27] Within the revolutionary circumstances of the time, van Swinden invoked the achievements of science as indicative of what human effort could accomplish, of what revolutionaries could effect.

The revolutionary mystique of science cannot easily be excised from the Western imagination. The metaphor of there having been a "scientific revolution" eventually became both a conceptual resource widely used in the historiography of science and a trope integrated into the histories of political transformation. The uses to which the history of science was put throughout the eighteenth century bespoke the revolutionary. Eventually by the 1780s, as I. B. Cohen has shown, the phrase itself, a scientific revolution or a revolution in science – thanks to Bailly – became fashionable.[28] It served the needs of secularism just as it affirmed the revolutionary tradition so basic, we still believe, to the creation of Western democracy.

Nowhere in Europe did the revolution in science find greater admirers than in France during the 1790s. The task was to translate the central premises of the French Enlightenment into the life of the entire nation. In 1794 the abbé Grégoire told the National Convention in Paris that "*Les Savants* and the men of letters carried out the first coups against despotism. . . . If the career of liberty has opened before us, they were the pioneers."[29] Just as had befitted free-born Englishmen of the seventeenth century, science now adorned the inheritance of late eighteenth-century French democrats and republicans. Not surprisingly, the French Revolution, coupled with its seventeenth-century antecedent in England, held the key to why heroic science – and the closely related concept of revolutions in science, politics, and industry

---

26 J. H. van Swinden, *Redenvoering en aanspraak ter inwijding van het gebouw der maatschappij Felix Meritis te Amsterdam* (Amsterdam, 1789), pp. 24ff.
27 For the truly heroic, see the opening of J. H. van Swinden, *Oratio de philosophia Newtoniana* (Franeker, 1779). Note the influence of Koenig (p. 40).
28 I. Bernard Cohen, *The Newtonian Revolution: With Illustrations of the Transformation of Scientific Ideas* (Cambridge: Cambridge University Press, 1980), pp. 120–2. Cf. by the same author, *Revolution in Science*.
29 Quoted in Luc Rouban, *L'État et la Science. La politique publique de la science et de la technologie* (Paris: Éditions du Centre National de la Recherche Scientifique, 1988), pp. 26–7.

− became so dominant and pervasive, particularly in Anglo-American historiography.

We all know that in the first instance Copernican science also provided one source for the notion of there being "revolutions" in matters of state. Indeed, the language of astronomy, partly indebted to Copernicus's "revolutions of the heavenly orbs," provided vocabulary for the profound changes in England during the 1640s and 1650s. By 1660 the terms "revolutions and commotions" in the state had become commonplace. Under the impact of those midcentury events, both political and intellectual, revolutions in the state and in thought began to take on the modern meaning of progressive, irrevocable change, not simply a returning to a previous place or a revolt that occurs periodically with little to show for the trouble. In 1651 Robert Boyle himself used the term "revolution" to describe the progress he expected in philosophy and divinity as a result of the civil wars: "I do with some confidence expect a Revolution, whereby Divinity will be much a looser, & Real Philosophy flourish, perhaps beyond men's Hopes."[30] For Boyle and his friends the expected revolution had millenarian associations; by the 1690s a more secular understanding of time had become commonplace in England. So too had the use of the term "revolution" to describe the events of 1688-9.

Largely because of seventeenth-century events, in the eighteenth century in English, French, and Dutch the term "revolution" became working linguistic capital.[31] It could function as both political and cultural currency. In 1766 Josiah Wedgwood, speaking about what would come to be known as the Industrial Revolution, wrote to a friend: "Many of my experiments turn out to my wishes, and convince me more and more, of the extensive capability of our Manufacture for further improvement. It is at present (comparatively) in a rude, uncultivated state, and may easily be polished, and brought to much greater perfection. *Such a revolution, I believe, is at hand,* and you must assist in, [and] profit by it."[32] It is a myth of recent origin, perpetuated by proponents of the so-called new economic history, that early industrialists experienced the changes they themselves wrought in manufac-

---

30 Jacob, *Robert Boyle and the English Revolution*, p. 97, citing the manuscript in the Royal Society of London.
31 In Dutch the more authentic term is "omwenteling" but by the mid-eighteenth century "revolutie" was also in use.
32 *The Letters of Josiah Wedgwood, 1762–1772*, ed. Katherine Euphemia, Lady Farrer (London, 1903), p. 165 (emphasis added). Once again in following E. Hobsbawm and others, I. B. Cohen placed the usage of this term in an industrial setting too late; cf. Cohen, *Revolution in Science*, p. 264.

turing as something gradual, almost imperceptible. Having available the notion of revolutions born out of profound political transformations and out of the history of science, early mechanists and industrialists like Wedgwood possessed the vocabulary to describe unprecedented transformation effected not by the sword or the air pump but by the machine.

As in most things modern, the French Revolution gave extraordinary circulation both to the word and, more important, to the concept of revolution as a sharp and irrevocable break with the past. Almost predictably by the 1790s the term "revolution" began to be applied by the French to economic phenomena, particularly to what was happening in British industry. A spy first identified as revolutionary the events about which Wedgwood had privately mused. In 1794 Le Turc wrote home with the following description of British technological development: "[When traveling in England] I saw with dismay that a revolution in the mechanical arts, the real precursor, the true and principal cause of political revolutions was developing in a manner frightening to the whole of Europe, and particularly to France, which would receive the severest blow from it."[33]

The spy knew whereof he spoke. He was addressing his urgent comments on British industry to the revolutionary ministers charged with the task of improving French manufacturing and mechanization. At that moment the leader of the effort was the minister, François de Neufchâteau, and to stimulate technological innovation he created a system of national exhibitions where craftsmen and women would come from throughout the country to display their skill and ingenuity. At the opening of the first exhibiton in 1798, François de Neufchâteau spoke about science and technology – and about history – to the thousands assembled in Paris for *ce spectacle républicain*. He explained that in the old regime "la technologie ou la théorie instructive des arts et des métiers[,] cette science était presque entièrement ignoreé." It was only Francis Bacon, and later Diderot in his *Encyclopèdie*, who saw that the mechanical arts were essential, a branch of philosophy based on the assumptions now championed by the new regime, that "l'industrie est fille de l'invention, et soeur du génie et du goût."[34] In the mind of the revolutionaries, industry, invention, genius, and taste

33 MS U 216 Le Turc to Citoyen, 14 Nivoise An 3 [December 1794], Conservatoire des Arts et Metiers, Paris. Le Turc was born in 1748 and in the 1780s as an engineer and spy he had traveled extensively in England, describing techniques and recruiting workers. I owe this splendid quotation to the kindness of the late J. R. Harris, and to my knowledge it is the earliest use of the term to describe industrialization.
34 Printed in *Reimpression de l'ancien moniteur*, vol. 29, 1847, pp. 402–3, no. 1, 1 Vendémiaire, Year VII, September 26, 1798.

were of a piece, and science and technology were exemplars of all that genius and invention could achieve. To the leaders of the revolution like Jean-Marie Roland, Girondist and the minister of the interior in 1793, responsible for financing the elite academies of science, the true genuises of science formed academies, not the reverse. "Rousseau, Bacon, Newton, Euler, Bernouilli and a multitude of other celebrated *savans* . . . were not *savans* because they were named to a seat (in the academy)."[35] The academicians, once necessary when science was less mature, should now concentrate their energies on giving public courses, in effect to imitating the culture of practical science that had been in place across the Channel for much of the eighteenth century. Roland's suspicion of the academies foreshadowed their demise. At the Terror they were ruthlessly purged, with 25 percent of their members executed and exiled.[36] In their place Roland wanted institutes and lycées; he and his compatriots wanted to "créez une immense aristocratie professoriale, vous établissez dans la république un sacerdoce scientifique."[37]

By the late eighteenth century heroic science provided a cherished model for revolutions, both political and economic. Helen Maria Williams put the linkage well in 1790 when she urged Britons to accept the French Revolution: "Why should they not be suffered to make an experiment in politics? I have always been told, that the improvement of every science depends upon experiment."[38] The coupling of science and politics had not been invented by Williams out of thin air; it derived from the way Westerners understood their political and scientific history. More than any other body of culture, science released the revolutionary imagination, helped to develop its fantasies, to eliminate doubt about what human beings could accomplish. After 1700 radicals and moderates on both sides of the Channel, and across the Atlantic – even conservatives interested in piety and politeness – could embrace the imaginary of revolutions whether in cotton or in regimes.

---

35 Jean-Marie Roland, *Compte rendu à la convention nationale de toutes les parties de son département, de ses vues d'amélioration & de prospérité publique*, January 6, 1793, BN, Lf 132.3, p. 225, Bibliothèque Nationale, Paris; "Je ne dirai pas que Rousseau, Bacon ne furent d'aucune académie; car on me répondroit aussitôt que Newton, Euler, Bernouilli et une multitude d'autres savans célèbres en furent; mais je dirai que ces derniers ne furent pas savans, parce qu'ils furent appelés au fauteuil académique. Au contraire, c'est parce qu'ils étoient savans qu'ils y furent appelés avec beaucoup d'autres qui ne l'étoient pas."
36 Dorinda Outram, "The Ordeal of Vocation: The Paris Academy of Science and the Terror 1793–95," *History of Science* 21 (1983): 251–73.
37 Roland, *Compte*, p. 226.
38 Helen Maria Williams, *Letters Written in France 1790* (London, 1790; reprint, Oxford: Woodstock Books, 1989), p. 220.

The histories of science that they told were filled with exaggeration and hyperbole, with assumptions that we would now characterize as naive. But the first believers in science had also allowed for, indeed celebrated, the possibility of rapid, irreversible transformations. In societies encumbered by hierarchy, blood, and birth, they had imagined, and experienced, significant intellectual changes.

We may find it fashionable now late in the twentieth century to cast doubt on the very notion of there having once been a "scientific revolution." But by 1720, with the exception of Fatio, not one of Newton's close followers, not Desaguliers, or Pemberton, or Clarke, or even the millenarian Whiston, or the student of the Temple of Solomon, William Stukeley, could have understood the master's alchemy.[39] They hung on his every word, but I think it to be the case that not one of them could have explicated his alchemical texts. Such a rapid shift may justly be imagined as revolutionary. It took historians like Dobbs, Westfall, and Karin Figala years of hard labor to penetrate a mind-set that had disappeared within one generation. By 1750 few among the literate in northern and western Europe and the American colonies would have found it remarkable that alchemy had become obscure, esoteric, and, the enlightened said, ignorant. The historiographical notion of there having been a revolution in Western thinking about nature makes for an inheritance that cannot be erased so easily. Bringing it down will entail dismantling a set of interrelated mental structures that support beliefs as basic to Western thought as the value of technological development, industry, human freedom, the rule of law, and the possibility of progress. Saying that it never happened cannot alter the gulf between Newton and his first generation of followers. We may be better served both as historians and people by a finer honing of our historiographical legacy, not by attempting its wholesale deconstruction.

39 For Stukeley's manuscripts on the Temple of Solomon, see the collection at Freemasons' Hall, Great Queen St., London.

INDEX

Abraham the Jew, 213
Académie Royale des Sciences, 220
Accademia del Cimento, 160, 223
Accommodation, Principle of, 71
active principles, 231n, 232
Adam, 192, 195, 195n, 210
aether, Newton on, 273–4, 279, 283
afterlife, 79, 200, 200n; Boyle on, 200, 200n; and knowledge of nature, 79
Agricola, Georgius, 120
air pump, 158, 323; Boyle on, 19, 156, 160, 161, 161n, 163, 163n, 170; More on, 154, 164, 167–8, 168, 179
al-Battani, 67
Albert the Great, 218n
Albumazar, Galbrion, 150
alchemy, 6, 14, 16–17, 32, 34–6, 42, 51, 52, 113, 120, 129, 131, 203n, 208–9, 217–9, 236, 318; in Boyle, 21, 202–4, 206n, 217, 219, 319; Digby on, 91, 97, 106, 108–9, 111–13, 115; historiography of, 51, 53, 137, 202, 203n, 217, 218n; Newton on, 4, 5, 14, 15, 17, 21, 36–8, 41, 52–3, 117, 183, 201–5, 215–17, 219, 234-6, 236n, 255, 277, 319, 332
Alexandria, Clement of, 286
al-Farghani, 67
Alsted, Johann, 160, 177–9, 177n
anatomy, 21
ancient wisdom (*prisca sapientia* or *prisca theologia*), 191, 210, 221, 244, 283; Boyle on, 193, 195, 210–11; Kircher on, 230–1, 243; Newton on, 15, 192–3, 215–16, 235, 242–3, 279, 283–5, 196, 210
angels, 35, 165–6, 194, 199n, 215
Apian, Peter, 67, 73
Apocalypse, 267, 306–7
Apollonius, 75–7, 78n
Appelius, Henry, 132
Aquinas, Thomas, Saint, 52, 100
Aratus, 67, 283, 285

Arianism, 54, 186n, 187n, 277n; in Newton, 38, 53–4, 91, 293–4
Aristotelianism, 11, 20, 21n, 44, 61, 72, 115, 172, 174, 233; and alchemy, 217–18; on atomism, 194; rejection of, 4, 33, 21, 45–7, 100, 110, 157, 165, 171, 322; on the void, 153, 176
Aristotle, 73–4, 153, 177n, 194, 202, 207
Arius, 187n, 277, 293
Arminianism, 101, 101n
artisans, 123–4, 127, 321
Ashmole, Elias, 220
astrology, 17, 137–8, 139n, 140–2, 145n, 147, 147n, 148n, 148–52, 152n, 234, 327; Melancthon on, 65–6, 66n; theological objections to, 138–9, 144
Athanasius, 241–2; Newton on, 262, 262n
atheism, 316
atomism, 154, 165–6, 166n, 194
Aubrey, John, 221
Augsburg Confession, 65, 68n
Augustine of Hippo, Saint, 285–6

Bacon, Francis, 33–4, 44, 116, 185, 224, 322, 324–5, 330–1
Bacon, Robin, 207
Bailly, Jean-Sylvain, 328
Barker, Peter, 6
Barrow, Isaac, 26, 245n, 298
Bassi, Laura, 27
Baxter, Richard, 173n
Becher, Johann, 32
Becker, Carl, 27
Beeckman, Isaak, 97
Bellarmine, Robert Cardinal, 50
Bellers, Fulk, 144
Bengel, Johann Albert, 309
Bentinck, Charles, 325
Bentinck, William, 325
Bentley, Richard, 287, 295–6
Bernal, J. D., 28

Bernoulli, Johann, 52, 320, 331
Biagioli, Mario, 20
biblical interpretation: Digby on, 98; Lutherans on, 70–1; Newton on, 267, 299–308, 303n, 310–11; Simon on, 299–303, 301n, 305, 308, 311; Spinoza on, 299–303, 305, 308, 311
Biot, Jean-Baptiste, 256–7n
Blackloists, 93
Blount, Charles, 298
Boas Hall, Marie, 157n
Boerhaave, Hermann, 26, 32, 325
Böhme, Jacob, 129
Booker, John, 143, 144, 149
Book of Nature, 96, 98, 128, 232, 241, 257, 259, 295, 353
Boyle, Robert, 3, 18, 26, 31, 33, 107, 118, 131, 153–79, 184–5, 195, 224, 235–6, 238–9, 251, 253, 318–19, 323–4, 329; on air-pump experiments, 19, 156, 160, 161, 161n, 170; on alchemy, 19n, 117, 184, 189, 201–7, 206n, 207n, 209–10, 213–19; on ancient wisdom, 196, 210–11; on angels, 199n, 215; on Aristotelianism, 172, 177; on the Bible, 194, 216; on Descartes, 173–4; and experiment, 19, 49, 171, 184, 209–10, 217; on Hermes Trismegistus, 211–12; historiography of, 183–4; on *horror vacui*, 164, 170, 172; on Kircher, 223, 224n; on limits of reason, 188–91, 194–5, 197–200, 198–9n; on the mechanical philosophy, 155–6, 169–70, 176, 213; and More, 22, 154–6, 158, 168, 169, 174; on Moschus the Phoenician, 194, 196–7; on the Philosophers' Stone, 207n, 214–16, 220; on providence, 159, 159n, 170, 259; role in the Scientific Revolution, 176–7; on the Spirit of Nature, 169–71, 173–4, 179; on spring of the air, 161, 163, 163n; Stalbridge Conference on, 184n, 194–5; theology, 155n, 156n, 159, 174, 183–4, 188–9, 194, 196; on the Trinity, 188–9, 189n; on void, 154, 159, 170, 172–79; voluntarism in, 155n, 168, 173, 173n, 184; on weight of air, 171
Boyle Lectures, 264, 323
Boyle's Law, 326
Brahe, Tycho, 3, 60, 80, 325
Brewster, David, 36–7
Brougham, Henry, 257n
Bryan, Margaret, 323
Burnet, Thomas, 50, 117
Burns, William E., 8
Burtt, Edwin A., 19, 28, 45
Butterfield, Herbert, 20, 29, 42, 251; on the Scientific Revolution, 11n, 30–1, 41–2, 157, 247

Cajori, Florian, 292
Calvin, John, 190
camera obscura, 237n
Cartesianism, 315–16, 324, 325n; More on, 166; in Newton, 273n, 279, 281; on void, 153
Casaubon, Isaac, 229
Cavendish, Charles, 97
Cellini, Benvenuto, 129
Chalcidius, 279–80
Chaldeans, 195
change, intellectual, 6–9, 22, 44–5, 272
Charles I, 91, 91n, 95–6, 137
Charles II, 137, 144, 220
Charleton, Walter, 102n, 107, 107n, 253, 259
Charlotte, Princess of Wales, 323
Châtelet, Emilie du, 325–6
chemical philosophy, 21
chemistry, 18, 31, 218n
Childrey, Joshua, 148, 148n, 149
Christina, queen of Sweden, 235
Cicero, Marcus Tullius, 286
Clairaut, Alexis, 26
Clarke, Samuel, 54, 253, 263, 323, 325n, 332
Clavius, Christopher, 239
Cleiphorus, Mystagogus, 208
Cockburn, William, 26
Cohen, H. Floris, 9, 9n, 13
Cohen, I. Bernard, 4, 28, 288, 295–6, 317, 317n, 328, 329n; on the concept of revolution, 12, 43, 45; on the Scientific Revolution, 25, 247, 247n, 251
Coke, Edward, 94
Colie, Rosalie, 298
Collegio Romano, 223
comets, 72–3, 141, 151
Comte, Auguste, 11n
Condemnations of 1270 and 1277, 100, 100n
Conway, Anne, 168–9
Copernicanism, 4, 62, 71, 138, 148n
Copernican Revolution, 30, 329
Copernicus, Nicholas, 3, 10, 18, 21, 25–6, 31, 32, 34–5, 63, 66–7, 69, 80–81, 86, 318, 325
Cornford, F. M., 269
Coste, Pierre, 290–1
Council of Constantinople, 187n
Council of Nicea, 187n, 277
Coxe, Daniel, 220
Craig, John, 241
Crawford, Charles, 309
Cromwell, Oliver, 146, 149
Crosland, Maurice, 31–2, 36, 51
Crouch, John, 150
Cudworth, Ralph, 287, 293, 298

Culpeper, Nicholas, 143–46, 149
Cunaeus, Petrus, 299
Cunningham, Andrew, 11n, 20, 60n
Curry, Patrick, 137–8, 139n, 148n
Cusa, Nicholas of, 82

d'Alembert, Jean, 34–5
Dalton, John, 32, 315–16, 315n, 320–24
Damascene, John, Saint, 286
Dampier, William, 44
Daniel, Book of, 139
Davis, Edward B., 183
Dear, Peter, 239
Debus, Allen G., 31–2
Dee, John, 35, 177n, 233
deism, 183, 255
Demiurgos, 282
Demosthenes, 69
Desaguliers, Jean, 315–16, 323, 332
Descartes, René, 3, 33, 35, 47–8, 97, 102–3, 102n, 173, 222, 236–7, 240, 254, 275, 321, 326, 327
design, argument from, 21, 81, 85, 156; Newton on, 253–4, 256, 259, 264
Despaigne, Jean, 145n
determinism in astrology, 140
Dickinson, Edmund, 220
Diderot, Denis, 325, 330
Digby, Everard, 91
Digby, Kenelm, 22, 89–118, 220
Digby, Venetia, 89–90, 109, 118
disciplinary boundaries, 4, 8, 11, 12n, 17, 17n, 60–1, 71, 85
Dobbs, B. J. T., 4, 22, 41, 43, 46, 48, 117, 137, 159, 183, 201, 209–10, 213, 215, 316, 319, 332; on alchemy, 51, 53, 111n; on Digby, 93n, 113; on Newton, 251, 253–4, 248n, 249, 258, 268–73, 273n, 279, 283–4, 286, 288, 293; on Newton's alchemy, 15–16, 51, 59, 214, 250; on Newton's Arianism, 91, 278; on Newton's theology, 16, 54, 216, 216n, 277; on the Scientific Revolution, 5, 51, 59, 118, 122, 218–19, 247, 249, 251–2, 258, 269–70, 272, 317
Duchesne, Joseph, 108–9
Duclo, Gaston, 218n
Duiller, Nicholas Fatio de, 208, 285, 308, 320–21, 332
Dyck, Anthony Van, 89

eclipses, 141–7, 147n
Eco, Umberto, 221, 225, 226n, 245–6
Edict of Nantes, 93, 93n, 94n
Einstein, Albert, 28, 202
English Civil War, 138–9, 150
Enlightenment, 50, 225, 318
Epicureanism, More on, 168

essentialism, 9n, 22; about alchemy, 202; and the history of science, 12; in natural history, 18
Eucharist, 61–2, 116n
Euler, Leonhard, 331
Eusebius, 196
Evans, Arise, 144
Evelyn, John, 151
Everard, John, 212
experiment, 19, 22, 48–9, 123, 158, 209, 237–9; Puy de Dôme, 171n
experiments, air-pump, 158, 161; More on, 154, 164, 167–68, 179

fact, concept of, 37n, 38
Faraday, Michael, 28
Feingold, Mordechai, 33
Ferguson, James, 324
Figala, Karin, 201, 332
Findlen, Paula, 8, 20, 27, 318
Flamel, Nicholas, 213, 218
Fludd, Robert, 31, 116
Fontenelle, Bernard Le Bovier, Sieur de, 26–7, 220
force: concept of, 48, 53, 225, 289; Newton on, 225, 272–3, 289
Force, James E., 16, 183, 190, 311n
Franklin, Benjamin, 328
free will, 101n, 189; Digby on, 94, 101–2, 104
French Prophets, 308
Frisius, Gemma, 80–1
Froom, Leroy E., 310
fundamentalists, 300, 309, 311
Furet, François, 29
future contingents, 189

Gadbury, John, 147, 147n, 148n
Galilei, Galileo, 3, 10, 13, 21, 26, 35, 44, 46–9, 223, 232, 295, 318, 320, 323–4, 326; and the Church, 34, 62, 63
Gassendi, Pierre, 17, 98–9, 102n, 103n, 118, 280
Gataker, Thomas, 140, 142, 145n, 148, 148n
Gaule, John, 146
Geber, 213, 218, 218n
Gilbert, William, 231
Geminus of Rhodes, 69
Glauber, Johann Rudolph, 119–33
Glauber's Salt, 119–21, 133
Glorious Revolution, 329
God, 103–4, 166; as geometer, 82, 84, 86–8; Newton on, 260, 287–8, 291–3, 296; relationship to the creation, 22, 282
Goethe, Johann Wolfgang von, 227, 228n
Goldish, Matt, 298

Index

Gravesande, Willem Jacob van s', 315, 324
gravity: cause of, 279; Newton on, 234–5, 283–5, 291–2, 294
Greatrakes, Valentine, 159n
Green, John, 146
Greene, R. A., 156–7
Grégoire, Henri, 328
Gregory, David, 285, 292, 322
Guericke, Otto von, 160, 323
Gunpowder Plot, 91

Hale, Matthew, 153
Hall, A. Rupert, 28, 251, 256
Hamann, J. G., 309
Hampson, Norman, 254, 269
Hanckwitz, Godfrey, 209
Harflete, Harry, 147
harmony of the spheres, 34, 285
Hartley, David, 309
Hartlib, Samuel, 126, 127n, 132
Harvey, William, 18, 20, 21, 26
Harwood, John T., 179
Hauksbee, Francis, 315, 324
Heinrich, Johann, 236
Helmont, Jean Baptiste van, 21, 31, 124, 193–4
Henry, John, 93, 107, 158–9
Hermes Trismegistus, 13, 211, 218, 230, 243, 245
Hermetic philosophy, 13–14; in Kircher, 229–30
hieroglyphics, Kircher on, 230, 230n
Hilderich von Varel (Edo Hildericus), 69–70, 86
historiography, 17, 22, 29–30, 41, 59–60; of science, 7, 17, 177, 177n, 321–4, 326, 328–9, 331–2
Hobbes, Thomas, 19, 97, 103n, 107n, 116, 161
Hobsbawm, E., 329n
Homberg, Wilhelm, 220
Homes, Nathaniel, 139, 151
Hooke, Robert, 237
horror vacui, Boyle on, 164, 170, 172
Horsley, Samuel, 309
Hume, David, 268
Hunter, Michael, 19, 138n, 183
Hutchison, Keith, 107, 107n, 245n, 246n
Huygens, Christiaan, 21, 26, 160, 222, 225, 256n

idolatry: Boyle on, 196; Kircher on, 243; Newton on, 192, 241–2, 244, 261, 262n, 264, 283
Iliffe, Robert C., 269
Industrial Revolution, 318

inertia, principle of, 47
instrumentalism, 79
instruments, scientific, 49
Interregnum, 137, 152
Islam and science, 60n

Jacob, James R., 318
Jacob, Margaret C., 5, 8, 33
James I, 95, 101
Janacek, Bruce, 22
Jenkins, Jane E., 22
Jesuit science, 34n, 44
Job, Book of, 108n
Josephus, Flavius, 139, 142, 299
Judaism and science, 60n

Kant, Immanuel, 13
Kaplan, Barbara Beigun, 159n
Katz, David, 311n
Kepler, Johannes, 3, 5, 17, 22, 26, 33–5, 46, 59–60, 62–88, 148, 231, 318–19, 325
Kepler's Laws, 272
Keynes, John Maynard, 226
Kircher, Anasthasius, 8, 221–46, 318
knowledge, 22; Boyle on, 189–91, 194–5, 197–8, 198–9n, 200; of God, 71; Newton on, 19, 193, 197–8
Koenig, Samuel, 324–5
Koyré, Alexandre, 19, 28, 32, 45; on Newton's theology, 38–9
Kuhn, Thomas S., 28–9, 32, 63–4, 82n

Laplace, Pierre-Simon, Marquis de, 269
Lateran Council, 99
Laud, William, 91, 91n
Lavoisier, Antoine, 31, 32
law, natural, 47, 83, 83n
laws of motion, Newton on, 285, 291–2
laws of nature, 47; Boyle on, 185n; Newton on, 267
Lead, Jane, 308
LeClerc, Jean, 301
Leibniz, Gottfried Wilhelm, 21, 26, 127n, 220–3, 227, 254, 260, 263, 288, 320, 326
Leibniz-Clarke Correspondence, 326
Leiden, 325
Lémery, Nicholas, 220
Leo X, 99
Lilly, William, 140–2, 141n, 144, 147, 147n, 149, 151
Lindsay, Hal, 300n
Linus, Franciscus, 161
Lipsius, Justus, 286
Locke, John, 33, 187, 214, 235, 254, 290, 298, 300, 326–7; and alchemy, 208, 219; on argument from design, 264; on bib-

lical prophecies, 264; and Boyle, 220; on miracles, 264n
Lucretius, Titus Carus, 316
Lull, Raymond, 236, 236n
Luther, Martin, 63–5
Lutheranism: and Copernicanism, 62, 64, 82, 86; on design, 88; on natural light, 82; on providence, 61, 82; on real presence, 61–2

McGuire, J. E., 15–16, 183, 210
Mach, Ernst, 10, 10n
Macrobius, Ambrosius Theodosius, 284
McVaugh, Michael, 36
Maestlin, Michael, 78, 80, 87
magic, 34, 234; mathematical, 234; natural, 14, 245n, 246n; Yates on, 13
magnetism: Kircher on, 231; Newton on, 234–5
Maimonides, 195–6
Malebranche, Nicolas, 26
Manuel, Frank, 229, 242, 255–6, 260, 308–9
Marsh, Bishop of Dublin, 299
Marsh Library, 299
Martin, Benjamin, 316
Martyr, Justin, 286
Marwick, Arthur, 25, 44
materialism, 323
mathematics, 239–40
matter, 22; Boyle on, 213; Gassendi on, 17; Newton on, 278, 280–2, 293; passivity of, 97, 156
Mauskopf, Seymour, 36
Maxwell, James Clerk, 28
mechanical philosophy, 6, 10, 21, 42–48, 254; and chemistry, 18; Digby on, 92, 102, 105; and theology, 97, 107
mechanics, 46
Melanchthon, Philip, 59, 65–6, 69, 82–4, 87; on astrology, 65–6, 66n; on astronomy, 63, 67–8; on Copernicanism, 63; on university curriculum, 64–5, 68
Menasseh ben Israel, 196
Mencke, Johann Burckhardt, 222
Mersenne, Marin, 85, 97–8, 223
Merton, Robert K., 18
method, 48, 78n, 323; Aristotelian, 72–3; Boyle on, 185; in Copernican astronomy, 81, 82n; Kepler on, 83n, 84; Kircher on, 236; Newton on, 236–7, 268; Paracelsus on, 128
microscope, 326
Millen, Ron, 107
millenarianism, 27, 250n, 299, 300n, 329
Millennium, 121, 127, 299; Boyle on, 329;

Glauber on, 120; Newton on, 265–7, 300n
Miller, William, 309
miracles, 102n, 326; Locke on, 264n; Newton on, 102n, 263–4, 267
Mirandola, Pico della, 212
Mishnah, 299
Moore, Paul N., 300n, 311n
More, Henry, 107, 159n; on air-pump experiments, 22, 154–8, 161, 164, 168–9, 171, 179; on atomism, 166, 166n; on concept of God, 166, 286; on Descartes, 165; on providence, 159; on the soul, 165–6; on the Spirit of Nature, 161, 165, 167, 168, 173, 179
Morelli, Henri, 298
Morhof, Daniel Georg, on Kircher, 223–4
Moschus the Phoenician, 194, 196–7
Moses, 192–3, 196, 210, 301
motion: Aristotelian conception of, 46; Galileo on, 46–7; Newton on, 288, 290
motion, laws of, Newton on, 285, 291, 292
Motte, Andrew, 292

natural history, 18, 20
natural light: Kepler on, 83–4, 86; Kircher on, 232–3; Melanchthon on, 82–3
natural philosophy, 17, 21, 241
natural theology, 323, 326
nature: concept of, 155, 195; mathematization of, 11, 13, 47, 279–82, 290, 295–6; mechanical view of, 11, 46; uniformity of, 13
Nedham, Marchamont, 150
Neoplatonism, 82, 115, 228–9, 233
Neufchâteau, François de, 330
Newcastle Circle, 97
Newman, William R., 202
Newton, Humphrey, 185
Newton, Isaac, 3–4, 8, 14, 16n, 18, 21–2, 31, 33, 35, 49, 117–18, 221, 226, 229, 231n, 234–5, 241, 245–6, 297n, 299, 318–27, 331; on aether, 273–5, 278–9, 283; on alchemy, 4–5, 14–15, 52–3, 117, 183, 201–5, 207–9, 214–16, 218–19, 234–6, 236n, 255, 277, 291, 332; on ancient wisdom, 15, 192, 196, 210, 215–16, 235, 242–3, 279, 283–5; on the Apocalypse, 267, 306–7; Arianism of, 54, 91, 186–8, 186n, 187n, 188n, 192, 216, 241–2, 244, 260–2, 277–8, 284, 293–4, 300–1; on Athanasius, 262, 262n; on biblical interpretation, 8, 50, 216, 265, 267, 297–301, 301n, 302, 303–8, 303n, 310–11; on biblical prophecies, 256–7, 264–5, 266n, 307, 311; on the concept of force, 53,

Newton, Isaac, (cont.) 225, 272–4, 289; on creation, 279–81, 289; and Descartes, 273n, 275, 279, 281; on design argument, 253–4, 256, 259, 264; Dobbs on, 15, 16, 250, 252, 254, 268–9, 272, 273n, 278–9, 286; on experiment, 209, 237; on God, 216, 253, 260, 282–3, 286–9, 291–6; on gravity, 193, 234, 278–92, 294; on Hermes Trismegistus, 211–13, 217–18, 243, 245; historiography of, 17, 36, 157, 226, 253–4; on idolatry, 192, 242, 244, 262n, 264, 283; influence of, 48, 309–10; on Kepler's Laws, 272; on laws of motion, 285, 288, 290–2; on laws of nature, 264, 267; on limits of reason, 184n, 186, 186n, 191, 193, 197–8; on mathematical physics, 8, 52; and mathematics, 14, 15, 186, 228, 239–40; on the mathematization of nature, 289–90; on matter theory, 216, 245, 275–82, 293; and the mechanical philosophy, 48; on method, 184n, 236, 240, 268, 294–6; on the Millennium, 250n, 265–7, 300n; on miracles, 102n, 263–4, 267; on optics, 14, 49, 240; on pendulum experiments, 272–5; on predestination, 190, 190n; on providence, 54, 257, 259, 263, 264n, 266–8, 275; on the Prytaneum, 193, 283; reputation of, 27–8, 222, 224–5, 227, 230, 233–4, 328; role in the Scientific Revolution, 14, 16, 29, 48, 250, 253, 255, 255n, 257–8, 269–70; on science and religion, 255, 255n, 257, 263, 267, 271, 283, 294; on the soul, 262, 262n; on space, 275, 279–82, 284, 285, 287, 291; Stoicism in, 112n, 286; theological manuscripts, 300n, 308, 310; theology, 4, 14–16, 16n, 38, 54, 60n, 183, 216, 241, 254–5, 260–65, 268, 271, 275–7, 283, 290; on the Trinity, 91, 186–8, 186n, 187n, 188n, 192, 216, 241–2, 244, 260–2, 277–8, 284, 293–4, 30–1; on the void, 274–5; voluntarism in, 257, 260–61; Westfall on, 15–16, 250, 252, 254, 265, 269
Newton, Thomas, 309
Newtonianism, 11, 315–17, 320–21, 323–6, 325n
Newtonian style, 294
Nieuwentyt, Bernard, 327
Noah, 192, 242, 283
Nollet, Jean Antoine, 324
nominalism, in Boyle, 175n

occult qualities, 107, 107n, 225, 228, 245, 246n
occult sciences, 226

Oldenburg, Henry, 127n, 148n, 208; on Kircher, 238–9
Origen, 196
Ornstein, Martha, 28
Osiander, Andreas, 67, 78
Osler, Margaret J., 102n

Paley, William, 323
palingenesis, 92, 108–9
Palissy, Bernard, 129
Paracelsianism, 17
Paracelsus (Philippus Areolus Theophrastus Bombastus von Hohenheim), 21, 26, 31, 120, 124–5, 127–9, 135, 177n
Partridge, John, 152n
Pascal, Blaise, 26; experiments, 49, 153, 172n
Patrizi, Francesco, 280
patronage, 20
Paul, Charles, 27
Paul, Saint, 283, 285
Peiresc, Nicholas, 223
Pelagianism, 103
Pell, John, 97
Pemberton, Henry, 332
Pena, Jean, 74
Pentateuch, 301
perception, 47
Peucer, Caspar, 68, 79, 87
Philadelphians, 308
Philalethes, Irenaeus, 52, 202, 205, 211, 213
Philo Judaeus, 286, 288
Philoponus, John, 280
Philosophers' Stone: Boyle on, 207n, 214–16, 220; Kircher on, 236; Newton on, 214
physics, 6, 21
physiology, 20
Piccartus, Michael, 69
planetary models, 75–7
planets, order of, 70
Plato, 243, 277, 282
Platonism, 110–11, 166, 286–7
Platonists, Cambridge, 38; see also More, Henry
pneuma, 111–12, 112n, 286–7
politics and science, 318
Popkin, Richard H., 8
Porta, Giovanni Battista della, 231, 234
positivism and the Scientific Revolution, 10, 10n, 11
predestination, 95, 101–2, 101n, 140, 186, 189–90, 261
Priestley, Joseph, 227, 228n, 309
Principe, Lawrence M., 21, 184
prisca sapientia, see ancient wisdom

prisca theologia, see ancient wisdom
Proclus, 279–80
prodigies, 139–42, 141n, 148–9, 150–1, 152n
progress, idea of, 27, 321, 324, 327, 331
prophecies, 264, 309, 311; Newton on, 256–7, 264–5, 266n, 307, 311
Protestants, 65, 66n
providence, 17, 34, 61, 82, 139, 156, 159, 254, 258, 263; Boyle on, 159, 159n, 170, 259; Kepler on, 84–5; Newton on, 54, 257, 259, 263, 264n, 266–8, 275, 279
Prytaneum, Newton on, 193, 283
Ptolemy, Claudius, 75–6, 85
Puerbach, Georg, 66
Puritanism, 91, 138
Pythagoras, 243, 269
Pythagoreanism in Kepler, 17, 197

Rachovian Catechism, 187–8
Ranelagh, Katherine Boyle, Viscountess, 209
Rattansi, P. M., 15, 183, 210, 283
Ray, John, 18
realism, 80
reason, limits of: Boyle on, 189–91, 194–5, 197–200; Newton on, 184n, 186, 186n, 191, 193, 197–8
Redi, Francesco, 224
Redondi, Pietro, 115n
Reformation, 59
Regiomontanus, 73, 325
Reinhold, Erasmus, 66–9, 79, 87
Restoration, 138
resurrection, Digby on, 92, 106–11
Revelation, Book of, 139
revolution, concept of, 4, 25–6, 30, 42–4, 272, 329–30
Rheticus, Georg Joachim, 64, 66n, 67–8, 78, 80–81, 86
Riccioli, Giovanni Battista, 224, 239
Ripley, George, 211, 218
Rohault, Jacques, 325n
Roland, Jean-Marie, 331
Rothmann, Christopher, 80–1
Rousseau, Jean-Jacques, 325, 331
Rousset de Missy, 326
Royal Society, 138, 165, 220, 223, 225, 238–9, 258–9, 259n

Sabellianism, 187n
Sabra, A. I., 6
Sacrobosco, 66
Saint-Evremond, Charles, 298
Salvation Army, 309
Sarton, George, 28
Scaliger, Joseph Justus, 177n, 242

Scargill, Daniel, 53
Schaffer, Simon, 19, 33, 158
Scheiner, Christoph, 239
Schiebinger, Londa, 27
Schmitt, Charles B., 33
Scholasticism, 11, 35, 46
Schöner, Johann, 67
Schott, Gaspar, 160
science, popularization of, 321, 325–7, 325n
science and religion, 49, 50, 254n, 269–70, 283, 318, 326; in Newton, 254, 294; Westfall on, 247–9, 253–4, 256
Scientific Revolution, 8–9, 10n, 41, 43, 46, 48, 217–19, 246, 317–19, 324, 326, 329, 331–2; Butterfield on, 31, 247; concept of, 29, 31, 33–4, 44–5, 47, 55, 137; Dobbs on, 25–39, 122, 219, 247, 249, 258, 269, 272; historiography of, 3–5, 10–13, 17–19, 22, 25, 28, 225, 251, 325, 328, 331; I. B. Cohen on, 25, 247, 247n, 251; influence of, 48, 331; Newton's role in, 14, 16, 29, 48, 225, 247–8, 250, 253, 255n, 257–8, 269–70; Westfall on, 122, 219, 247, 249–50, 258, 269, 272
Seidelmeier, Suzanne, 68
Sendivogius, Michael, 207n
Sennert, Daniel, 213, 218n
Seventh Day Adventists, 310
Shapin, Steven, 8, 19, 19n, 33, 158, 158n
Sheppard, Samuel, 150
Simon, Father Richard, 297, 299, 301; on biblical interpretation, 300, 301n, 302–3, 305, 308, 311
Singer, Charles, 63, 64
skepticism, 254, 255n
Smeaton, John, 320
Smith, Pamela, 7
Snobelen, Stephen David, 269
Socinianism, 186n, 187–8, 187n
Solomon, 210, 283–5
Sorbière, Samuel de, 126, 131
soul, 17, 21–2, 79, 98–104, 108–9, 165–6, 262, 262n
space, Newton on, 275, 279–85, 287, 291–2
Spargo, Peter, 215
Sperling, Otto, 126
Spinoza, Benedict de, 238, 297, 301, 326; on biblical interpretation, 299–300, 302–3, 305, 308, 311
Spirit of Nature, 170; Boyle on, 155, 171, 173–4; More on, 161, 165, 167–8, 173, 179
Sprat, Thomas, 207, 253, 259
spring of the air, 163, 163n, 169
Stahl, Georg, 32

Starkey, George, 52, 131, 205, 211
Steno, Nicolas, 232
Stillingfleet, Edward, 299
Stoicism, 280, 283, 285–8; Boyle on, 194–5; in Digby, 97, 112, 115; in Newton, 112n
Stukeley, William, 225, 332
sunflower clock, 222
Swan, John, 141n, 145n
Swift, Jonathan, 152n
Swinden, J. H. van, 328
Sydenham, Thomas, 26
sympathetic medicine, 224
Synod of Dort, 101

Talmud, 299
Taylor, F. Sherwood, 204
Thales, 193, 285
Thomas, Keith, 137, 138, 140
Thomson, Thomas, 36
Thorndike, Lynn, 67, 252
Thornton, Alice, 144, 145
Timaeus, 277, 280, 281, 282
Torricelli, Evangelista, 49, 160, 172
transubstantiation, 62, 91, 115–16n
Trinity, 91, 186n, 187n; Boyle on, 188–9; Digby on, 102, 114–15; Kircher on, 243–4; Newton on, 91, 186–9, 186n, 187n, 188n, 192, 216, 241–2, 244, 260–2, 277–8, 284, 293–4, 300–1
Tullius, Cicero, Marcus, 286
Tyacke, Nicholas, 101n

universities, 33–4, 64–5, 316n

Valentine, Basil, 211
Valeriano, Piero, 229
Vesalius, Andreas, 3, 26, 32
Villiers, Frances Coke (Lady Purbeck), 94, 94n
Virgil, 286
void, 22, 153–4; Alsted on, 178; Aristotelianism on, 176, 178; Boyle on, 170, 175–6, 178; concept of, 159, 170, 174–5, 178; Newton on, 274–5, 321
Voltaire, François Marie Arouet, 227, 228, 325, 327
voluntarism: in Baxter, 173n; in Boyle, 155n, 168, 173, 173n, 184; in Newton, 16, 184, 184n, 257, 260, 268, 290
Vondelius, Johannes, 230

Wallis, John, 21
Warr, John, 179
Watt, James, 320–3
weapon salve, 92, 92n, 105
Webster, Charles, 18
Wedgewood, Josiah, 329, 330
Westfall, Richard S., 4, 7, 20–22, 117–18, 138, 201, 241, 332; on alchemy, 137, 209, 217–18; on Newton, 249–50, 253–6, 255n, 256n, 258, 293; on Newton's alchemy, 15, 16, 17, 183; on Newton's theology, 266n; on patronage, 20; on science and religion, 247–8, 25–4, 256; on the Scientific Revolution, 5, 8, 11, 17, 41–56, 122, 217–19, 247–8, 249, 252, 258, 272, 317–18
Westman, Robert S., 7
Wharton, George, 144, 147n
Whewell, William, 13
Whiston, William, 54, 253, 332, 363–4
White, Andrew Dickson, 62–4
Whitehead, Alfred North, 28
Whitla, William, 297, 309
Wilkins, John, 234, 253; on providence, 259
will: divine, 283, 289, freedom of, 101n, 189
Willey, Basil, 254n
William IV, 326
Williams, Helen Maria, 331
Williams, Perry, 11n
Willughby, Francis, 18
witnessing, 238
Wittenberg astronomers, 66–7, 69
Wojcik, Jan, 210–12
Wolf, Abraham, 44
women in science, 27, 323
World War II, impact on history of science, 30
Wren, Christopher, 21

Yahuda, Abraham Shalom, 310
Yarworth, William, 208
Yates, Frances: on the Hermetic philosophy, 14n, 216; on the Scientific Revolution, 13–14, 14n, 225

Zeno the Stoic, 195
Zosimos of Panopolis, 202

www.ingramcontent.com/pod-product-compliance
Ingram Content Group UK Ltd.
Pitfield, Milton Keynes, MK11 3LW, UK
UKHW032325190125
453752UK00011B/163